Supporting small-scale farmers and rural organisations:
Learning from experiences in West Africa

A handbook for development operators and local managers

Supporting small-scale farmers and rural organisations: Learning from experiences in West Africa

A handbook for development operators and local managers

Scientific editors: Dr Sylvain Perret and Dr Marie-Rose Mercoiret (CIRAD)

Protea Book House, IFAS & CIRAD

Pretoria 2003

First English edition, 2003, published by:

Protea Book House
PO Box 35110
Menlopark
0102
Republic of South Africa
protea@intekom.co.za

Institut Français d' Afrique du Sud (IFAS)
PO Box 542
Newtown
2113
Republic of South Africa
www.ifas.org.za

and

Centre de Coopération Internationale en Recherche Agronomique pour le Développement (CIRAD)
Avenue Agropolis
TA 283/04
34398 Montpellier Cedex 5
France
librairie@cirad.fr

Typography and design by PrePress Images, Pretoria
Cover design by Tienie du Plessis
Reproduction by PrePress Images, Pretoria
Printed and bound by Interpak, Pietermaritzburg

ISBN Protea: 1–919825–92–4
ISBN CIRAD: 2–87614–505–7

© 2003 Protea Book House and CIRAD

© 1994 French text *Editions Karthala and Ministère français de la coopération*

French edition

The French version (first edition of this book) was published in 1994, under the co-ordination of Dr Marie-Rose Mercoiret from CIRAD (Centre de Coopération Internationale en Recherche Agronomique pour le Développement), with the support of the French Ministry of Co-operation, and the contribution of several French developmental and co-operative institutions.
Original title:
L'appui aux producteurs ruraux: Guide à l'usage des agents de développement et des responsables de groupements, Karthala, Paris, France.

This Edition

Scientific editors: Dr Sylvain Perret and Dr Marie-Rose Mercoiret, from CIRAD

Advanced translation and edition: Dr Sylvain Perret, Sue Burdairon (translator and editor).

Associate scientific editor: Juliana Rwelamira, from the University of Pretoria.

Basic translation: Jean Pierre Akima Mavian (University of Pretoria) and Dr Roger Pambu (Department of Agriculture, Gauteng Province, South Africa).

All sketches were drawn by Jacques Mercoiret.

Financial support from the French Embassy in South Africa, through IFAS (Institut Français d'Afrique du Sud), has made this edition possible.

Table of Content

Foreword 7

Introduction 9

Part one: Procedures 17

Chapter   I.   Local planning 19
Chapter   II.   Diagnosis 39
Chapter   III. On-farm experimentation 69
Chapter   IV.   Monitoring and evaluation 91

Part two: The Tools 115

Chapter   V.   Training of farmers 117
Chapter   VI.   The farmers' organisation 141
Chapter   VII. Contracts between role-players 161

Part three: The Fields of Intervention 179

Chapter VIII. Extension services and farm management advice .. 181
Chapter   IX. Natural resources management 205
Chapter   X.   Product management 225
Chapter   XI. Management of collective assets and facilities . 243
Chapter   XII. Financing local development 271
Chapter XIII. Women and development 291
Chapter XIV.   The non-agricultural sector 307

About the editors

Dr Sylvain Perret is an agronomist and agricultural economist at CIRAD, where he has specialised in farming systems and rural development. He has been an associate professor at the University of Pretoria (South Africa) since 1998, in the Department of Agricultural Economics, Extension and Rural Development, and conducts research, educational and outreach activities in various developing rural areas of South Africa.

Dr Marie-Rose Mercoiret is a sociologist at CIRAD. She has specialised in rural institutions and developing small-scale farmers organisations, and is based at the CIRAD scientific headquarters at Montpellier (France). She has carried out research and outreach programmes in many developing countries, especially in sub-Saharan Africa.

CIRAD

The Centre de Coopération Internationale en Recherche Agronomique pour le Developpement – CIRAD – is a French organisation specialized in agricultural research for the developing tropical and subtropical areas.
CIRAD has a mandate to contribute to rural development in southern and developing regions of the globe through research, experimentation, training operations in France and abroad, and dissemination of scientific and technical infonnation.
Its fields of competence and operations include agronomy and crop sciences, veterinary and animal sciences, forestry, food processing and technology, environmental sciences, economics and agricultural policy.
CIRAD cooperates with over 90 countries worldwide.

Head Office:
42, rue Scheffer, 75116 Paris, France
Tel: +33 1 53702000

Montpellier Research Hub:
Avenue Agropolis, 34398 Montpellier Cedex 5, France
Tel: +33 46761 5800

The French Institute of South Africa (IFAS)

This French institute includes a cultural and a research section, concerned with human sciences. It is one of 25 institutes maintained by the French Ministry of Foreign Affairs.
Under the guidance of its scientific committee, IFAS promotes and supports joint research programs and partnerships between French and French-speaking academic and research institutions and their counterparts in South Africa and, more generally, the sub region. The institute offers bursaries and subsidies to individual researchers and encourages exchange across the North-South division. It offers a library service with internet facilities to visiting researchers.
IFAS publishes a series of occasional papers, *Cahairs de l'IFAS*, of which three issues have yet been released. IFAS also maintains a website.

66 Wolhuter Street, Newtown Cultural Precinct, Johannesburg
PO Box 542 Newtown, 2113
South Africa

Tel: + 27 11 836 05 61
Fax: + 27 11 836 58 60
ifas@ifas.org.za
www.ifas.org.za

Foreword

The translation of this book into English comes at a very appropriate time as the government in South Africa is trying to come to terms with the challenges of rural development and rural poverty in particular. An Integrated Sustainable Rural Development Strategy (ISRDS) was launched early in 2001 to create a broadbased approach to improve the quality of life of the rural population of South Africa. Many of the principles discussed in this book have already assisted the thinking behind the design process of the ISRDS and we argue that the success of the implementation of the strategy will, to a large extent, be assisted by some of the tools and principles discussed in this book. There are also a number of the other initiatives, such at the Kellogg Foundation's Integrated Rural Development Programme, which are also trying to extend the message of integrated development and decentralised/localised planning to other countries in southern africa. Kellogg's IRDP is a true example of the application of planning and diagnostic processes (some of the aspects discussed in Part One of the book) at local level to assist the implementation of programmes to improve the conditions of communities in specific localities in six countries in Southern Africa.

The translation of this book into English has opened the way for many students in East and Southern Africa to learn from the wealth of experience with rural development, farming systems research and various interventions in rural communities. As such this effort will certainly make a major contribution to capacity building in this part of the world.

Professor Johann Kirsten
Head of the Department of Agricultural Economics, Extension and Rural Development
University of Pretoria
Pretoria, South Africa

Introduction:
Local development vs. globalisation: seeking the match for southern Africa's rural areas

In the context of southern African rural areas, key issues face the local people, development operators, as well as decision and policy makers.

Firstly, a number of these issues are inescapable:
- Southern Africa's rural areas face severe backlogs in infrastructure, services, human capacity and self-confidence, access to information, markets, and so on (owing to former dispensations), often along with poor access to natural resources.
- Globalisation is taking place with its national and local implications: the state's withdrawal from its former commitments and controls, the liberalisation of markets, decentralisation, the transfer of competencies to local management and decision structures. Local government structures or emerging private management structures are seldom prepared for this quick handover process.
- Policy makers are struggling to strike a proper balance between a pure rights-based, gap-filling approach, and a productivity oriented approach.

Secondly, the rural environments undoubtedly bear the marks of diversity, complexity and dynamics:
- Activities and livelihood systems are diverse as are land use patterns, people's strategies, farmers' practices and so on. The so-called homogeneity is a myth. Addressing and understanding this diversity is a prerequisite to any step towards the further development of rural settings.
- Human relationships (at household, community and small region levels) are complex, as are local institutional arrangements (laws, rules, socio-cultural norms), and the entanglement of institutions and decision levels acting upon local settings.
- Dynamics exist, as all the systems developed are always adapting to constraints, taking advantage of new opportunities, and then shifting to new strategies or activity systems.

The globalisation process as described above is likely to reinforce these traits. It is crucial to create awareness and to support the stakeholders with proper tools, methodologies, examples and illustrations.

Relying only on notions such as income maximisation, optimum decisions, normative economic theory and the like appears highly illusive and unrealistic. The idea is to avoid further normative, homogenous and ultimately blind initiatives for rural development (failure to extend the Green Revolution to most rural areas was a serious warning).

The authors of this handbook support the idea that the rural environment and circumstances have to be taken into account for development purposes, be they planning at regional level, policy making, targeting research efforts, designing training and advice for capacity building, or promoting technical change, just to mention a few.

Such concern is derived from the following facts:

- Rural people are increasingly likely to become the main participants of their own development.
- However, they do not have the information needed to make proper decisions. Furthermore, they often lack the skills and self-confidence to make change happen.
- Their activities and desires are limited by poor access to most markets (land and resources, product markets, credits, inputs, and so on), and often by the absence of actual local democracy (so-called participation and empowerment are still more often talked about than really implemented).
- Nevertheless, rural households often show great flexibility, and develop a number of objectives and activities.

This handbook was initially based on West African experiences. It has been decided to translate it into English, and to adapt it, as it is likely to prove useful to most development operators and local rural managers throughout sub-Saharan Africa, as it addresses key issues in a very open, simple, comprehensive and convenient way.

Dr Sylvain Perret
CIRAD / University of Pretoria

Presentation

1. Objectives

In the specific area of local rural interventions and support to small-scale farmers, this handbook strives to achieve two objectives:
- It attempts to summarise the basic components (whole or part) of one field of intervention with the aim of providing the most suitable support to the farmers' needs; each chapter corresponds to one of these components (diagnosis, monitoring and evaluation, training, organisation, management, financing, and so on),
- For each component in a chapter, the handbook strives to underline the main issues to be borne in mind (specific objectives, principles, methods, tools) when working with small-scale farmers and rural people.

2. Target market

The handbook has been produced for two types of people:
- Firstly, it targets *development agents, extension officers and the rural area external operators* who, while working with the rural people, wish to adapt and upgrade their working methods and support in order to address the needs expressed by the farmers and other rural people. It also targets the technical staff from the public sector, NGOs, parastatals, research and co-operative organisations who are involved in projects.
- Secondly, it targets *stakeholders in charge of rural and agricultural organisations* who organise and manage the support to their members. Although such organisations are not very common in sub-Saharan Africa, some do, however, already exist, and their leaders have made requests regarding operational methodology. The recent tendency towards the professionalisation of agriculture linked to the emergence of rural organisations and to the local and regional dynamics (decentralisation and the emergence of local governance systems) should result in the proliferation of professional agricultural organisations, which would take charge of diverse mechanisms of support to the farmers (water users' associations, small-scale co-operatives for marketing support or access to inputs, local agribusiness, and so on). The number of this calibre of professional is expected to increase in the years to come.

3. New context leading to specific choices

This handbook has been produced during a period of tremendous change in sub-Saharan Africa. Although these changes are well known (state disengagement, decentralisation, transfer of responsibilities, changes in the development organisations), their operational consequences have not yet been analysed. In fact, while most African governments are disengaging from their previous functions, other role-players are emerging. Rural organisations are being created or reinforced, private economic role-players are now operating publicly, and local public authorities are emerging or are gaining more autonomy in the management of local affairs. Each of these organisations has its objectives, its own strategies and is defining itself as playing, as is right and fair, a role in the development process.

However, this raises some questions regarding the new relationship between these different role-players: What will be the role of each in future? How will this role be defined? How will the tasks be shared and what collaborations may be established?

The authors of this handbook are of the opinion that the answers to these questions cannot be dictated from outside. If they are expected to be appropriate, such answers must arise from consultation between role-players.

Constructive negotiation (which will take into account the expectations and interests of everybody) is necessary. It must allow convergence to a consensual compromise accepted by all the role-players involved.

This will necessitate a review of the role of the extension officer, who might appear as both an advisor and a mediator in
- supporting the establishment of contracts (by analysing and explaining the objectives and the expectations of the farmers during the negotiations), and
- facilitating access to information and to training.

4. Proven tools and methods

This handbook cannot be considered a package of propositions for new types of local interventions in rural environments. While reading it, one may gain, for good reasons, the impression that the contents have been said or heard before. However, two important points have to be mentioned on this subject:
- The handbook strives to summarise the different aspects of a successful intervention in the field; in this respect, the authors have considered the existing literature and their own experiences. Whenever certain principles, or the proposition of methods and tools are considered as crucial, the risk of repeating them has been taken, especially when they have not yet been translated into common practice.
- The proposals (approaches, methods, tools) herein formulated have already been applied here and there in real situations. Therefore, they are not vague intentions to be tried out but actual operational indications whose reality and efficiency have already been proven. As an illustration, some examples are provided to stimulate the creativity of the field role-players, who are always confronted with specific realities and who must therefore adapt their methods and forge their own tools.

5. Contents

The handbook is divided in three major parts:

- **PART ONE: Procedures**

This first part comprises four chapters that correspond to four permanent components of support to the farmers:

I. Local planning

Local planning is the means by which the local role-players (smallholders, artisans, rural role-players, etc.) define firstly their priorities regarding the question of development, and secondly the multi-sectorial action plans in the short, medium and long term. The role-players have to discuss these with external partners (technical services, NGOs, financial partners, administration) located on other geographical scales or other levels of decision making.

II. Diagnosis

A sound knowledge of the starting point is a prerequisite to the support being adapted to the expectations of the farmers. While diagnosis is a preamble to action, it continues during action, allowing different role-players to acquire the knowledge necessary for action.

III. On-farm experimentation

In most agricultural African situations, technical changes are necessary. From the results obtained in agronomic research stations, experimentation in rural environments may help to design some adapted techniques.

IV. Monitoring and evaluation

Monitoring the actual running of scheduled actions, and frequently evaluating the results are two important functions necessary for the permanent adaptation of these actions and for addressing the needs and constraints role-players face.

- **PART TWO: Tools**

The second part of the handbook comprises three chapters corresponding to three main concerns that should be permanent in all programmes dealing with support to the farmers.

V. Training of farmers

Training farmers makes it possible for them to acquire the necessary skills for the implementation of their increasing responsibilities. This chapter discusses the information, technical training and management training respectively.

VI. The farmers' organisation

For the farmers to take charge of their economic activities (input supply, credit, marketing, processing of products, etc.) and of their increasing role in activities of general interest (advice, management of natural resources, etc.), they very often need to improve the level of their professional organisation. This is crucial to enable them to defend their interests, and to increase their negotiation power.

VII. Contracts between role-players (discussing contracts)

Smallholders cannot do everything, nor is it advisable that they do everything. It is therefore important for them to enter into contracts with other economic role-players (artisans, shopkeepers, tradespersons, industrialists, agribusiness role-players, etc.) and other institutions (technical services, NGOs, local authorities, etc.).

- **PART THREE: Fields of intervention**

The third part consists of seven chapters that correspond to several key areas of rural development.

VIII. Extension services and farm management advice

How does one communicate innovations? Extension services, which are usually used to propagate simple and uniform themes, do not respond satisfactorily to this question. Is it possible to make some improvement by using new means of information and technical training, or should one encourage extension services to evolve towards a farm management advice system capable of helping farmers to take decisions that are adapted to their particular situations?

IX. Natural resources management

How does one create conditions for a sustainable agriculture that will use valuable resources without destroying them? Is it possible to define, with the farmers, some regulations related to the use of the environment, based on the maintenance and regeneration of natural resources, so that the entire system may remain compatible with both their short-term needs and their constraints?

X. Product management

The modernisation of agriculture increases the number of products to manage and the channels through which they must go. The quality of product management does impact upon production. Inputs must arrive on time, products must find a good market and they should be sold at the right prices in order to generate a profit, and so on.

XI. Management of collective assets and facilities

Social collective assets and facilities are put in place in the villages. They include water points, sanitary facilities, and so on. Village people also acquire equipment such as motor pumps, mills. How should such facilities be maintained or replaced? This chapter will attempt to answer such questions and provide some general methodological orientations, while also developing a case study.

XII. Financing local development

Local development always requires financial capacity and the sound management of financial resources. There are many discussions around the diverse modality and nature of such management. This chapter is an attempt to present the key issues, and from diverse experiences arrive at some methodological orientations.

XIII. Women and development

Women have always played an important socio-economic role in the rural society. Although in the past many interventions for development considered women as nothing other than mothers and spouses, their economic role has become progressively evident.

Some kind of specific support for women is necessary. However, this does not mean that the approaches leading to segregation based upon sex are valid. The support for women must always be taken in a general context where men and women have specific but interdependent roles to play.

XIV. The non-agricultural sector

Supporting farmers cannot be limited to agricultural and agribusiness sectors. Rural people are not only involved in agriculture, but also in some other sectors (like small, micro and medium enterprises and industries, local markets, etc.) that exist or may be developed. Specific approaches are necessary, but which ones? How should these activities be connected to the cropping system, animal production, and so on?

6. How to use the handbook

The handbook does not provide answers to all the questions. As already mentioned, it is definitely not a collection of formulas and recipes, which in any case do not exist anywhere.

The diversity of situations encountered makes it impossible to develop one single method which might be applied everywhere.

Moreover, the debate has been left open on many issues (extension, saving and credit, collective management of productive equipment or privatisation, etc.) It has been necessary to report on existing controversy here and there. Who is right? Who is wrong? The differences in the reality are so great that no one can ensure that what has succeeded in one place will also succeed elsewhere. It is therefore advisable for the reader (e.g. extension officer, development operator) to reconsider and analyse his or her own practice in order to create his or her own methods of development and come up with some original approaches which will be well adapted to the situations he or she faces.

The handbook was not produced for reading from cover to cover. The reader is advised to select the most appropriate chapters, those related to the genuine questions arising from the nature of the work being carried out, its progress and the field of interest.

The following table serves as a reference to the 14 chapters:

PART ONE: PROCEDURES

| I. Local planning | II. Diagnosis | III. On-farm experimentation | IV. Monitoring and evaluation |

PART TWO: TOOLS

| V. Training of farmers | VI. The farmers' organisation | VII. Contracts between role-players |

PART THREE: FIELDS OF INTERVENTION

| VIII. Extension services and farm management advice | IX. Natural resources management | X. Product management | XI. Management of collective assets and facilities |

| XII. Financing local development | XIII. Women and development | XIV. The non-agricultural sector |

A sketch with comments is given at the beginning of each chapter. It refers the reader to the other chapters related to the issue in question. For example:

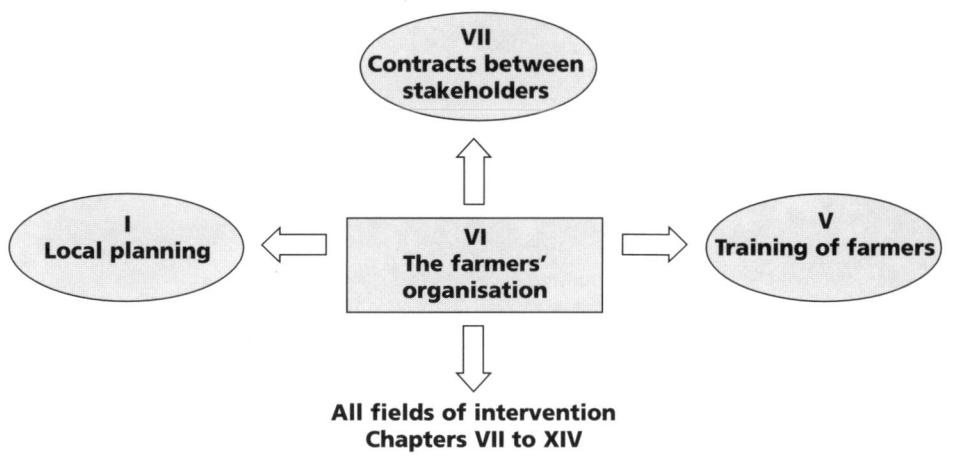

PART ONE:

PROCEDURES

I.   Local planning
II. Diagnosis
III. On-farm experimentation
IV. Monitoring and evaluation

Chapter I: Local planning
J. Berthomé, J. Mercoiret

Chapter I may usefully be connected to the following chapters:

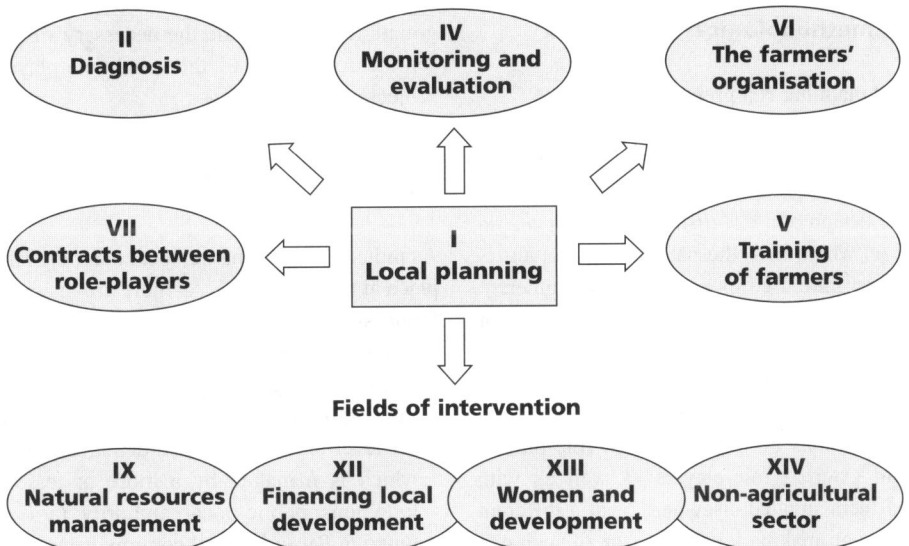

Nowadays planning is sometimes perceived negatively, since many past practices have contributed to its negative image. Planning has sometimes been imposed on the population, and bitter memories of this still remain. Indicative planning has not been more convincing and is equally disregarded by people because of its top-down approach.

Is the current context of economic liberalisation a reason to replace planning with spontaneous initiatives from the farmers and private economic role-players?

For several years, there has been the occasional interest in local planning associated with the concept of local development, for instance when there have been local decentralised authorities (e.g. a rural community in Senegal, a village in Rwanda). This is also the case in certain strong co-operative rural organisations that have a meaningful social and territorial basis (e.g. in Burkina Faso, Senegal; see Chapter VI: The farmers' organisation). Similar cases exist with regard to some developmental projects whose objectives imply a long-term impact, for example, the projects of local territories' management in Burkina Faso (see Chapter IX: Natural resources management).

Thus, in various cases, planning has been re-evaluated. Planning here means the *way in which the local role-players formulate a local developmental project so that it will serve as a guideline for the project in the short and long term* and will also be a reference for the co-ordination of the activities performed in different sectors.

> In Senegal for instance, the Ministry of Planning has recently been removed, and at the same time, one can notice a proliferation of local planning initiatives:
> - For many years, a federation of NGOs (FONGS), which includes most farmers' organisations and associations (see Chapter VI) has organised the training of rural managers in local planning.
> - A large farmers' group (Amicale des Agriculteurs du Walo, ASESCAW) in the delta of the Senegal River has developed a very ambitious triennial plan.

1. A methodological framework

Local planning is a process:
- It involves local role-players (farmers, shopkeepers, rural entrepreneurs, small-scale industries, etc. living in a given territory) in the design of a *mobilising project for the future*, where both the natural and human resources are used in a sustainable manner.
- This helps people to translate projects for the future into action plans and programmes they can manage.
- It strives to create conditions conducive to *negotiations* between the local role-players and external role-players as well as with partners situated elsewhere or at a different decision-making level. The role of such negotiations is to provide the necessary *support* for the realisation of the local role-players' projects and programmes.

1.1 The importance of local level

As indicated by the name, local planning takes place at the local level. As there are many different situations, each case should be considered individually:
- Local planning may take place within an administrative entity corresponding to the lowest level of administrative demarcation and which is managed by a more or less autonomous public local authority (a commune in Rwanda, rural community in Senegal, municipality in South Africa). In this case the context is both a precisely delineated geographical area and an easily identifiable administrative area.
- Local planning can take place within rural organisations involving many federated farmers' groups. The territory is then defined according to the level of involvement of the different villages in the organisations; its dimensions change depending on defections or new memberships. It is an economic territory with a developmental area, the boundaries of which are defined at a later stage, depending on the social basis of the organisation (see Chapter VI: The farmers' organisation).
- When these two structures do not exist, planning can take place within a small region. The identity of this small region may be based on geography (e.g. a catchment

> If the intervention is to be applied to a wide area (a large agro-ecological or economic region), it would be of value if the process of planning was implemented at first only in smaller units resulting from a delineation that takes into account the agro-ecological, economic as well as socio-cultural criteria.
>
> In most cases, it is preferable for the planning process to start at the village or community level. This may be determined sociologically or for legal and administrative reasons (as in the case of plot management in Burkina Faso). Grouping may be organised later, in order to discuss inter-villages issues (livestock management, land management, marketing and economic issues).

area, a valley): in this case, resources or common interests may serve as the basis for a planning process, which will in turn contribute to promoting them. A small region may also be based on historical or socio-cultural characteristics: in this case it is a cultural entity composed of localities, values and social practices that are recognised "as theirs" by the inhabitants even if there is no agro-ecological homogeneity. These cultural entities may become developmental areas and the planning process may contribute to this development.

Generally speaking, local planning must take place in a designated geographic and social context that is small enough for local role players to understand and in which they are able to manage and carry out actions efficiently.

1.2 Who can take initiatives?

1.2.1 Two possible cases: a demand or a proposal

- The initiative may come from a local source emanating from elected local representatives, the leaders of a rural organisation, or similar. Although still rare, this does occur. It happens when there are well-trained and informed local leaders who are interested in long-term actions; their initiative, in this case, consists in demanding external support (be it methodological and/or financial).
- Or, it can be an external initiative: a technical or administrative service, a NGO, or similar, can suggest to a local authority or to a rural organisation that it conduct a planning process in order to organise a time frame for its actions. In this case, the external role-players make a proposal to the local role-players.

In both cases the planning process can only reasonably be conducted when those in charge of its implementation have made a *preliminary analysis of the support proposal or of the support demand.*

The following questions may facilitate the analysis:

- Who is expressing the demand for support? What is the social, economic, administrative or political position of the group or the persons formulating the demand?
- What are the circumstances surrounding the demand; how did the demand come about?
- What are the reasons for the demand and the key issues involved? Is demanding local planning a way of adopting an existing project? Is it a way of realising the population's or its leaders' aspirations which have been dormant up to now? Is it a way of obtaining external funding?

The above analysis has many objectives. It makes it possible

- to identify the different local role-players who are (or will be) involved either in developing the local plan or in its future implementation.
- to evaluate the gap that can exist between the local leaders and the people. The local representatives and leaders of rural organisations may be fully aware of the necessity to define the orientation of work in the medium and long term. However, the people at grass-roots level may be preoccupied with the day-to-day problems (digging a borehole, obtaining a motor pump, building an area for drying products, or food supply).
- to see whether or not there are possibilities for real participatory work. (For instance, does the political and administrative context leave enough room for the local role-players to think and take real initiative?)

1.2.2 Co-ordinating the process

Co-ordinating the planning process is more often the role of those who initiated it:

- If the process of planning emanates from a rural initiative, it is the leader and the staff

members of the organisation who will oversee the process with, if necessary, external support.
- If the initiative is from outside, in most cases it will be the agents of those institutions who will co-ordinate the planning process. It is however advisable for those agents to seek the support of the local workers who might need to be empowered through training.

1.2.3 Building collaboration

The external support's first contact will be the local representatives or the leaders of a rural organisation, if the initiative started with them or is targeting them. If the context of the work is not institutionalised, the permanent contacts will be the officially recognised representatives (village chiefs, local leaders or any person delegated to this job).

However, in all cases, as many people as possible should be involved in the action without marginalising any social groups (women, youth) or other social minorities (small-scale entrepreneurs, shopkeepers, crop or livestock farmers, etc.). The permanent contact persons are the mediators between the support team and the population. It is really important for the population of the area to play an active role in the planning process because this will determine their future involvement in the implementation of the planned actions.

The other local role-players must also be associated with the action from the beginning to show, on the one hand, that their knowledge is valued, and on the other, to create an environment where their future contribution will be welcome. In fact, one should avoid the presentation of a fully developed and completed plan to the population, without any inputs, collaboration and validation from them. There are generally many potential local partners: local technical and administrative services, developmental companies, local offices of agronomic research, NGOs operating in the area or private economic role-players are just a few that can be mentioned.

In some cases, the planning process takes place in an area dominated by a small town. It will be necessary to take into account the town/village relationship and to involve those who work in the town but still maintain contact with the rural areas. These will include entrepreneurs and traders who sell agricultural products or secure supplies; small-scale industries and sometimes consumers.

1.3 Characteristics of local planning

The local planning process is not simply a technique for obtaining better knowledge of an area. It is meaningful only when it takes place in the context of a genuine decentralisation policy, for which the following conditions apply:
- The local level must no longer be considered simply as a basis for implementing developmental plans decided upon by external operators (natives or foreigners) on the basis of both external analysis and macro-economic objectives defined at national and regional level.
- It must be recognised that the local community and its different social and economic role-players have the right to take over the initiative, whatever its origin, in order to define the methods and orientations of the economic, social and cultural development as discussed with the external role-players. This is possible only when a genuine decentralisation of certain decision-making powers is guaranteed by the legal and administrative dispensations in a democratic state.
- Local role-players must have enough leverage to take economic initiatives resulting in the emergence and development of profitable and sustainable projects. This requires an open market and liberal economic context in which, however, the state may act as a regulator.
- To implement their projects, local role-players must have access to new financial resources (credits, subsidies). This requires suitable financing mechanisms (see Chapter

XII: Financing local development) and the decentralisation of public resources to local authorities (tax reform, autonomous management).

If these conditions are created and/or if there is a political will and dynamic to promote local development, the planning process can then be driven from a local level and become a means of
- defining an overall goal for the future, which should be built up according to the different role-players' interests, the existing convergences and the negotiated settlement;
- having a pluri-annual programme, which should be negotiated with the decision makers and the external partners (see Chapter VII: Contracts between role-players).

For the planning process to be efficient, it should have the following characteristics:
- It has to be based on dialogue, whereby the population and its different groups may find the opportunity to analyse the local situation, define its own developmental objectives and identify the actions it should undertake and the resources it should mobilise.
- It must go hand in hand with a multi-sectorial development approach. This implies, firstly, that all the local role-players must be involved in the process, and secondly, that analysing and developing the plan of action cannot be limited to the agricultural sector only (see Chapter X: Product management and Chapter XIV: The non-agricultural sector).

It should be noted that because there are so many rural people and because they are the least prepared to conduct local negotiations with external partners, they need a particular kind of support. They should not be seen only as small-scale farmers and/or resource-poor people, but rather as *economic role-players* (capable of producing, purchasing, selling, transforming, signing contracts, etc.) as well as *social role-players*, citizens, members of a rural community, with their own strategies, diversified needs and numerous aspirations.

The local planning process cannot be compared to a withdrawal, or to the quest for an illusive self-sufficiency or autonomy. It is rather an opportunity to create the basis for contractual relationships between community members, other local role-players and role-players situated elsewhere geographically (region, national level and foreign partners), or with other centres where economic and political decisions are made.

In fact, it is not a question of focussing solely on local rural communities. The planning process is about helping the latter to discuss their own future with other involved external role-players, with institutions that can help in the realisation of a developmental plan (e.g. research, technical services, local and regional administration, agricultural credit, etc.) and with sponsors (see Chapter VII: Contracts between role-players).

The planning process is about promoting thoughts and actions in the short, medium and long terms:
- The starting point for the process is the situation experienced by the local role-players and an analysis of the external and internal causes of such a situation.
- From this initial and consensual diagnosis, the planning process will identify some immediate actions (priority programme, emergency programme, motivating actions) which can in the short term solve the immediate problems; this in turn, will give some credibility to the process undertaken.
- The implementation of a priority programme makes it possible to maintain rural people's involvement and to define local developmental issues in the long run, from which an action plan for the medium term (four or five years) can be defined.

Co-ordinating a local planning process involves helping rural people and the most disadvantaged members of the community to

overcome their shortcomings and grievances (see Chapter V: Training of farmers). Co-ordination and support consist of
- giving methodological support and the necessary information so that rural people can analyse the socio-economic reality and see how it fits in with other geographical levels (relations with the town, the region or the province),
- helping rural people to establish a coherence between the problems they are intending to solve, the possible solutions, the resources available or to which they can have access. This means evaluating the agro-ecological resources, identifying the internal and external human resources, the existing and potential economic assets, the existing or accessible financial resources, and so on.
- helping rural people to define realistically the priorities as dictated both by the urgency of the problem experienced and the real chances of solving them (What means do they have? What kind of alliance can they organise with the external partners? What are the existing opportunities? How can they best be used?).

In other words, the process of local planning strives to answer the following questions:
  In view of the prevailing problems (based on the analysis and diagnosis) and the long-term objectives and issues, and given the possibilities and constraints related to the ecological environment, the economic, institutional and political context
- which actions should be undertaken immediately?
- which action plan should be established for the five years to come?
- which sectors should be involved in the action plan?
- which relationships and connections should be established between the different role-players?
- What are the human, organisational, material and financial means that should be mobilised inside and outside the local society?

2. General procedure and tools

In practice, the implementation of a local planning process can take different shapes, but whatever the case, the following general steps are advisable for setting up a local plan:
A. Preparation of the process
B. External diagnosis
C. Consensual diagnosis
D. Researching solutions, their means and conditions, definition of a priority programme.
E. Elaboration of the developmental plan, implementation of the priority programme

2.1 Preparation of the process

Preparation is generally divided into two parts:

2.1.1 Informing all the partners involved
This includes all the villages and rural organisations concerned (farmers' associations), technical and administrative services and all those who interact with the area regularly.
The themes discussed are the following:
- The origin and reason for the process
- The objectives (e.g. better co-ordination of development programmes, management of natural resources, reinforcement of farmers' organisations, local empowerment)
- The framework (i.e. time frame, steps, progress)

2.1.2 Training of local operators
These can be either community development agents already in place in the village, leaders of farmers' associations, or local development workers and so on, who will play an important role in the implementation of the process. The training can be about
- how to get an overview of the local situation;
- the objectives and steps of the process;
- indications of the possible relations to be established between community members and other role-players, and so on.

2.2 External diagnosis
(see Chapter II: Diagnosis)

The external diagnosis deals with collecting and collating the existing information, identifying the major problems and formulating hypotheses in order to prepare for the next step, the consensual diagnosis.
A number of methods can be used:
- Valuation of the existing information on the area, documentation, maps, interviews with key informants residing in the zone (e.g. teachers, development officers, shopkeepers)
- Observation of the landscape, local activities, the market, and other economic variables
- Agro-ecological zoning
- Identification of the socio-cultural characteristics and the social organisation (ethnic diversity and homogeneity, modes of existing organisations, sexual partitioning of work, etc.)

It may be necessary to conduct a short survey (the results of which will be used immediately) among the local leaders, farmers and other social and economic role-players.

2.3 Towards a consensual diagnosis

A consensual diagnosis aims to ensure the participation of community members in the analysis of their situation in order to come up with a diagnosis that truly reflects their situation. Such an appraisal should indicate
- the principal characteristics of the area, the changes observed and the current and foreseeable consequences;
- the problems and constraints confronting the farmers, the resources at their disposal, and the identification of the possible competition between activities, the conflicts concerning resources, and so on.

This will be fundamental in the search for solutions.

2.3.1 Reporting back the external diagnosis
This stage creates the opportunity for the farmers to start analysing their situation and seeking consensual solutions.

For the report back to be efficient, it must follow certain rules:
- Information reported must be processed and organised.
- It must be presented in an attractive and understandable manner to all local role-players, especially when the majority of local people are illiterate.
- In the villages, the report-back meetings should be organised for different audiences: extended groups (village, area), specific groups (e.g. women, local leaders, associations' leaders), homogeneous groups (e.g.

Boards with illustrations

Such boards may constitute a pedagogical support during the report back / appraisal to the community and other local role-players. The information must be organised in homogeneous topics, each topic being reflected on one or several boards (e.g. the demography of the area, its evolution and the consequences, or the existing projects in the area).

The illustrations must be clear (visible from afar), simple (without superfluous details; only the information to be represented must be shown), univocal (one idea per picture), adapted to the local situation (and therefore made locally). The colours used with moderation should facilitate the understanding of the illustration. A detailed pedagogic sheet must be prepared before the report-back meeting; this sheet should define all the objectives of the meeting and should recall the previous steps. This document should provide all comprehensive information on the diagnosis, while the illustrations on display should be only pedagogic support.

elders, men, youths, farmers, rural small-scale entrepreneurs). Report-back meetings allow for greater discussion with fewer constraints. A combined general meeting may also make it possible to report back the major points discussed in the small groups.

During a report-back meeting, the role of the developmental agent consists of
- presenting the diagnosis/appraisal of the situation by using illustrations. At the beginning, the agent must take time to get the public used to this technique and help the people to understand the illustrations. At a later stage, it is advisable to remind them, even briefly, of the connection between the idea and the picture;
- underlining the main issues pertaining to the situation presented (its origins) for each topic discussed and highlighting the causes related either to the rural people or to the natural and economic environment. The following should be done:
  - Encourage discussions about the accuracy of the analysis presented; seek out additional information, amendments, corrections or classifications.
  - Help prioritise the problems identified.
  - Initiate a search for solutions or ideas.
  - Provide information on the general economic environment (product prices, structural adjustment impact and available services, etc.).

2.3.2 Consensual diagnosis

This is initiated when a report back on the external diagnosis is made to the community and is followed by analysis meetings. These meetings take place in accordance with the modalities defined in consultation with the community members. They must allow the various groups and other social categories to express themselves concerning their perceptions of the situation, the analysis of the problems that confront them (causes, consequences) and the possible solutions (conditions, expected results).

2.3.3 Involvement of other institutional and economic role-players

Villages cannot survive in isolation, independent of structures and services existing in the region. The support of other institutions is indispensable to the villages. That is why structures and services must be involved in the report-back meetings and the analysis, as often as possible, in order to
- understand the real situation of the farmers and the logic underlying their decisions and acts;
- make an input or at least discuss the relevance of the analysis carried out by local people;
- inform the community about their own perception of the situation.

Development workers and extension officers must be familiar with the local analysis process. This will then help them in the preparation of and search for solutions. Moreover, some surveys must be conducted among the services and developmental structures in the area so as to identify
- the human, technical and financial resources at their disposal which can contribute to village development;
- the results available from technological and agricultural research regarding the agro-ecological zones;
- the actions already taken or earmarked for development, improvement and research in the area concerned;
- the proposals that can be made (as alternatives to local community solutions) to solve the dominant problems prevailing in the zone.

Community members also need to build or to strengthen their relationships with the local economic role-players. It not advisable for the latter to participate in all analysis meetings, but some surveys can be conducted among them in order to prepare for future meetings.

The consensual diagnosis made at this stage is not likely to exhaust all the subjects. It needs to be kept relatively short (somewhere between a few weeks and a few months). It should be finalised with a synthesis of the issues raised by small groups. This synthesis should also be presented to the whole community, in the presence of other local institutions and economic role-players.

2.4 Definition of a developmental plan and action programme in the short term

The orientations towards a strategic developmental plan and short-term action are defined following the presentation of results from the analysis meetings. Two results are expected at the end of this phase:
- A concerted definition of the long-term developmental framework (the master plan): this presupposes that the propositions, reservations and suggestions from institutions and community members expressed during the diagnosis have been effectively taken into account.
- The design of an immediate, short-term action programme: this is necessary to keep the community and other role-players thinking, because it provides the operational character of the ongoing process that rapidly results in concrete actions.

The content of this programme depends on the community priorities and on the available resources (available solutions, existing means, etc.) It is obvious that one should prioritise actions that have more chance of succeeding.

In the agricultural sector, such actions can be crop trials conducted in rural areas (see Chapter III: On-farm experimentation) with the purpose of testing the efficiency of a new technique or the coherence of a combination of techniques. This presupposes a discussion of the trial's objectives with the institutions involved (agricultural research). The priority may be an improvement in the supply of inputs or in marketing, which would require negotiations with agribusiness stakeholders, traders (urban wholesalers and retailers) and possibly consumer associations or representatives.

In some cases, priority actions concern the improvement of water supplies (which may re-

It is about participation, without sidelining any social group.

quire negotiations with public services, private experts in pumping, a company, etc.) or of the sanitary conditions in the village (which may require contracts with health services or a NGO for instance).

2.5 Implementation of the priority action programme and the definition of a concerted mid-term developmental plan

Realising an immediate, short-term action programme (priority programme) must come before any other consideration in order to reinforce the credibility of the actions undertaken. It is necessary to
- identify the tasks to be performed (division of the programme into specific tasks);
- identify collaborations or other contracts and negotiate them;
- divide tasks between the different role-players involved;
- give support to the farmers' organisations and their training programmes so that they are empowered enough to manage the tasks they have been assigned;
- establish a concerted monitoring-evaluation system that can make any necessary readjustment.

The implementation of a priority programme results in the creation of conditions conducive to the concerted elaboration of a mid-term developmental plan because it generates self-confidence among the local role-players and highlights the possibility of local-level alliances.

Building up a development plan means shifting from general objectives (e.g. improving livestock management, increasing staple food production) to sustainable action projects. This requires
- well-defined priorities (Where should one start? How can the operations planned follow on from one another smoothly?);
- actions being compatible with one another (e.g. compatibility between crop production improvements and sustainable livestock husbandry).

2.5.1 Defining priorities

Two factors must be considered when selecting priorities:
- The needs expressed by the community (the actions that they consider as priorities): for example, what they ideally want and what can be realistically achieved
- The resources, assets and external constraints (soil conditions, rainfall patterns, existing markets, opportunities offered by a trader or local entrepreneur, etc.) which de-

---

Example 1

Some community members want to increase out-of-season vegetable production for the market because the income generated might counterbalance the low yields recorded in dry land crops. This requires new cropping techniques, new investments, increased water supply for irrigation, water-supply equipment, fences, implements, input, and technical training.

However, there are many similar demands coming from other villages; the local market is limited and marketing is already a problem. What can be done?

- Either this priority must be discarded and one must focus on another activity (e.g. fruit trees, fattening pastures) if external conditions are suitable;
- or one must strive to link the increase in vegetable production to better use of this production by organising the supply chain (seeking new markets, establishing contracts with shopkeepers and traders, setting up some processing units and so on).

> **Example 2**
>
> In many sub-Saharan countries, the priority for farmers is often household food security because they have repeatedly faced food crop failures and shortages. Increasing cereal production then becomes a priority objective.
>
> However, by analysing the ways and means of realising such an objective, the difficulty in intensifying the production of one solely staple food is highlighted. As a matter of fact, the necessary equipment and inputs will be acquired and made profitable only through cash crop production (e.g. groundnuts, cotton).
>
> A more holistic approach is then required: How can the productivity of dry land crops be improved? What costs (in money and in labour) are involved in the suggested changes? What effects can be expected? (Agricultural calendar, cropping systems, mechanisation, plant-health measures.)

> **Example 3**
>
> In the lower Casamance area in Senegal, groups of women experienced problems in marketing their vegetable production. The presence of hotels in the surrounding area made them think of exploiting this market.
>
> However, the conditions set by the hotels were for them to be supplied from the beginning of the tourist season (November–December). The women collected rice right until December/January and it was impossible for them to start vegetable production before November. Thus the objective of improving vegetable production challenged that of improving rice production, seen also as a priority. Three solutions were possible:
> - Discard one priority. This was totally unacceptable for the women.
> - Some women could specialise in the early growing of vegetables. This was a problem for the married women.
> - Put the early vegetable production in the hands of the youngsters (girls and boys).

termine at a given time what is possible or reasonably likely to be achieved.

This does not mean making choices for the community or other local role-players, nor trying to convince them about the validity of a choice. It rather means talking seriously with them about what is wanted and what is possible and about what conditions must be fulfilled in order for the wishes to become achievements.

At this stage, mediation is often indispensable to establish a dialogue between the different local role-players. It should help challenge preconceived ideas, highlight complementarities and points of convergence, and establish negotiation and compromises (see Chapter VII: Contracts between role-players).

2.5.2 Looking for coherence between actions

To find solutions to the problems encountered, one is often obliged to adopt a sector approach: for example to seperate dry land crops, out-of-season activities, livestock husbandry, or health care. Some conflicts may then occur: different objectives, competition for time and labour and so on.

Above and beyond coherence per sector (which is the logical organisation of all the ac-

> **Other examples**
> - A forge, initially set up to manufacture light equipment for vegetable production (e.g. watering cans, shovels, rakes), can be used to maintain larger implements for dry land crops (ploughs, hoes, and so on).
> - An input store, initially designed for large-scale commodity farming, may develop a section dedicated to small-scale horticulture or gardening.
> - A system of savings and credits, initially devoted to agriculture and livestock farming, may finance some food-processing or marketing activities.

tivities scheduled for one field of activity), one should check the coherence between the various activities pertaining to different sectors.

This coherence may also be sought in the progressive expansion and creation of diverse activities within a given service at local level.

2.6 Implementing the plan

Implementation implies the use of specific methods and tools. This will be discussed in different chapters of this handbook. At this stage, the idea is only to underline some points.

2.6.1 Financing a plan (see Chapter XII: Financing local development)

A local development plan must be worked out. The cost of the different actions planned within a local development plan must be assessed, taking into account
- the direct contribution of community members (savings in the form of subscriptions, a realistic evaluation of the human capital investment),
- credit for the productive activities and equipment, be they individual or collective,
- the subsidies that are considered necessary and which should be used, if possible, only for heavy investments (small dams, infrastructures) related to public intervention.

It is crucial to raise funds for the implementation of the plan as this maintains the dynamics and fosters the enthusiasm generated by the planning process.

Some general indications:
- The principle of communal financial contributions (self-financing) to the project must be discussed as the plan is being drawn up.
- Contacts with the credit institutions must also take place during the design phase of the plan so that a basis of collaboration may be created in time. It should be noted that the specialisation of certain credit institutions that provide loans only for some activity sectors is a serious drawback.
- The time frame needed to obtain grants from private or public donors, national or international institutions is often too long. Comprehensive and early information often eases and speeds up the grants.
- Diversifying financial sources is sometimes a good idea as it limits the dependence on one exclusive sponsor, who may use this situation to exert pressure. On the other hand, by diversifying financial sources some complex co-ordination problems may be generated.

2.6.2 Programme planning

Programme planning consists of taking stock of (1) the actions that will be carried out during a given period and (2) the tasks related to these actions, in order to adapt the modalities of their implementation to the available human, material and financial resources.

A. Timing

An annual work plan is essential to avoid delays, overloading, unexpected peak work times, and sometimes the postponement of actions. The plan must take into account the constraints related to certain activities (cropping schedule, climate), but also social constraints (rural communities are sometimes immobilised for many weeks owing to certain ceremonies, e.g. initiatory rites).

The annual programme can later lead to quarterly, monthly and even weekly work programmes. It should be noted that the closer the period concerned, the clearer the programme planning must be (tasks to be done, persons involved, identified means).

The capacity to translate an annual programme into monthly programmes that are effectively carried out is an indication firstly of the existence of an efficient developmental strategy and secondly of the mastering of the programme by the local leaders. Conversely, a succession of spontaneous monthly programmes shows that efforts are dispersed and that external opportunities or proposals are very influential.

Delaying certain actions from one month to another must always lead to a thorough analysis. Those tasks that have been scheduled but not implemented are often a sign that there is a discrepancy between what is desired and what is realistic.

B. Programme organisation

Programming cannot be a solitary exercise carried out by external operators, development agents or the leaders of a rural organisation. If this is done, then programming becomes technocratic or even bureaucratic, since only a few people will be committed. This may also nullify all the efforts made during the planning process to have local role-players and more specifically the community take the initiative.

The initiative for programme planning may be taken by the organisations' leaders (if the organisation is well structured, significant and representative enough) or external operators supporting the realisation of a local plan. How-

Planning is about helping people to overcome a mere series of grievances, and to establish a structured plan to solve their problems.

ever, these initiatives should be made in consultation with the people concerned and who will participate in the scheduled actions.

The participation of local role-players in the programming process is important for several reasons:
- It helps avoid unrealistic forecasts and expectations since the community members will constantly bring to the fore any constraints and will require commitment from themselves and other stakeholders.
- It helps the preparation of the action and enables information to flow.
- The members learn autonomy through the multiple tasks that are to be performed in order to reach an objective or to overcome obstacles.

This supposes that programming is undertaken at different levels: wards, villages or communities, specific interest groups and also at regional level for those actions involving many villages.

C. General programme planning

General programme planning refers to the overall orientations, objectives and estimates of a local plan and takes account of appraisals from previous actions conducted. Programming is possible only if this appraisal has been done and if the actions to be carried out have been identified (in consultation with partners or beneficiaries).

The general programme takes stock of all the actions scheduled. Each action is broken up into major steps, which must take place at suitable times.

One or several persons must be specifically identified to take responsibility for each scheduled action and each step required for its implementation.

Experience has shown that the more precise the programme (Who should do what? When? With whom? With what means?), the greater the chances that the scheduled actions are performed effectively.

The development of a programme
- makes it possible to specify how work is to be organised;
- provides a common framework (platform) for all partners and vis-à-vis external observers. It also acts as a memorandum for all operators involved, so that the preparatory operations may take place in due time;
- will highlight peak work periods and establish the compatibility of actions planned for the same period with regard to human resource availability.

The general programme must be optimal. It is the result of a compromise between "what

Example

If a programme to intensify agricultural productivity is planned, and this requires the acquisition of materials and inputs, it is advisable to start the countdown from the day when all conditions are in place:

If rainfall is forecast for early July, it is advisable to have all materials and inputs ready from mid-June.

The inventory of all needs should be completed by March, so that all the ordering is done in time and the goods are delivered in June.

This means that the discussions with the farmers, in February, must include the modalities for acquisition of equipment (credit for instance), transport up to the villages, as well as the contracts with the traders, shopkeepers, local entrepreneurs and transporters.

would be nice to do" and "what is really possible to achieve", given the available resources. It is useless to design ambitious programmes if the resources required to implement them are hypothetical. This can only result in discouraging the different partners and discrediting the promoters.

It should be noted that the general programme must also include intermediary actions, for example informing the administration, making contact with the hydraulic services, requesting the support of an agricultural training centre.

D. A sectoral programme planning
Based on the general programme, each and every action co-ordinator must prepare a detailed and specific programme stipulating the exact tasks that are to be done, their order, the people involved, and so on.

A particular effort must be made to define, link up and organise the elementary operations. It is often useful in the beginning to help the co-ordinator of an action to identify the tasks to be performed, the time it will take to perform them and the required resources.

A programme is only efficient if a co-ordinator is responsible for mobilising, when necessary, the various operators.

Developing a budget helps make a programme realistic. It enables various expenses and benefits, such as those related to the implementation of actions and the recurrent capital depreciation expenses connected to certain investments, to be taken into account.

2.6.3 Programming and monitoring institutions

It is important that the major partners associated with the planning process form a programming team that meets regularly. This team is efficient only if it can monitor the actions conducted and be involved in their evaluation. This presupposes that a real climate of trust and an effective partnership between role-players exists.

In the daily implementation of the planned tasks, the nominated co-ordinators must enjoy total autonomy. Nothing is more discouraging than falsely delegated responsibilities, interferences, and so on.

This does not preclude the need for operational field co-ordination, which firstly ensures that the co-ordinators nominated are doing their job and secondly, helps solve unforeseen problems and arbitrates in cases of competition between actions.

2.6.4 Adjustment of the plan

The plan represents a reference for developmental actions for several years. However, it must not become a constraint, suppressing initiatives and opportunities that are in keeping with general orientations. The plan must remain open and flexible and may be adjusted.

Many factors can lead to the local plan being adjusted:
- The external constraints of the environment (e.g. drought, markets), institutional constraints (slow progress in the negotiating of a contract with an external partner), financial constraints, etc.
- Initiatives emanating from innovative local role-players (e.g. the development of a local agro-industry).
- Institutional opportunities (e.g. a health-care programme decided on at national level) or equipment opportunities, etc.

Every adjustment of the plan must, however, be carefully thought about in terms of the diagnosis and defined general orientations. It must also be discussed thoroughly with the different local role-players.

The monitoring-evaluation system provides the necessary information for adjusting the local plan (see Chapter IV: Monitoring and evaluation). It must enable the following to be identified:
- Those actions that were effectively performed and the practical problems encoun-

tered; the factors explaining success and failure; the observed internal and external dysfunctions, etc.
- The effects of the actions conducted and the possible discrepancies between objectives and results.

If there is continuous monitoring, evaluation must intervene at the end of each major phase (half-yearly, annual).

Monitoring and especially evaluation must be concerted, which means that
- external quantitative and qualitative evaluations are carried out by the local development agents and their external supports;
- the results of this external evaluation are reported back to different role-players involved in the programme.

This consultation makes it possible for the local role-players to continue in the diagnosis of their local reality and its environment. It also encourages internal debate and enables resources to be adapted to the targeted objectives.

2.7 Appraising the plan

At the end of the period initially defined for the implementation of the plan, a final appraisal must be carried out and reported back to the different role-players.

This appraisal must highlight
- the discrepancies between the actions planned and the actions performed, as well as the internal and external causes for this;
- the qualitative and quantitative results obtained;
- the lessons learned from this experience.

From this appraisal, which follows the procedure used in the development of the plan, a new plan can be defined for several years. This process (upward spiral) is possible only if the concerted evaluation/planning process is based on well-established results.

3. An example in Senegal: local planning and programming in CADEF, a Senegalese farmers' organisation

The Action Committee for the Development of the Fogny area (CADEF) is a farmers' organisation of 40 federative groups in the administrative sector of Sindian, region of Bignona, in Senegal.

Created in 1983 by villagers, and supported by some inhabitants residing in Dakar for the first two years, CADEF conducted, with its own resources, some sectorial actions in response to the priorities expressed by its members (development of market gardening, boreholes, health care, technical training).

3.1 Establishing a development plan

In 1985, CADEF requested external support from a NGO to help it establish a development plan.

Illustrations were used during the consultation process with the communities, as visual supports to the diagnosis, and to help define in a concerted manner the orientations for a local development plan.

3.1.1 Problems encountered in market gardening
CADEF farmers' groups grow market gardening crops on small plots (blocks), but there are many problems:
- Wooden fences are fragile and easily destroyed by animals that trespass thus causing serious damage to the crops.
- Little cropping equipment is available and the farmers have to wait their turn before using it.

- It is difficult to get sufficient good seeds (in quality and in quantity).
- Farmers are not well trained in market gardening production; they are totally defenceless when confronted with diseases. This results in low yields.
- It is difficult to market vegetables, because the commercialisation is not well organised.

3.1.2 Analysing hypotheses and solutions

Before attempting to improve production, it is important to consider the various possible outlets. The farmers are considering the following possible solutions:
- The creation of a weekly local market
- The introduction of conservation (drying) and processing techniques
- Planning production collectively in the different farmers' groups, so that not everyone produces the same thing at the same time
- A progressive replacement of market gardening by fruit trees as the fruit market is wider and the products less perishable
- Increased vegetable consumption

3.1.3 Possible technical improvements

Improvements can be expected in various sectors:
- Building new and improved boreholes
- Setting up new water-pumping systems
- Manufacturing of small agricultural equipment by trained local blacksmiths
- Planting of hedges with rows of *Jatropha*, *Anacardia* and *Citrus* being used as fences.
- A seed supply organised by CADEF
- Organisation of technical training for farmers, with the support of a training organisation, which will also provide specialised training to some development agents

3.1.4 Problems arising in the rice fields

The rice fields are affected by five factors: lack of water, salt, decrease in soil fertility, sand invasion and the presence of iron. These result in a dramatic decrease in productivity, which is aggravated by the fact that the cultivars are not adapted to wintering reduction. Manual ploughing (with the use of *cayendo* or *fanting*) can only take place when wintering is well in place. Rice is planted and rainfall stops before the seedlings reach maturity.

3.1.5 Possible solutions for rice fields

The challenge is to restore fertility in the rice fields (organic fertilisers and chemicals). Some early varieties of rice are adapted to the reduction of wintering. However, they are demanding in terms of fertilisers, they need deep ploughing and their productivity is related to efficient hoeing.

Building dams is envisaged to maintain water security in the valleys. This, however, requires preliminary and thorough investigation. A forest-planting programme in sloping areas is required to fight sand invasion.

3.2 Implementation of the development plan

This has resulted in
- the development of sectorial programmes that enable ideas to be translated into actions;
- a borehole programme (45 boreholes have been put in place);
- a market gardening improvement programme (creation of a forge, a fence-making unit, a shop for input supplies, technical training, organisation of marketing, drying trials and so on);
- a rice improvement programme (sub-programmes for rice field improvement, dams and intensification, e.g. research on adapted cultivars, resources management, agricultural equipment, motorised ploughing, management of water and soil fertility, management of the sloping lands).

External supports have been sought (inside as well as outside Senegal) in the technical and methodological fields (national agronomic research (ISRA), training, support to the organisation) as well as financial sources (local and international sponsors).

An internal support system has progressively been put in place with certain farmers with technical skills (management/bookkeeping, dams, seeds, plant health programme, credit) acting as local development agents.

3.3 Assessment of the plan

In 1989, the plan was assessed and discussed with the community during meetings organised at grass-roots level. The outcome and final achievements were compared with the expectations at the outset.

Debates and discussions highlighted the factors leading to setbacks or delays, unexpected problems or unplanned actions. This analysis resulted in another action plan and in an internal reorganisation of CADEF.

The work done by CADEF and its partners shows clearly
- the value of a planning process that can help to achieve clear operational objectives approved by the community;
- the complexity of the problems to be solved in order to reach the objectives;
- certain unavoidable mistakes owing to the fact that it is not easy to find the right solution first time round or the right way to implement the solution once found.

4. Conclusion

When a local planning process starts, it sometimes gives the farmers the impression that it is too theoretical, whereas they are confronted with very serious concrete problems that need to be solved. This may be true if the local planning is conceived externally and implemented as a formal exercise. On the contrary, when it is a participative process, it creates a vision, opens perspectives and results in initiatives. In order to achieve this, it is important to
- accompany the local planning process with immediate and motivating actions which correspond to local people's needs and thus give it credibility;
- obtain the necessary means to implement the plan (financial resources, efficient organisation, training, etc.).

This means that the administration, the public and parastatal technical services must recognise that rural people are competent and that they can play a role in programming developmental actions. Practices must be in line with official statements. The ability of official institutions to become partners with rural people and farmers' organisations (e.g. to the extent of making contact with them) is undoubtedly facilitated by the existence of local and decentralised authorities that are able to be mediators between rural people, other local economic role-players and the state.

Contractual relationships between rural communities and other local private stakeholders are often easier to establish, even though rural people need some support to carry out negotiations in conditions that are not too unfavourable. Once again, the creation or reinforcement of local authorities may be very beneficial.

5. Recommended literature

Anandajayasekeram, P. et al. (2000) *Agricultural project planning and analysis*. University of Pretoria, FARMESA, Pretoria, South Africa.

Bendavid-Val, A. (1998) *Rural area development planning: principles, approaches and tools of economic analysis*. FAO series: Training materials for agricultural planning. Rome, Italy.

Berthomé, J. & Mercoiret, J. (1993) *Méthodes de planification locale pour les organisations paysannes d'Afrique sahélienne*. L'harmattan, Paris, France.

Corbridge, S. (editor) (1995) Development planning. In: *Development Studies, A Reader*. pp. 64–77.

Cusworth, J.W. & Franks, T.R. (editors) (1993) *Managing projects in developing countries*. Longman, London, UK.

De Beer, F. & Swanepoel, H. (2000) *Introduction to development studies* (2nd edition). Oxford University Press, Cape Town, South Africa.

Holtz, U. (1995) Towards a new development paradigm. In: *Development and Cooperation*. Vol.6, pp. 4–6.

Mangin, J. (1989) *Guide du développement local et du développement social*. L'Harmattan, Paris, France.

Osborne, D. (1988) *Laboratories of Democracy*. Harvard Business School Press, Massachusetts, USA.

Pycroft, C. (2000) Integrated development planning and rural local government in South Africa. *Third World Planning Review* 22 (1): 87–102.

Selener, D. (1997) *Participatory action research and social change*. Cornell University, Ithaca, NY, USA.

Van der Ploeg, J. & Long, A. (editors) (1994) *Born from within: Practices and perspectives of endogenous rural development*. Van Gorcum, Assen, The Netherlands.

Chapter II: Diagnosis
P. Jouve, M.-J. Dugué, M.-R. Mercoiret, S. Perret

This chapter presents methods and tools used in the study of farming modes and conditions in developing rural environments. Methods and tools for the diagnosis of other aspects of rural society are discussed in Chapters VI, IX, X, XI, XIII and XIV.
For reference purposes, Chapter II may be connected to the following chapters:

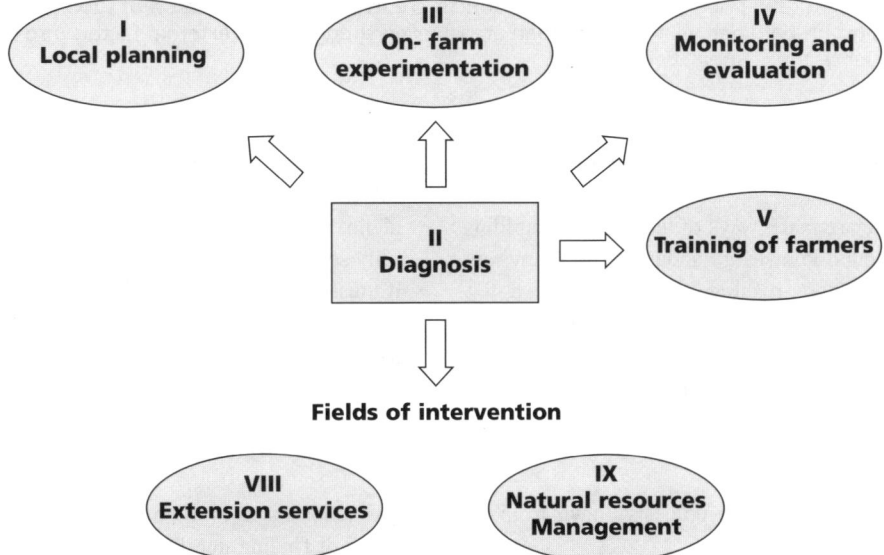

1. Important points

1.1 Knowledge as a prerequisite to action
The well-proclaimed objective of a developmental action, whether launched by public services, a NGO or a rural organisation is always to "improve the existing situation". Experience has however demonstrated that this "improvement" seldom occurs.

In many cases the failures, or short-term or limited successes are the result of unsuitable proposals made to the farmers. This mismatch may be caused by two factors:
- Those proposing innovations to farmers are unfamiliar with the environment in which they operate[1]. Consequently, they propose new techniques or new forms of organisation that are not necessarily what the farmers want. As a result, all the proposals miss their points.
- They do "know" local people and their practices but they do not take into consideration the economic context in which the community lives.

---

1. or act as if they are, as in the case of extension officers who keep preaching certain techniques because they have been told to do so, even though they are fully aware that these will not work.

Two scenarios are common:
- The farmers do not react as expected (or desired), they are reluctant to adopt the technical, economic or organisational suggestions made to them.
- The farmers adopt the new propositions but unexpected problems then arise, for example, they are granted credit but fail to pay back properly; they buy sowing equipment but do not use fertilisers; they build a local health care centre but cannot keep it operational for long.

This is, for instance, the case with some NGOs that have targeted a certain village or a group of farmers, but have shown little interest in the rest of the socio-economic environment (e.g. prices of agricultural products, credit system, organisation of supply, relationships between villages and urban areas), even though these external factors also influence the farmers' behaviour.

1.2 Learning about the rural environment

There are many ways of learning and building up sound knowledge about the rural environment before rushing into action. For a few years now, the word "diagnosis" has taken over from the previous "field studies", "preliminary analysis" or "monographs". Dictionaries usually define "diagnosis" in two ways, namely:
- The action of determining a disease based on its symptoms
- A prediction or hypothesis drawn from information or signs

Diagnosis is therefore from the medical field, referring to the act that guides the physician in firstly establishing the causes of a disease and secondly, prescribing the most appropriate treatment.

In a developmental context, diagnosis is a way of preparing for action:
- It must enable one to identify some of the problems the farmers are confronted with, the major constraints they are faced with, and the available resources that can help solve these problems. These problems can be identified based on certain signs and carefully selected indicators.
- It must be conducive to the identification of work lines and to the definition of priorities (see Chapter I: Local planning).

1.3 Proposed framework

The diagnosis is not just an exercise, or a simple way of finding out things, it is a prerequisite, a means of preparing for a specific action or an action programme.

To be efficient, it needs to be *reliable* (providing the most accurate possible image of the reality), *rapid* (so that the action may begin as quickly as possible) and *concerted* with different role-players concerned with the action.

1.3.1 A reliable diagnosis

A reliable diagnosis must be based on approved data collection methods and tools. The information collected must be processed and interpreted accurately, even if some of it is contradictory to the ideas previously contemplated.

The following procedures contribute to the reliability of the diagnosis:
- Describe[2] as objectively as possible the current situation with emphasis on the relationship that exists between the different observations made and the different issues noticed.
- Question the origins of the issues noticed and formulate hypotheses with regard to the factors and mechanisms underlying them.
- Test the explanatory hypotheses by making other observations (for instance, by comparing two opposite situations), and present them to the rural people to see their reactions.
- Formulate the final diagnosis, in other words, draw some conclusions (even provisional) in order to define the plan of action. The diagnosis should not only be descriptive but also analytical, both approaches being complementary to each other.

A diagnosis is reliable only when it takes into account different geographical levels and the interactions between them. In a rural environment, one can distinguish four of these levels:
- *The plot*, the cropped field: It is at this level that the farmer makes technical decisions such as how to plough when the land slopes, the planting density, and so on.
- *The production unit*, also called the farming system[3]: It is, undoubtedly, at this level that the farmer has to make the most decisions: the importance attributed to the agricultural sector, to livestock farming and to other activities; the priority given to certain cultures on the cropping schedule, choice of equipment, and the management of labour, and so on.
- *The village or the community*: This is often an important geographical level in Africa, as certain collective decisions are taken at this level, for example grazing camps and cropping areas, water supply and anti-erosion measures, relationships with non-agricultural sectors, and such like.
- *The region*: It can be defined as a social, geographical and cultural space where resources are used according to practices and regulations often resulting from history. This level may be important in development because certain decisions concern many villages (resource management, e.g. hillsides, inland valleys, forests, water resources, etc.) (see Chapter.I: Local planning).

Diagnosis must strive to describe the practices at each geographical level corresponding to a different level of decision making (Who is the decision maker? What is he/she supposed to decide?) and to the interactions between the levels.

For instance, the farmers make a number of decisions inside their own production units but these decisions are not entirely independent of other decision-making levels. The choice made may be influenced or determined by the existence of higher decision-making centres (the village, the region). These units represent a geographical delineation of the space. In fact, it is often useful to associate a geographical space with the way the land is used in order to define a scale of work.

Lastly, it is important to remember that farmers and the geographical space in which they live are all part of a more extended eco-

---

2. Methods and tools are discussed further.
3. The term "farm", as defined in a western agricultural context, refers to the existence of a unique decision maker and the strict co-existence of dwelling, production and consumption units. Compared with the African family rural organisation, this is not always the case, that is why the term "farm" has to be used cautiously, with accurate definition (see further in this chapter, and also Chapter VIII: Extension services and technical advice; and Chapter XIII: Women and development).

nomic and social system (the region, the country or even the global environment). The farmers depend on decisions made at these levels (national and regional policy, international markets) and these may have a favourable or unfavourable impact upon their behaviour (product prices, fertilisers, conditions of credit, etc.).

1.3.2 A concerted diagnosis

Consensus is the second condition for an efficient diagnosis.

The perception one has about a given reality is always influenced by personal references resulting from experience, training and one's nature and identity.

Therefore, it is not sufficient for an external development operator to make a diagnosis. His/her analysis and his/her opinion must be compared with those of the local people so that the gap between the external and internal impressions is reduced.

At this point, one should distinguish between

> **An example**
>
> In a Sahelian region it is observed that women spend more and more time collecting firewood. They have to collect it from afar since good quality firewood is scarce in the vicinity and the situation is worsening from year to year. The farmers have noticed a tremendous decrease in the size and number of trees (aerial photos prove it).
>
> - First hypothesis (favoured by the farmers): drought has increased tree mortality (the dead wood has been quickly consumed), and has compromised vegetation regeneration.
>
> In fact, further observations in the area (made on the ground and from aerial photos) show that, even in the lowlands (which should normally be less sensitive to drought), the tree cover has dramatically decreased.
>
> More extensive investigations are therefore required and it is clear that some other parameters which are aggravating the problem must be taken into account, such as:
>
> - The cultivated areas have increased (as a response to demographic growth and reduced land productivity).
> - The grazing areas have decreased in surface. They are now far away from the village; moreover, they are less productive because of drought and overgrazing.
>
> Because of the above factors, the pressure on the tree cover has increased (severe browsing on forage trees at the end of the dry season, young shrubs and shoots destroyed or damaged by animals) with two major consequences: tree mortality remains high and regeneration is impaired because rainfall is hardly sufficient for certain plant species.
>
> Lastly, in the cultivated fields, the regeneration is difficult as young shrubs are often chopped by hoeing.
>
> Depending on whether the first hypothesis has been considered or the diagnosis has been further extended, the action proposals will differ.
>
> In the first case, one would be inclined to import tree species that are less demanding in terms of water (and which are possibly not appreciated by livestock) while the problem is, in fact, wider than that.
>
> - One should improve the fodder plan to reduce the pressure on existing vegetation.
> - Regeneration should be supported (identification of shoots, protection against animals).
> - The location and type of reforestation should be discussed. (Scattered or limited on plots? On which kind of soil?)

- *external diagnosis*, which is made by one or more agents who are external to the system investigated, is based on data collected using specific tools (surveys, observations) and is analysed in comparison with other realities (representations);
- *auto-diagnosis*, which is established spontaneously or stimulated by external operators, with the role-players being involved in the system themselves (meaning the local people).

Generally, the conclusions from each side differ.

> This is particularly true when it comes to subjects like natural resources management (e.g. cutting of wood). The short-term objectives of the farmers (fulfilling their needs for firewood or wood cutting) may still be satisfied and therefore lead them to a quite optimistic diagnosis. A long-term study on the other hand may depict an alarming imbalance and demonstrate that there will be a severe shortage in the resources in the long run if corrective measures are not taken.

This confrontation leads to a *concerted diagnosis*. It means
- presenting the external diagnosis to the local people;
- discussing the external diagnosis in an open manner (true dialogue), the aim of which is neither to convince local people that the external analysis is right, nor to accept without criticism the views of the local people;
- discussing together a concerted (consensual) diagnosis, which will be the starting point for prospective work.

1.3.3 A quick and permanent diagnosis

It is important to distinguish the initial diagnosis from the more thorough one that will take place continuously throughout the plan or programme.

The initial diagnosis must be made immediately, before any action. This is to make sure that
- the actions planned will still be valid and relevant when they are carried out (rural societies are, for many reasons (climate, demography, economic context), in perpetual evolution; a diagnosis is a reflection of a situation at a given time and its results are valid only for a limited period after it has been made);
- local people are not discouraged by endless surveys, the results of which are not forthcoming.

The duration of the initial diagnosis will vary according to the objectives and the size of the field of intervention. Depending on the case in question, the duration will vary from a few weeks to a few months.

The diagnosis must continue throughout the implementation of the action.

All actions conducted in an intervention context (e.g. experimentation, technology transfer, new organisation) give indications which can help to increase one's understanding and diagnosis (appearance of new constraints, expression of new demands, resistance, etc.).

The data required for in-depth and permanent diagnosis will be collected through monitoring and evaluation (see Chapter IV). That means that monitoring and evaluation, information systems and data processing mechanisms must be in place from the beginning.

Several methods and tools are proposed later in this handbook for the study of modes and conditions of agricultural production in rural areas, from regional to farming household level.

Methods and tools for the diagnosis of other aspects of rural society (types of organisation, natural resources management, product management, collective equipment, the place and role of women in production and rural society, the non-agricultural sector) are discussed in Chapters VI, IX, X, XI, XIII and XIV.

Some useful definitions about agricultural and rural systems with regard to diagnosis (compiled by S. Perret):

- **Crop management sequence:**

"A logical and methodical combination of cropping techniques applied to a given plant species to manage the environment in order to achieve a given production objective" (Sebillotte, 1978).

- **Cropping system:**

"The range of technical methods used on fields treated in an identical manner. Each cultivation system is defined by two criteria: the nature of the crops and the cropping sequence (which used to be called crop rotation), and the crop management sequences applied to these different crops, including the choice of crop varieties" (Sebillotte, 1990).

- **Livestock production system (livestock system):**

"A range of interacting elements and technical methods organised in such a way that resources are managed to gain different products from animals (milk, meat, skin and leather, traction force, manure, etc.) or for other objectives" (Landais, 1992).

"A group of techniques and practices carried out by a community to exploit the plant resources in a given space by the animals in conditions compatible with the objectives and constraints of the environment" (Lhoste, 1985).

- **Farming system (agricultural production system):**

"The coherent combination, in time and space, of the means of production (land, labour force, equipment, capital) dedicated to plant and/or animal production" (Dufumier, 1987).

"A structured ensemble of the means of production ... combined together in order to provide crop and/or livestock products with a view to satisfying the objectives of those responsible for production" (Jouve, 1992).

"A goal-oriented system encompassing the family and open to its environment" (Ruthenberg, 1981).

"A unique and reasonably stable arrangement of farming enterprises that the household manages according to well-defined practices in response to a physical, biological and socio-economic environment and in accordance with the household's goals, preferences and resources. These factors combine to influence output and production methods. More commonality is found within the system than between systems. The farming system is part of larger systems, e.g. the local community, and can be divided into subsystems, e.g. cropping systems." (Shaner et al., 1982).

"A specific farming system arises from the decision taken by a small farmer or farming family with respect to allocating different quantities and qualities of land, labour, capital and management to crop, livestock and off-farm enterprises in a manner which, given the knowledge the household possesses, will maximise the attainment of the family goals" (Norman, 1980).

"The total of production and consumption decisions of the farm-household including the choice of crops, livestock, off-farm enterprises and food consumed" (Byerlee et al., 1980).

> - **Agrarian system:**
>
> "The association in space of the products and the techniques used by a society in order to satisfy its needs. It expresses in particular the interaction between a bio-ecological system represented by the natural environment and a socio-cultural system, through practices stemming principally from technical experience" (Vissac, 1979).
>
> "A historically constituted and sustainable manner of utilising the environment, a system of forces of production adapted to the bio-climatic conditions of a given space and responsive to the social needs and conditions of the moment. The internal coherence of the manner of utilising the environment is related to the broader technical, economic and social conditions of production" (Mazoyer, 1979).

2. Diagnosis of the conditions and modes for agricultural production and their effects on the environment: from regional to plot level

Based on experiences, new strategies on development have shown the necessity for diagnosis as a prerequisite for any intervention in a rural area. The recognition of this necessity has resulted
- firstly in increased attention regarding the study of modes and conditions of agricultural production in rural areas;
- and secondly, in the development of methods and procedures which have in turn been debated to such an extent that the practice of diagnosis itself has been questioned, especially when it takes too long or when it is not yet finalised.

It should be remembered that a diagnosis is a judgement made during a short period of time on a situation or a state in order to guide an action. It is in this context that it will be discussed here, even if, in real terms, the time required for diagnosis in a rural area may, in certain cases, correspond to the agricultural year (cropping campaign) which can be considered as a reasonable duration.

This is quite understandable if one considers that rural areas are complex environments that can be studied in different ways depending on which viewpoint is adopted (agronomic, geographic or sociological) and especially on the purpose of the diagnosis.

The result is that one cannot suggest methods for rural area diagnosis without specifying this objective and the major priorities. The choice that we have made here is to present a diagnostic procedure capable of guiding rural development interventions. By this we mean that the end result of such a procedure will be the improvement of production patterns and the development of a rural area (i.e. improving people's living conditions and satisfying their needs). Such objective can be achieved by
- defining the prioritised intervention themes in the context of supportive programmes for farmers;
- formulating technical and managerial advice in order to provide supportive programmes with a technical content adapted to the farmers' needs;
- orientating applied research programmes in order to avoid such programmes being steered from the outside and showing no correlation with local and regional realities;
- proposing economic, social and institutional measures to ensure a more satisfactory utilisation of both the rural area and its resources.

The point of view that we have adopted is a result of both observation and options. Role-players in rural areas are first and foremost farmers and members of rural communities to which the farmers belong. This option is not solely a reflection of the new orientation of agricultural policies that are being implemented and that are characterised both by state disengagement and increased autonomy for rural entities, but also refers to the fact that the success or failure of any intervention in a rural area depends on the decisions made by farmers and their level of compliance with the suggestions made to them.

From these choices and observations a number of issues need to be addressed:
- We will attempt to diagnose the way in which farmers manage the natural resources, how they mobilise such resources, the means required for agricultural production and its effects on the environment, as well as how they manage animals and crops, and so on.
- To achieve this we will study the conditions of agricultural production and its effects on the environment, be it physical or human.

However, we will not limit ourselves only to these conditions as they are insufficient to assess agricultural production. Following this we will concentrate on the modes of agricultural production in a given context. From this stage, we may make judgements and recommendations on the ways in which farmers manage their resources and perform both crop and animal farming.

3. Methodological orientations

3.1 Agricultural practices: main targets for diagnosis

The ways in which farmers manage their productive means and activities are called agricultural practices. These practices will be at the

Studying agricultural practices first requires thorough observation and description.

centre of the diagnosis. Agricultural practices are quite diverse. How should they be studied? How should one draw a rural development diagnosis based on such practices?

3.1.1 Steps in the study of agricultural practices

A. Observation and description

The study of agricultural practices first requires one to observe and to describe these practices. It is an elementary precondition which is far from being the rule. What is known about the social management of resources such as land, water and grazing areas? What do we know about the successive cropping systems carried out by the farmers, the displacement of livestock and the way it is fed? Most of the time not much is known. Yet, these are precisely the practices that the developmental interventions are intended to transform. The consequence of this ignorance is that, rather than studying the proposals for innovation from the realities observed, one often determines them in advance according to the following considerations:

- General and normative, for example that the land must be ploughed before sowing. Whereas, if one observes the state of the environment, and mobilises technical background information on ploughing (e.g. impact, reasons, constraints, costs), this operation may be avoided in certain instances. Another example may be that millet must be sowed at a density of 10 000 seed holes per hectare. In reality, the Sahelian farmers' sowing practices are very varied and depend on the soil type, the soil fertility, and the distance between plots and households. A general norm has little meaning for them.
- Subjective, resulting from the external operators' personal experiences, which lead them more often to think by analogy, rather than to take into account the specificity of each situation.

B. Understanding and evaluating

Once the agricultural practices have been observed and described, one has to analyse them. This analysis can be made in two complementary ways.

The first way consists of analysing, from the farmers' point of view, the reasons that lead them to decide on such or such a practice. The aim of this is to find out the farmers' rationale behind their agricultural practices. In this way, the constraints impacting on the farmers' decisions, the opportunities they take and finally the objectives and strategies they follow may be put on record.

On the other hand, a study of the way farmers work (especially women farmers, since in most developing areas women are the ones who perform the majority of farm work) will reveal their skills and the immense capital of empirical knowledge accumulated over a long time. It will be in the interest of the development support services to make good use of such knowledge. This internal analysis of practices documents the qualitative factors of a technical, economic, social, cultural and historical nature. It is a demonstration that innovation and changes in agricultural practices are not merely about the transfer of technology.

> One can evaluate the performances of different cropping practices by analysing their effects on the yield, which is basically an agronomic analysis. One can also evaluate this practice from an economic point of view by comparing the labour productivity, the added value and the income thereof. Other points of view may be adopted in evaluating agricultural practices. If protection of the environment is the major concern, then one will be bound to study the impact of these practices on the conservation and renewal of natural resources.

> Thus, when the farmers are faced with adverse production conditions, be they climatic or economic, their objective is more to minimise these hazards than to strive to maximise production. This risk aversion approach of farmers' strategies is particularly important in arid areas and for farmers who do not benefit from a protected market. It explains their reluctance to embark on intensification programmes suggested to them by development services.

The second way of analysing practices is to evaluate the results and their efficiency. Such an evaluation can only be made in comparison with an objective.

Once agricultural practices have been identified, it is necessary to combine the internal analysis, which provides the role-players point of view, and the external analysis, which enables the efficiency of these practices to be evaluated. The evaluation can lead to contradictory conclusions as to the validation of these practices, depending on whether the short or longterm is taken into consideration and also depending on the interest and objectives of the various role-players (farmers, owners, renters, local authorities, state, sponsors).

3.2 Structuring diagnosis

To structure the diagnosis of the modes for utilising the agricultural environment, we will start from the premise that the practices are the result of implicit or explicit decisions made by the farmers. The analysis of these decisions helps to distinguish between the different levels at which production is organised in the rural environment.

The different levels of organisation that are worth mentioning are:
- the region: this is where most developmental interventions are defined[4];
- the village, or its equivalent (rural community): this corresponds to the human and territorial community utilising one space and the resources therein;
- the farm, or the production unit: this, despite the large diversity of its organisation, remains an important decision-making centre, its diagnosis is indispensable in evaluating how production systems adopted by the farmers work;
- the farming plots and grazing areas: it is at this level that the diagnosis will be carried out to evaluate the efficiency of the farmers' technical practices and to identify the possibility of improving such practices.

These levels of organisation are interdependent. The way a village operates depends partially on the way the farms operate. Conversely, the way a farm operates depends on a certain number of rules and characteristics common to the villages where it is located and particularly in as far as the management of natural resources is concerned (e.g. grazing areas, forest, water).

Diagnosis in rural areas must therefore take into account the different levels of organisation, and the different levels of analysis. But must it proceed from local levels to upper levels or the reverse?

Generally, the diagnosis of a rural area starts at the regional level and finishes at plot level, passing through village and farm level. This pathway is unavoidable for obvious rea-

---

4. Beyond the region, there are other levels of organisation e.g. national entities, bigger agro-ecological zones, global markets, etc. Although these levels also have an influence upon the functioning of the lower levels, we will not take them into consideration in order to limit ourselves to the spaces and systems corresponding to the usual level of interventions in rural areas. The upper levels are considered as an external environment to the systems studied.

sons related to the necessity to stratify different levels of studies. Thus, the diagnosis of farms cannot be done on all the farms within a region. It is necessary to make a selection. Sampling can be done randomly, although a large number of farms will be required to cover the range of farm types. On the contrary, the number may be reduced if sampling is stratified according to the diversity of the organisations at upper levels (region, village), seeing that the degree of interdependence between levels and the differences in the environment or in the type of village imply differences in the way farms run.

Before discussing the methods, it is important to bear in mind that each level of analysis corresponds to a particular objective of the diagnosis, and that it is by combining these different objectives that one can reach the overall diagnosis of the rural environment. In the same way, each level of study, be it the zone, the village or the farm, requires specific tools and methods for diagnosis. In the context of this presentation, it is not possible to detail all these tools and methods. More information on tools and methods can be obtained from the documents mentioned in the bibliography.

For each scale of analysis, the objectives of the diagnosis will be discussed and the general procedures to follow indicated with an emphasis on the tools and methods for investigation that are most suited to the specificities of the sub-Saharan countries of Africa.

4. Procedures and methods for diagnosis

4.1 At regional and village level

The region, as a geographical entity, represents a space delimited by a combination of physical, historical and ethnic factors. This space can extend across many different countries (a large river valley that crosses borders, for example). At this level, the final aim of the diagnosis will be to prepare a master plan for development or improvement, reinforced by a decentralisation policy. Diagnosis must lead to the determination of strategies and main lines of intervention, taking into account the physical and human characteristics of the environment as well as the farming practices in the area.

4.1.1 Stratification and zoning of the area
Generally the area is heterogeneous. It is therefore necessary to take into account this heterogeneity when developing the intervention plans. This requirement leads to the concept of zoning the area. However, "zoning" means nothing if we do not define what is going to be zoned.

The aim of the first zoning will be to characterise the conditions in which resources are utilised for production purposes in a given area. One will strive therefore to stratify the environment according to characteristics that are supposed to impact upon agricultural production. Among these characteristics, one can distinguish those related to

- the natural environment (soil, vegetation, climate), which determine not only the choice of production systems, but also production techniques that play a role in improving the environment to make it more favourable to agricultural production;
- human factors, that is population density and dynamics, social and administrative organisation and support structures;
- the agrarian system, which highlight the partitioning of production means, especially the land tenure systems, between the different farms;
- the socio-economic environment of the agricultural sector, including communication networks, transport, markets, agro-industrial plants, agriculture support services, and the like.

The combination of these different characteristics will help to design a preliminary stratification of the intervention zone which corresponds to an "agro-socio-ecological" delimitation. This stratification makes it possible to identify areas where it can be assumed that the production conditions are relatively homogeneous.

4.1.2 Regional diagnosis on the modes of utilisation of the environment

At this stage, zoning cannot help make a diagnosis on the utilisation of an area. It is necessary to look at how the area has been utilised since the same area can be used in many ways, depending on the history of its inhabitants, its social organisation, and its socio-economic environment. It is also advisable to study the practices and modes used by the farmers to exploit an area. This should be done within a reasonable time frame stipulated when processing a diagnosis.

The concept which is particularly suited to making a diagnosis at regional level is the agrarian system. This can be defined as a result of the association of production and techniques developed by rural society to utilise its area, manage its resources and satisfy its needs. Establishing a regional diagnosis based on the agrarian system is aimed at understanding the running and dynamics of the social, technical and economic organisation set up by rural society over time to utilise and manage its environment and resources. The development of such a diagnosis raises a number of practical and methodological questions. Firstly, how can an agrarian system be defined, and what is its territorial delimitation?

Looking at the landscape facilitates this identification. Agrarian landscapes are the visible expression of the way a rural society has organised, utilised and occupied its environment. This occupation may not be apparent, as in a region where the primary productive activities are characterised by nomadism and gathering (desert area). Conversely, occupation is extremely dense in very populated areas (e.g. central Africa, south east Asia) where the natural landscape has been replaced by a totally man-made agrarian landscape.

At regional level, describing the agrarian landscape in cultivated areas enables one to identify the major types of agricultural systems adopted by farmers. This identification is an essential step in characterising the modes for utilising an environment.

Another means of identifying an agrarian system consists of investigating the social, technical and organisational regulations that apply to all production units and determine the agricultural exploitation of an area. These "rules" that are generally implicit and inherited from the technico-cultural patrimony of a rural society concern the management of natural resources, in particular land, as well as the social organisation of labour, equipment, animal husbandry, or crop farming practices.

The territorial expansion of an agrarian system will correspond to a space that displays a certain homogeneity in terms of agricultural landscape, rules and practices, and modes of production. As agrarian systems are of historic and social construction, it is not surprising to see that their expansion in a number of underdeveloped countries is strongly correlated to the geographical distribution of certain ethnic groups. However, the reality is usually more complex and the utilisation of a given area more often results from the association of different ethnic or social groups.

4.1.3 Steps for regional diagnosis

The first step consists of identifying the elementary territorial and human unit, meaning the level at which the characteristics of the land use modes and the local agrarian system will be analysed and identified. In many regions of the world this unit is the village. It can also be a valley in a mountainous area (Andes, Nepal, Atlas heights), a hillside in Rwanda or a hydraulic sector in a large irrigation scheme.

Thereafter, by using the pre-established stratification of the conditions for agricultural land use in the region where different agricultural situations have been identified, a sample of villages (or any elementary analysis unit) will be taken in an attempt to study regional diversity and modes of land use in the area. The number of villages chosen will depend on how much time and what tools are available to carry out the survey. Experience has demonstrated that a survey of 20 villages is sufficient to cover such diversity.

At the same time, a questionnaire adapted to agricultural situations and to the territorial elementary units chosen can be used for the investigation. In the case of a village, this questionnaire is generally structured around the following themes:

- The history of the village, its population and agriculture (dates and reasons why certain crops or technical innovations disappeared or appeared)
- The spatial organisation, the different soil types, the plots, distribution of land, gardens, housing (schematic plan of the village or a transection)
- The social organisation of the village and production units, including the different authorities, the traditional and professional organisations
- Production and natural resources management modes (labour, land, water, forest, grazing area)
- The production systems, that is cropping systems, livestock husbandry systems, fishing, gathering
- The village's socio-economic environment

Since the diagnosis is made from a development perspective, the themes will be completed by a survey on the community's problems and wishes.

The aim of this survey is not to make a monograph of the village but to have an overall and synoptic understanding of its operational mode as well as the way farmers utilise the environment at village level.

In this process, the village is not merely the sum total of the farms and households that it groups together, but a human and territorial entity with its own identity and its own coherence. In an effort to transfer this methodological choice to the countries where this process was applied (Niger, Togo, Senegal, etc.), the concept of a village agro-system was used to underline in a synthesised way the operational mode of the villages investigated.

In the context of a regional diagnostic, village surveys must be short (2 to 3 days per village visited, if possible, twice). As in any diagnostic process, the survey progress relies on pre-selected criteria and indicators. This presupposes some minimal knowledge of the agrarian systems investigated and the formulation of hypotheses about the functioning of the village agro-systems surveyed. In order to achieve this, it is necessary to acquire information of previous studies conducted in the region and draw conclusions from them.

Surveys rely partially on the spoken word. The reliability of responses can be cross-checked with secondary information on the overall functioning of the village agro-system.

> In this way, youth emigration is usually related to pressures exerted on land. If emigration is high despite adequate availability of good land, one should investigate the causes of such movement (e.g. opening of mines).

Surveys require not only knowledge of previous studies conducted in the area, but real intellectual curiosity. That is why the implementation of this kind of survey cannot be delegated to inexperienced officers. On the other hand, surveys are mainly collective, and the diverse or even contradictory points of view that appear can be used to conduct deeper analyses.

A survey based on verbal information must be systematically associated with direct observation of the area, which means that one should walk across the village following one or several different transections to see the different types of spatial utilisation. The use of aerial photography, whenever possible, may facilitate the identification of the community territory and its occupation. This type of fieldwork is indispensable to complete and eventually to correct the information collected in surveys.

To facilitate the analysis of surveys at the level of the village chosen, certain indicators have to be selected to make a comparative study between the ways in which different village agro-systems work. These indicators may be quantitative (cropping and fallow land schedules, duration, percentage of farms using animal traction, etc.) or qualitative (presence of lowlands, of rural small-scale industries).

The step that follows the completion of field surveys is their interpretation with a view to analysing the nature and diversity of land use at regional level. To concretise this analysis of diversity, it is extremely useful to establish a typology of the village agro-systems (or of any other elementary unit of investigation).

This typology is based on the constitution of groups of villages, possibly subdivided into sub-groups having certain homogeneous features in their operational modes. The establishment of this typology is altogether a product of and an important step in regional diagnosis. It allows one to
- understand the geographical diversity and historical evolution in land use patterns and identify their principal determinants (population density, introduction of commercial crops, etc.) (From this point of view, one should note that in many regions of sub-Saharan Africa the analysis of regional diversity in agricultural land use gives one much information on its historical evolution, since, within the same region, different agricultural situations corresponding to different evolutionary phases of the regional agrarian system may coexist.);
- pursue the stratification of the area by providing an objective base for the selection of a sub-sample of villages that are representative of the main village agro-systems identified.

Then further diagnosis may be conducted on other levels of rural organisation, and in particular, the farms. This typology will also make it possible to select rationally a limited number of pilot villages where, in the context of development projects, experimental actions will be conducted in order to provide references for interventions at larger levels.

At this stage, the problem arises of how to extend geographically the results obtained from a sample of villages (limited to 20 or 30) to all the villages in a region (many hundreds). With regard to this problem of scale, two scenarios can be distinguished:

Either a good correlation is found between the different types of agro-systems and the agricultural situations identified during the preliminary zoning of the region. In this case, one considers that the geographical dimension of each type of agro-system is represented by the zone where it has been noticed. This situation highlights the fact that the characteristics selected for the preliminary zoning of the area strictly determine its land use. Such a situation, although possible, is unlikely to be systematic because the analysis of the village agro-systems usually results in the appearance of other determinants originally ignored or overlooked during the preliminary phase of the diagnosis.

Or conversely, there is no correspondence. In this case, the village agro-systems at regional level can be applied in the following manner:

For each type of system, easily identifiable discriminatory indicators are selected, such as the presence of a particular production system (e.g. pigs, palm oil), or particular incidental factors (e.g. vicinity of a market, existence of lowlands). After a quick enquiry in the villages of the region, the possibilities of a geographical application are identified. If all the villages belonging to the same type are geographically grouped, one can demarcate the region according to land use patterns instead of conditions. Otherwise, one can also limit oneself to classifying the villages according to the types of village agro-systems. In this case, it will not be possible to establish a map of different types. However, the classification of types based on village agro-systems will become a useful reference to orientate and plan development actions, this being one of the goals of the diagnosis at this level.

In conclusion, we must emphasise the importance of regional diagnosis. When diagnosis takes into account land use patterns in an area, some amazing and realistic developmental orientations appear. In many development projects, where this process has been put into practice in Africa, evidence has shown that the major lines of intervention have been identified after such a diagnosis. This result would probably not be the same in countries with a controversial and unequal agrarian system (Latin America, northern Africa, southern Africa).

4.1.4 Village diagnosis

From the previous observations, two lessons are to be learned. Firstly, that diagnosis at regional level is not only essential, but that it also saves time. Secondly, the village, or its equivalent, in many regions represents the core level of analysis and intervention and thus justifies the new development strategies based on the development of villages and homesteads (see Chapter IX: Natural resources management). However, one should bear in mind that this level of intervention, as relevant as it is for some developmental actions, is not relevant for all of them. In particular, it is not adapted to some activities (nomadic livestock) and resources (forest), the management of which may go beyond the areas covered by intervention at village level.

In projects where intervention strategies are based on community development, diagnosis on this level is obviously indispensable. The general process that can be adopted is not significantly different from that prescribed for analysis in the context of regional diagnosis. In this case, as the village is not considered a basic unit of analysis nor as the regional level for land use, its diagnosis should go deeper into details. Here the diagnosis, it could be complemented by quantitative data concerning the population, production, farms, and the like. One should also make a point of recognising functional diversity among different production units within the village.

Most interventions at village level tend to organise their actions into plans for village development. The diagnosis of how a village operates is the cornerstone for the development of such a plan, but it must also take into account the community's suggestions and problems. This participation of the village people in the development of the most appropriate action plan for their village requires both real dialogue and real negotiations between development officers and the community. Thus, this concerted diagnosis, indispensable to the success of the village developmental plans, takes more time than the quick diagnosis of a village made during regional diagnosis (see Chapter I: Local planning).

4.2 Diagnosis at small-scale farm level

Regarding household agriculture, it is at this level of utilisation of the environment that most decisions dealing with technical, economic and other changes are made, hence the necessity to diagnose at this level in order to understand the operation of farms. However, the latter is partly conditioned by external factors depending on higher levels of organisation (village, region, country, etc.).

The aim of this diagnosis is firstly to analyse the farmers' decisions, which determine the overall running of their farms and, through the analysis, to understand their motives and strategies. Secondly, one should ask questions about the evolution of the farms studied, and more particularly about their future and the conditions for sustainability are asked. Lastly, as in all diagnoses, an attempt should be made to highlight the problems and constraints met by the farmers in the running of their farms.

The diversity of operational modes among farms should also analysed with the aim of establishing a typology at order to adapt the recommendations of the diagnosis to the different types of farms. At this level the diagnosis is particularly useful at the following stages:
- In the preparatory phase of the intervention which aims at transforming and improving the production systems;
- In operations aiming to give technical and managerial advice to farmers;
- In the context of support programmes and technical advice, in order to adapt propositions to the diversity of farms.

4.2.1 What is a small-scale farm in a developing country?

When a diagnosis at farm level is undertaken, the first thing to do is to define the exact nature of the farm-household system. Indeed, if in developed countries the identification of farms does not cause problems, it is totally different when it comes to developing countries. This is particularly so in Africa, where a dwelling unit, a consumption unit, a production or even an accumulation unit may be distinguished.

> For instance, in the Serer area in Senegal, the dwelling unit, the *mbind*, generally includes many consumption and production units called *ngak*. On the contrary, the matrilineal type of accumulation is kept out of the farm.

Although these traditional organisational family unit structures tend to be simple, a number of peculiarities still exist, particularly the limited economic autonomy of women and young men. In the context of a diagnosis based on local agricultural land use, priority is given to the production unit as a unit of analysis, while still keeping in mind the relationship that exists with the other units mentioned above.

> This production unit is more often identified from the existence of a collective cultivated crop field, known as the *gandus* in the Haussa tribe in Niger, the production unit authority being the *may-gandu*. To this production unit can be attached some individual field crops or the *gamanas* which belong to related members of the family, women, sons or younger brothers.

4.2.2 Sampling the production units

Owing to the number of farms, or production units, it is obviously necessary to take a sampling for diagnosis. At this level, it is also preferable to take a stratified sampling according to the territorial subdivisions established at regional level (agro-ecological zoning of village agro-systems). Carrying out this sampling is twofold. Firstly, a limited number of survey sites is selected (villages, wards, communities) according to the diversity of the regional land use. Secondly, a limited number of production units is selected within each of the sites.

This second selection is made in order to reflect the internal diversity inside each site. This sampling is generally based on structural characteristics that are easy to identify: family size, level of equipment, and so on. In Africa, the frequent lack of statistics in agriculture requires one to make a quick inventory of the production units in order to establish the ones to be investigated.

4.2.3 Methods and procedures for diagnosing production units

The diagnosis must be based on an overall, synthesised and dynamic understanding of the way in which farms operate as production units. To achieve this, the farm is considered as a system, an organised entity. The production system could be defined as a structured framework of production means (work, land, equipment, labour) combined together to ensure plant and/or animal production aimed at satisfying the objectives and the needs of the farmer (or the head of the production unit) and those of his/her family.

The general research process should determine the nature of the production system (characteristics), its structure, and study its running and dynamics.

A. Structural characteristics of the production systems

a) Family group (composition by age, gender and responsibilities)

In Africa, agricultural activities are generally divided according to gender and age, which makes the overall evaluation of the workforce meaningless. The extra-agricultural activities of different members of the family group should also be recorded. In most cases such activities bring additional income which is necessary for the viability and the replication of production units.

b) Means of production

- The family and outside family workforce (salaried or informal support): the latter, although not well defined, is mostly used by farmers facing heavy workloads.
- Land: it is difficult to know the actual area owned and/or cultivated if land is not a scarce resource and is not related to monetary exchange. One should investigate the layout of the plots and the quality of the land in relation to its local designation. Farmers usually attach a different value to land and designate local territories according to their fertility (soil quality, accessibility, etc.). Such information can help highlight their choices in as far as the modes of utilisation of the different plots are concerned.
- Equipment: it is generally limited and therefore easily identifiable. The tools and modes of traction have to be investigated because they have distinctive roles within production systems (use of a machete or of a traditional manual hoe, animal or mechanical traction).
- Capital: this can be generated by the production units through possible production surplus or income from it. This capital is usually difficult to quantify. In the context of a diagnosis, it is however important to identify carefully all the capitalisation assets, which may well be land, livestock or plantations, that can be mobilised in response to some specific needs.

> In this way, livestock, which is the most common capitalisation asset, may be utilised differently, depending on the nature of the animals utilised. In northern Africa for instance, small ruminants are usually sold to offset small expenses like seasonal cultivation costs. On the other hand, cattle units are sold only in the case of important expenses like the purchase of land or a tractor.

c) Productions

The following should be considered in order to determine the essential elements that characterise and differentiate production systems:
- For crop production, the types of crops are to be recorded (species, varieties), their proportion, their allocation on plots, what they are used for and their destination. Two survey tools are particularly useful to characterise crop production at farm level, namely plot status and a map.
- For animal production, the types of the stock, the numbers in the different herds, their roles and destinations are recorded.

It is possible to make a farm monograph by sticking strictly to these structural elements, but this does not allow one to make a systemic analysis of the farm. For such an analysis, the operation of the production systems must also be studied.

B. Studying the working of production systems

The production systems' operational features are the result of the following:
- The relations between its components: labour, land, equipment, livestock, and so on.
- The relationship with outside forces (credit, market, etc.)

The production unit operates in order to meet the objectives of the farmer or the head of the production unit.

These objectives are seldom explicit; they are known only at the end of the diagnosis, to the extent that they can be considered as the results of the diagnosis rather than as an initial parameter. Whatever the case, these objectives remain at the core of the system. They are determined by a number of internal and external factors which in turn depend on
- the constraints and advantages of the socio-economic environment (agricultural prices, markets, social rules and techniques imposed by the agrarian system);
- the soil and climatic characteristics;
- the nature and importance of the available means of production;
- the composition, needs and prospects of the family group.

The head of the production unit and those with whom he/she shares his/her power are inclined to make decisions according to these objectives. It is the analysis of these decisions that helps one to understand how the production system works. This analysis can be done by distinguishing
- decisions dealing with the organisation and mobilisation of the means of production (e.g. renting land, resorting to salaried labour), which determine how the farm is managed;
- decisions related to the technical production processes (e.g. choice of crops, cropping systems, modes of animal feeding), which characterise how the technical production systems are run.

By identifying and analysing the technical and managerial decisions, it will generally be possible to establish the farmer's and his/her family's objectives, which are usually far from being a simple quest for increased income. The analysis of these decisions will not just explain the farmer's objectives, but will also reveal the constraints and advantages of the physical and socio-economic environment of the farm. The same analysis will bring to light the reciprocal relationship that exists between the managerial and technical systems. For instance, land accessibility may lead to the use of more extensive technical systems. Conversely, the introduction of market crops into the plot allocation may result in fertilisers being purchased and salaried workers being employed.

As a result, if one wants to elaborate on sound propositions for the improvement of a farm, one must undertake a technical analysis of its operational features, meaning of its technical production systems.

However, as the structural characteristics of a farm determine the farmer's technical choices, improvements are adopted only if they are compatible with the means of production available. Therefore, diagnosis at farm level should simultaneously take into account both technical and socio-economic modes of operation.

Hence, the three fields for investigation at farm level are discussed below: the socio-economic situation and operation features, the technical operating modes, and the relationships between management systems and technical production systems.

a) Socio-economic status and operation at farm level

Once the means of production available have been investigated, one should analyse how these means are managed, by distinguishing between individual decisions, the rules, and the collective use that determines such management. The combination of these factors and the different adjustments that may result from it (renting or leasing of land, using salaried labour, mutual help, etc.) should be studied. The choices made by the farmers in realising such adjustments are strong indicators of their overall strategies.

In a diagnosis context and in the absence of all records, it is difficult to make economic calculations that enable a precise and reliable income statement to be drawn up. On the other hand, evaluating work and land productivity constitutes an essential diagnostic tool for establishing the economic status of the production systems. This evaluation makes it possible to compare the economic results of different types of farms. It also reveals the farmers' economic way of thinking which is generally concerned with making the most of the rarest or benefiting from the most strategic factors (in Africa for example, this is often labour).

Owing to limited savings and scarce monetary resources, the analysis (even simplified) of cash flow in production units by establishing a schedule of yearly financial flows may be very useful in understanding the rationale of certain farmers' decisions. These are not necessarily well explained by the overall economic analysis of the production units' results.

> For instance, in large irrigation schemes along the Niger River in Mali, farmers happily practise double rice cropping which is less profitable than the single cropping system, essentially because they need money and food at their disposal during certain critical periods of the year (hunger times).

b) Technical operating modes

The diagnosis of technical systems at farm level requires the analysis and understanding of the ways crops and animals are farmed.

- Cropping systems

The cropping system consists of a given combination between crops and fallow in space and in time. It represents land use patterns and a homogeneous mode of cropping in a certain area. On plots with the same cropping system, one can find
- the same kind of crops succeeding one another in a predetermined frequency and order,
- possibly the same type of association (millet-sorghum-cowpeas, in the Sahel for example),
- and comparable technical procedures for each crop.

In general, and particularly in traditional agriculture, a significant correlation is seen between cropping systems and soil types. This correspondence underpins the concept of local territory (*terroir*), which explains why the cropping system is an important factor in the characterisation of agrarian systems.

In many cases, farmers strive to divide their land into different groups of plots with certain soil features in order to diversify their production, spread labour and reduce the risks. Moreover, even when the farms are in a homogeneous environment, the distance between plots may dictate different cropping systems. This results in a diversity of cropping systems within any given farm.

The diagnosis of a technical system at farm level requires one to identify to which cropping systems the farm plots belong. This identification can be made quickly with a plan and a map showing the farm plots. As a first step, it can be considered that the plots with the same successive crops are all part of the same cropping system. If, in case of mono-cropping (in the case of rice for instance), this single criterion is not enough, one must resort to other criteria of differentiation between crop systems. Once this identification has been done, the operating systems are analysed.

The first objective of this analysis is to understand the reasons that explain the basic characteristics of each system:
- The nature of the crops cultivated
- The succession of crops (and of fallow) in time
- A possible association of crops on the same plots
- The cropping systems themselves (meaning the succession of practices and techniques that are applied)

It is simply a diagnosis of different practices applied to crop production. As in any diagnosis of this kind, one should not only understand the reasons for these practices but also be able to evaluate their results and their productivity.

- Livestock systems

Livestock systems are the result of relationships between the stock, the farmer, space and grazing resources used by the animals. In fact, just as in cropping systems, livestock systems can be identified at levels other than the farm. For instance, to understand how nomadic and semi-nomadic farming operates, it is necessary to consider the entire area concerned with the

A model of farming household operation

displacement of the animals.

At farmer level, and especially with regard to sedentarised livestock, the diagnosis of livestock systems (once the stock has been characterised) means analysing farmers' practices such as
- animal reproduction modes,
- their feeding modes (throughout seasons, according to animals' age, etc.),
- their utilisation.

The analysis of these practices helps one to understand the logic behind the livestock production systems, while at the same time it reveals the problems and constraints the farmers face. If this diagnosis has to be final, an evaluation of animal performance should be added to the analysis of farmers' practices. This entails resorting to specific tools and methods of diagnosis used in animal science.

Livestock systems are generally strongly related to cropping systems and this relation represents a very important aspect of the operational features in production systems at farm level.

An important analytical tool for studying the relations between crop and animal production is the fodder calendar, which helps to identify the different feeding resources available to each type of animal during the year. This calendar documents the contribution of crops (products and by-products) to animal feeding. It also enables one to identify the most critical periods and subsequently to design strategic adjustments which the farmers may resort to at such times.

However, the diagnosis of the relationships between agriculture and livestock farming in the overall production system requires one to transcend the problem of animal feeding and take into account the animals' effects on
- maintaining soil fertility, through their transfer from grazing areas to cultivated zones (grazing organisation, selling/exchanges of manure, kraal);
- transport, which can be a priority for certain farmers (donkey carts); this function can also generate some significant income;
- the mechanisation of crop farming through animal traction, the nature of the work (ploughing, weeding), as well as the effect of animal traction on the intensification of cropping systems depending on the local agricultural conditions (environmental conditions, population density, etc.) (A good analysis of these conditions may help one to understand why farmers' choices and their logic often differ from the recommendations of researchers and extension officers.);
- the provision of financial resources, since, in some dry African areas, stock farming is the main, if not the only, source of income;
- capitalisation, since the role of livestock is very important in contexts where land is not sold and saving money is not practised.

c) Relationships between management systems and technical production systems

The study of these relationships can be done by reviewing the availability of the production means (resources), production requirements and the overall running of the farm (assessment of labour, accounts, household food supply, etc.), this review being recorded in a sort of balance sheet. However, this type of procedure, which requires periodical records for at least one cropping year, is such a burden that it cannot be adopted in a diagnosis context. It should be borne in mind that such an assessment generally shows an equilibrium (balance), but may hide the problems the farmer is having in balancing his/her needs, his /her resources as well as the strategies he/she is adopting in an effort to reach this equilibrium.

That is why it is crucial to analyse these relationships with an emphasis on the most critical periods (work peak time, periods of food scarcity, etc.) in order to examine the means and adjustment strategies farmers resort to when faced with critical times. Unlike the balance sheet method, this diagnostic procedure is very efficient in revealing problems encountered by farmers while running their farming activities.

4.2.4 Diversity, typology and dynamics of production systems

Owing to the heterogeneity of environmental conditions and the unequal distribution of production means, among other factors, there is a diversity of production systems at village/community level. One of the results of the diagnosis of production systems diagnosis is to reveal this diversity, which can be formalised through typological studies.

It is commonly said that there are as many typologies as there are points of view and work objectives. From the point of view of farm productive activities, however, one can distinguish two main types of typologies depending on whether the emphasis is on the farmers' socio-economic status, or on their technical modes of operation. In the first case, typologies based on structural characteristics are established, that is on the organisational modalities and combinations of production means. In the second case, typologies based on functional characteristics are established, meaning that they are based on the analysis of the production systems.

If we refer to the overall model of how a farm operates, as presented earlier (see sketch), it can be seen that structural typologies mainly take into account the farm's managerial system and its means of production, while the functional typologies prioritise the technical production systems.

Depending on the point of view adopted, each typology can be legitimate and one can still establish much more, in accordance with the specific problems encountered (access to land, to markets, level of intensification).

With this in mind, it is advisable to estab-

In the village of Hrarda, in the Chaouia area in Morocco, where the distribution of the means of production is very unbalanced, El Hailouch (1982) established a structural typology based on the strategies for adjusting the means of production adopted by the farmers. Thereafter, he made a connection between these adjustment strategies and the farms' technical operations. This allowed him to define clearly differentiated types of production systems.

- Of the farmers 25% are landless. They sell their labour force to other farmers as part-time occasional farm workers. Those who possess animal traction plough a small plot of rented land on an annual-based contract. They are obliged to grow staple food crops.

- Of the farmers 33% possess less than 5 hectares, which is the farm-size threshold for survival in this region. Among them, those who are well equipped with animal traction and have sufficient salaried labourers rent more land in order to reach the survival threshold. Those who are not well equipped resort to animal traction co-operatives and occasionally sell their surplus labour force.

- Of the farmers 15% own land areas ranging from 5 to 15 hectares. Their means of production are well balanced. These farmers live relatively self-sufficiently and invest their benefits in small businesses or in livestock farming.

- Certain farmers in a position (*adoul, fqui*) that is not really compatible with agricultural activities rent their lands.

- Farmers who possess large areas (15 to 60 hectares) use mechanisation that they have either acquired or rented. Such farmers prioritise soft wheat rather than hard wheat, which is for self-consumption. They provide services for small-scale farmers and are involved in large stock commercialisation or in transport.

lish typologies that integrate the technical and socio-economic features of the operational modes of farms. This means that these features make it possible to distinguish production systems with the same problems, for which one can prescribe the same kind of ameliorating solutions (definition of recommendation domains).

At a given time in the farms' operations, the analysis allows one to envisage their diversity. However, this "picture" may be conducive to erroneous conclusions if one does not take into account their dynamics. This requires one to reconstitute the evolution of their structural characteristics and their technical production systems. Hence farm surveys must systematically integrate information concerning the farms' dynamics (history, factors for change).

It is from this analysis of the farms' dynamics and their production systems that the evolutionary trajectories are firstly drawn. Secondly, the conditions and modalities of the farms' reproducibility are analysed as essential components of the diagnosis at farm level.

While interpreting these elements in traditional agriculture, one should distinguish between evolutional factors linked to cyclical features within the family group (ageing, assets' transmission to descendants) and various conjectural (drought, agricultural price fluctuations, market opportunities) or structural factors (monetarised exchanges, technical progress).

4.3. Diagnosis on cropping systems

This diagnosis is concerned with evaluating the efficiency of the modes of cultivation, identifying and then classifying the causes of variations in productivity. As this makes it possible to explain the differences in the productivity between different plots, it is indispensable to all programmes based on technical advice and support to farmers and as an alternative to prescribing normative recommendations. It is also very useful as a means of orientating research programmes according to the real problems limiting production.

4.3.1 Agronomic bases for diagnosis and the choice of cropping situations

The diagnosis of the cropping system is based on a fundamental agronomic statement, which is that the relationship between techniques and yield is never direct. Most of the techniques applied by farmers (i.e. tillage, irrigation or fertilisation) are based on transforming the environment (structural and chemical state, and soil water content) and on making it more favourable to the growth and development of crops. *The diagnosis of the cropping system will therefore be based on the agronomic analysis of the relationships between techniques, the state of the environment, crops, and plant productivity.*

To carry out this analysis in a given situation, there must be some homogeneity of both the techniques and the environment. In commercial and mechanised agriculture, this condition is generally fulfilled at cultivated plot level, which is defined as a surface of land occupied by one or several cultivated crop species, all grown in a homogeneous manner.

In fact, management sequences for different plots are often the same for a given crop (meaning the logical and co-ordinated combination of techniques and operations, starting from the planting up to the harvest). Plots have often been demarcated to ensure a certain homogeneity in the environment.

In sub-Saharan Africa, things are quite different. Firstly, one should distinguish between the concept of a plot as defined previously, and the concept of a field, corresponding to a piece of used and cultivated land sometimes with many plots owned by a person or a group of persons. Secondly, recent clearing of the forest, common grazing practices on remaining crops, the presence of trees and many other factors tend to make the environment very heterogeneous and complex. Because of both manual cultivation and diverse operators, crop management sequences are usually variable within the same field.

Consequently, in order to perform an agronomic analysis in these conditions, one has to work in a portion of the plot where one is sure of the homogeneity of both the environment and the techniques, that is the so-called cropping situation. This situation can be defined as a homogenous area for agronomic observation with regard to the physical environment and the crop management practices.

In the context of an overall diagnosis of land use patterns at farm level, the realisation of a cropping diagnosis, meaning the agronomic analysis of the cropping situation, is an operation that requires time, method and discipline. Therefore, it can be undertaken only when it is necessary, for instance, to analyse significant differences in productivity between farms, or gaps between farmers' productivity and theoretical or potential productivity for techniques or management sequences prescribed by extension officers.

As soon as the decision for the diagnosis has been taken, it is necessary to select carefully which cropping situation is be studied. To increase the efficiency of the diagnosis, one selects a sample of the cropping situation by taking into account the variability at the highest levels (region, village, cropping system) of the area investigated and with regard to the hypothesis on the causes of productivity variation resulting from preliminary observations.

Another way of improving the efficiency of the diagnosis is to compare several cropping situations that have only one or a very limited number of variable factors (date of sowing, precedent crop, etc.) In this way, a field tool is developed, which, in a sense, is similar to an experimental apparatus (see Chapter III: On-farm experimentation).

4.3.2 The yield build-up model as a reference

The yield build-up model is used. This model constitutes a synthesised and explicatory model of relations between techniques and productivity for a given crop. The basic principles for the development of such a model are the following:

- The final yield of a given crop depends on a series of component factors, which develop according to specific mechanisms throughout the cropping cycle. For instance, the maize yield can be broken down in the following table:

Grain yield = [number of planting holes per hectare] x [number of shoots per planting hole] x [number of cobs per shoot] x [number of grains per cob] x [mean body mass of grain]

- The development of each and every component of the yield depends upon
  - a certain number of factors and environmental conditions related to the formation of the component factor, for example the number of seedlings in one planting location (population per location) depends on the sowing density, and germination and shooting conditions (temperature, humidity, soil structure);
  - the level of the previously developed yield component factors, which are caused by compensation processes occurring between components (e.g. too many shoots per planting hole generally generate weaker shoots).
- Environmental conditions and factors that determine the level of yield component factors are partially dependant on cropping techniques. In the above example of the seedling population per planting location, the germination and shooting conditions depend partly on the soil preparation mode, as well as on date and mode of sowing.

For most crop species, the process of yield build-up has already been investigated. It is therefore better to rely on the available literature before undertaking any cropping system analysis, and then to focus more on the local conditions of its operation (farmers' practices and techniques).

4.3.3 Procedure for analysing a cropping situation

This procedure, which is based on the analysis of the yield build-up, follows the steps mentioned below.

A. *A precise identification of the cropping operations applied to the cropping systems*

For precision sake, it is advisable to conduct surveys at different periods of the climatic cycle associated with the monitoring of the crop cycle.

A good compromise consists of making a series of observations and surveys during cultivation, for example at specific physiological stages (e.g. flowering), and later, shortly before harvest, at the moment when it is possible to evaluate all the yield components.

B. *The evaluation of the yield component factors*

This evaluation causes certain problems. The first, which has just been mentioned above, concerns the periods for observing crops and practices. The second concerns the survey and sampling procedures for evaluating these components.

Without going into detail, there are two types of methods for sampling:
- For some yield component factors, individual plants are selected randomly from the entire cropping system (e.g. for the evaluation of maize tillering).
- For other components, sampling only makes sense when related to a unit of surface; it is therefore necessary to count and make observations on certain elementary areas, such as small experimental plots which have been randomly located and demarcated within the plant population. This second method is necessary to evaluate components like population density (shoots) or cob density. If these counts do not have any damaging effects, they should be performed on the same elementary plots throughout the productive cycle.

The number of elementary plots, which is seldom lower than six, is determined by the statistical requirements and heterogeneity of the vegetation.

C. *Assessing the level (value) of the yield component factors*

To evaluate this level, some norms or references are necessary. These norms and references come from either the experiments carried out in the area, or from the best values recorded on the farmers' plots.

By comparing the level of these components to the norms and references, it becomes possible to make a judgement about the conditions of realisation of each component. All the same, there may be interactions may appear between components resulting in compensation phenomena.

Hence, in the case of cereals, a low number of seeds per hole (sowing density) can entail a high cob tillering. Interpretation based on one yield component factor must take into account the level of other components.

D. *Formulation of hypotheses explaining the differences between the level observed and the potential level of yield component factors:*

From the yield build-up model, it is possible to identify the environmental factors and conditions that explain these differences and thus to refer back to the cropping techniques which influence both these factors and the environmental conditions. These techniques may explain the variations of yield that are seen in the cropping situation. This analysis of the components will result in a number of hypotheses being made that concern the techniques-yield relationship, which may need some verification.

E. *Research enabling the explenation of the level of the yield component factors*
This research is concerned either with the direct observation of the environmental status (e.g. soil structural state, state of health of the crop population, weed rate) or the reconstitution of the cropping conditions that can influence the yield build-up process (e.g. nitrogen balance, water content).

When comparing cropping situations with a surveying apparatus designed to test particular technical effects, the emphasis is especially on analysing the environmental factors and conditions that impact directly on the level of components with regard to the techniques that are studied (see Chapter III: On-farm experimentation).

However, by orientating the observations and tests towards the environmental conditions and factors related solely to expected effects, there is always the risk of misleading observations and neglecting the unpredictable, but determining phenomena can help to understand the final yield (e.g. bad emergence, parasite attacks). This is why usually, it is a good idea to carry out a number of systematic controls of the environmental state throughout the production cycle (e.g. rate of weed, plant health situation).

4.3.4 Extending the diagnosis on cropping systems

In traditional agriculture, the variation in productivity (yield gaps) observed in crop fields is generally high. It is common to see yields ranging from 1 to 5 (e.g. from 0.4 to 2 tons of maize per hectare).

It is precisely this high variability that helps the diagnosis on cropping systems and allows significant results. This diagnosis is therefore a very useful tool for orientating technical advice and extension programmes for farmers.

However, although it is possible to identify which techniques need to be improved, the limited control and observations that are carried out in a diagnosis context do not always make it possible to explain the mechanisms that limit productivity (e.g. weed pressure, degradation of the soil structural state, etc.). If those mechanisms are to be explained, it is necessary to carry out experiments that will ensure a better control of the process of yield build-up.

Lastly, it is worth mentioning that the diagnosis on cropping systems cannot be confined to the technical factors limiting crop production. The conditions that determine the farmers' technical choices (i.e. lack of information or training, time constraints, limited availability of inputs, deliberate choice of an extensive agriculture, etc.) must also raise questions and demand investigation. To achieve that, it is necessary to interpret the result of the cropping diagnosis while, at the same time, taking into account the other level of production organisation, in particular, the constraints identified during the diagnosis of production systems and agrarian systems. It is through this process of using the knowledge obtained at different levels of organisation that the diagnosis will result in a complete description of the ways the environment is exploited, and in sound intervention strategies for agricultural and rural development.

5. Conclusion

It would be pretentious to consider that after many years of relatively successful experiences, efficient diagnostic methods have been unequivocally defined. Therefore, the propositions that have been made here can only be taken as methodological indications to be adapted to each situation. Most particularly, the importance that will be given to the diagnosis of the different levels of organisation will depend on the nature and objectives of the type of intervention. In the same way, in the projects concerning soil management at community level (see Chapter IX: Natural resources management), the regional diagnosis and the diagnosis of village agro-systems are of paramount importance. On the other hand, in the programmes on managerial and technical advice to farmers, the diagnosis must rely on the diagnosis of cropping systems in order to identify and prioritise the techniques needing improvement (see Chapter VIII: Extension services and farm management advice).

However, as the different levels of organisation and intervention in rural areas are interdependent, it is advisable not to limit the diagnosis to one level exclusively. This openness to different levels of analysis in the diagnosis of rural areas should make it possible to solve a major problem encountered by a number of rural developmental interventions, that is the issue of changing levels. How can one transfer knowledge and experience from the local level to the region? How can one extend actions conducted at village level to regional level and achieve a significant rural development impact? This problem is not easy to solve. The solution requires one to evaluate the representativity and relevance of local activities at a greater level. Proper rural diagnosis makes this possible, by taking into account the different levels of production organisation.

The methods and procedures presented above especially target the different field operators involved in rural development actions, whose concern is to base their action on a preliminary analysis of the realities which they are intending to transform and improve. This analysis, although necessary, is not sufficient. In fact, even though it gives a central place to the farmers' practices and takes into account the farmers rationalities, it remains the expression of realities analysed from an external viewpoint. That is why it is necessary to confront this external perception by analysing the farmers' own situations and objectives (see Chapter I: Local planning).

The linking of these two points of view is indispensable for establishing of a true dialogue between development operators and farmers. This will give the concept of concerted diagnosis its real meaning. By taking into account only one point of view, there is no room for dialogue. Moreover, it represents a number of risks and limitations. Focussing on external diagnosis only is surely running the risk of falling into a technocratic process, which has characterised a number of rural developmental projects over the past few decades.

Conversely, basing intervention on the exclusive expression of the needs and desires of the farmers represents a number of limitations as well. These are as follow:
- The information may be a combination of needs based on immediate problems or may arise from the nature of the person being interviewed and what one can expect from him/her;
- The difficulty of farmers to imagine alternatives other than those tried locally and also the fact that some constraints have been internalised to such an extent that farmers do not perceive them as problems with a solution;
- The legitimate concern of prioritising short-term survival techniques, rather than long-term actions, the effects of which may be delayed (e.g. management of natural resources), but remain the condition for maintaining the environmental productive capacity in the long run.

This necessary dialogue between internal and external viewpoints causes communication and negotiation problems that will be discussed later. Mastering these issues is essential to ensure the real participation of the farmers during the diagnosis phase and for undertaking developmental actions on a contractual basis, in line with the new developmental strategies that one wishes to promote (see Chapter V: Training of farmers).

Lastly, let us remember that diagnosis is not a finality as such, but rather a step in a dynamic process. It must therefore be extended by a monitoring-evaluation process that targets the actions undertaken and their impact upon the environment and the productive systems (see Chapter IV: Monitoring and evaluation).

6. Recommended literature

Overall perspective

De Beer, F. & Swanepoel, H. (2000) *Introduction to development studies* (2nd edition). Oxford University Press, Cape Town, South Africa.

Jouve, P. (1992) *Assessment of the rural environment: from region to field. A systems approach to agricultural exploitation of the environment.* CNEARC/CIRAD, June 1992, Montpellier, France.

Mettrick, H. (1993) *Development-oriented research in agriculture. An ICRA textbook.* ICRA publ., Wageningen, The Netherlands.

Sebillotte, M. (1994) Systems research and action. Interdisciplinary excursions. In: *Systems-Oriented Research in Agriculture and Rural Development.* International Symposium, Montpellier, France, 21–25 Nov. 1994, pp. 35–72.

At plot / crop / animal level

Aubry, C. et al. (1998) Modelling decision-making processes for annual crop management. *Agricultural Systems*, 56 (1): 45-65.

De Steenhuijsen Piters, B. & Fresco, L. (1995) Conceptualizing system diversity with examples from northern Cameroon. In: *Agricultural Research and Development at the crossroads. Merging systems research and social actors approaches.* A. Budelman (editor). Royal Tropical Institute, The Netherlands, pp 33–42.

Papy, F. (1994) Working knowledge concerning technical systems and decision support. In: *Rural & Farming Systems Analysis. European Perspectives.* Dent & McGregor editors., CAB International Publ., pp. 222–235.

Ruthenberg, H. (1981) *Farming systems in the tropics* (3rd edition). Clarendon Press, Oxford, UK.

Watson, D.J. (1976) The physiological basis of variation in yields. *Advances in Agronomy*, V (28): 101–145.

At farm level

Chambers, R. (1994) Participatory rural appraisal: the origins and practices of PRA. *World Development*, 22 (7).

Ellis, F. (1993) Peasant economics. *Farm household and agrarian development* (2nd edition), Cambridge University Press, UK.

Gilbert, E.H. et al. (1980) *Farming systems research: a critical appraisal.* Michigan State University, Rural Development, paper num. 6, East Lansing, Michigan, USA.

Landais, E. (1998) Modelling farm diversity: new approach to typology building in France. *Agricultural Systems*, 58(4): 505–527.

Low, A. (1986) *Agricultural development in southern Africa: a household economics perspective on Africa's food crisis*. Raven Press, Cape Town, South Africa.

Matata, J.B.W. et al. (1999) *Farming systems approach to technology development and transfer.* FARMESA, Harare, Zimbabwe.

Moock, J.L. (editor) (1986) *Understanding Africa's rural household and farming systems*. Westview Press, Boulder Co., USA.

Norman, D.W. (1980) *The farming systems approach: relevancy for the small farmer*. Michigan State University, Rural Development, paper num. 8, East Lansing, Michigan, USA.

Perret, S. (1999) *Typological techniques applied to rural households and farming systems. Principles, procedures and case studies*. University of Pretoria / CIRAD, working paper 99/2, Pretoria, South Africa.

Perrot, C. & Landais, E. (1993) Research into typological methods for farm analysis. The why and wherefore. In: *Systems studies in agriculture and rural environment.* Brossier et al. (editors) INRA publ., pp. 373–381.

At rural regional level

Deffontaines, J.P. et al. (1992) Managing rural areas. From practices to models. In: *Systems studies in agriculture and rural environment.* Brossier et al. (editors), INRA publ., pp 383–392.

Lhopitallier, L., Perret, S. & Caron, P. (1999) *Participatory zoning techniques, as support to rural development planning and management. Principles and procedures, a user's guide*. University of Pretoria / CIRAD, working paper 99/1, Pretoria, South Africa.

Whiteside, M. (1998) *Living farms. Encouraging sustainable smallholders in southern Africa.* Earthscan Publications, London, UK.

Chapter III: On-farm experimentation
A. Guillonneau

For reference purposes, Chapter III may be connected to the following chapters:

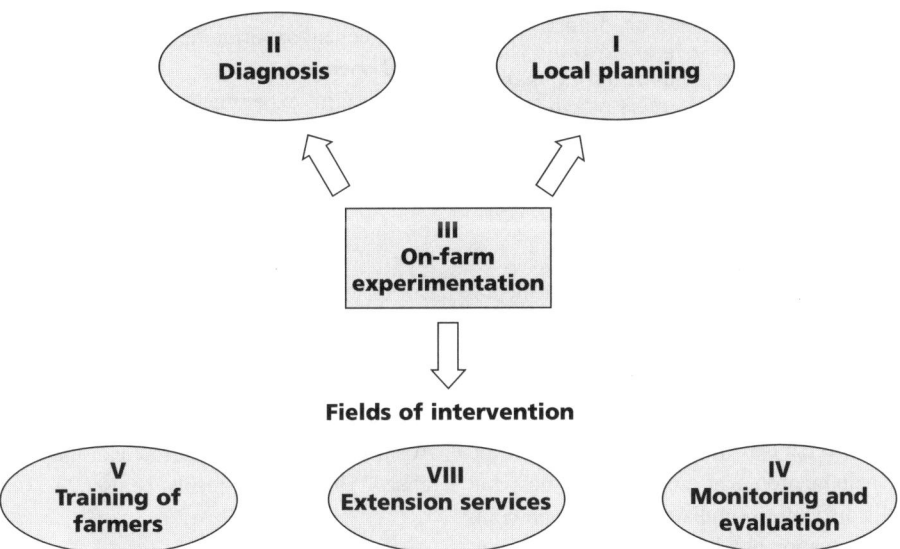

Farmers, almost everywhere, are faced with technical problems. Their cropping systems and traditional way of managing animal production usually become inadequate when confronted with new constraints or challenged by new strategies (e.g. intensification, access to a new market). Technical innovations are nearly always necessary in order to adapt to ecological changes (drought, declining fertility) and to socio-economic changes (reduction of land availability, decline in agricultural product prices, increasing input prices, new market requirements, quality, etc.).

Up to the beginning of the 1980s, the technical improvements proposed to farmers through extension systems were set up by agronomic research (see Chapter VIII: Extension services and farm management advice).

Since then, the question of experimentation with farmers in rural areas has been raised and implemented by certain teams in various contexts (see references). Research stations are no longer considered as the only centres for innovation, because most technical innovations developed by research and propagated through extension programmes have not been well adopted by farmers in sub-Saharan Africa, especially by the small-scale developing ones. Even though certain interesting results have been seen in specific contexts (irrigation schemes) or for certain crops (cotton), the overall results have been disappointing.

A number of factors explain the resistance or mistrust of the farmers towards the techniques they have been exposed to (see Chapter VIII: Extension services and farm management advice). One of the factors is that the technical innovations earmarked for the farmers were not always adapted to their expectations and their means.

> In the Ivory Coast, after several years of research, a high-yielding variety of rice, IRAT-13, was developed in the laboratory. During the extension process, this variety revealed a major defect: the hairs on the grain severely irritated the skin on the farmers' hands during manual harvesting. The farmers would have readily eliminated this variety if only the experiments had been conducted in the rural area before distribution.

This resulted in new methods of experimentation in rural areas being designed with the collaboration of both producers and development operators[1]. There are many justifications for this:
- The diversity of both the physical and the human environment are taken into account. It is difficult to simulate an actual and complex situation in a laboratory or in a station.
- Farmers' opinions are taken into consideration. This helps to prevent mistakes and enables quicker appropriate techniques to be used. The farmers' participation in experimentation guarantees efficiency and appropriation.
- Knowledge and know-how are exchanged. Farmers possess much information that they do not always share easily. They must have good reasons to give such information, either to support or contradict the researchers' arguments. On-farm trials, as a means of exchange, provide these good reasons.
- Collaboration facilitates the adoption of new techniques by farmers in the future.

On-farm experimentation demonstrates that other possibilities do exist. They can be carried out by farmers, on their farms, despite the harsh environmental conditions of the area.

Farmers are usually only convinced by what they can see, touch and evaluate. A farmer often trusts him/herself, which is normal given the fact that he/she is the one who will carry all the risks related to the innovation.

On-farm experimentation is training. Trials are excellent opportunities to train farmers and local operators.

On-farm experimentation incorporates extension and dissemination. On the one hand, trials create opportunities for farmers to meet and exchange viewpoints and experiences, and on the other hand they help spread information.

However, experimentation in rural areas has its regulations and constraints. With regard to problems encountered in the past, four main questions can be raised: What is the role of experimentation in a development programme? What are its objectives? How can they be realised? How can the results be used?

1. Pros and cons of on-farm experimentation in developing rural areas

1.1 Role of on-farm experimentation in the context of a development programme

On-farm experimentation should be discussed and then implemented as part of a whole development programme. The process includes four major steps:
- The concerted diagnosis, This makes it possible for farmers to identify the main prob-

---

1. In this chapter, as well as in the rest of the book, the term "development operators" refers to all operators involved locally in development-support activities to farmers and rural communities. It includes extension officers, researchers, technicians, consultants, NGO members, local stakeholders, etc.

lems and prioritise their actions. It must be quick and also permanent in order to allow in-depth knowledge to be obtained and actions to be readjusted. If the characteristics of the farms are varied, a typology may be useful: farmers that have the same type of constraints are grouped together in order to seek adapted solutions to their problems (see Chapter I: Local planning, and Chapter II: Diagnosis).

- Seeking solutions and developing technical improvements: This is done at research station at the level of farmers' plots or homesteads, and by introducing technical innovations are introduced to the farms.
- Wide dissemination of the developed technologies. This includes
  - the study of the agribusiness environment in order for farmers to master the changes (i.e. input supply, credit, commercialisation, training on how to negotiate with partners. See Chapter VI: The farmers' organisation, Chapter VII: Contracts between role-players, and Chapter X: Product management).
  - the dissemination of the new techniques on a large scale (see Chapter VIII: Extension services and farm management advice).
- Monitoring and evaluation (see Chapter IV): This makes it possible to evaluate the interest in the techniques introduced and to understand their effect on the evolution of the farms.

The more integrated on-farm experimentation is in the whole development process, the better its chances of being efficient.

1.2 Researchers must make choices

"There was no way to harvest the plots as the gravel road to the farm was cut off."

"The farmer has thinned the sorghum plot devoted to a trial on crop density without telling me."

"My fertilisation trial gave ridiculous yields compared with that of the farmer next to me."

"My trial equipment has been stolen (pegs demarcating small experimental plots)."

"The farmers did not plough the trial plot as planned in the protocol."

This is how agronomists who run trials in rural areas sometimes speak, since it is often easier to run trials at research stations than with farmers. The latter is often seen as risky agricultural research, although it does have some advantages.

As a final objective, the agronomist must find solutions to the farmers' real problems and not suggest ready-made solutions to them that eventually may or may not be adapted.

The content of his/her programme must take into account the immediate questions formulated by the farmers, but must also offer long-term development perspectives that integrate maintenance and the preservation of natural resources.

1.3 Questions arising

Why is it necessary to undertake on-farm experimentation? What are we looking for? The following sections investigate and discuss the different objectives of experimentation.

1.3.1 Acquiring accurate references on cropping systems

An example: The aim of the trial is to investigate the consequences of late weeding on the component factors affecting maize yield build-up.

This is an experimental procedure conducted by a researcher, the aim of which is to gain information and knowledge. It is not directly connected to a development-support programme. This case will not be discussed here.

1.3.2 Developing technical innovation

The objective is to create either a technique or a cropping system which does not exist among farmers (e.g. introducing a new crop, or setting up minimum tillage techniques), or to make sure that the available references are adapted to the local environment and/or adjust them (e.g. introduce new cultivars that are resistant to certain diseases or some well-known herbicides currently used in other countries).

The aim of experimentation is therefore to develop a *technique*. This does not mean that farmers should be excluded from the implementation of trials, from their conception and/or the development of new alternatives. The solutions from the experiments will either be as a result of research or the skills of the farmers themselves. This kind of trial will be called on-farm experimentation.

In this case, one must be very strict with experimentation. It is necessary to control a number of factors in order to have the means to explain and to analyse the outcome of the trials with the participation of all the interested people. The objective is to obtain reliable technical references, which are not only applicable to the experimentation plot. The trial also aims to provide the farmers with a full range of systems or techniques to enable them to choose. In this kind of experiment the researchers take the risks.

1.3.3 Studying the farmers' reactions to technical solutions, the adoption of this technique and its impact on farm operations

The target of this kind of experimentation is the *technique/farmer relation*. It will be called the on-farm test.

In this case there is little control of the technical factors. The key questions are what do farmers select and how do they adapt the technique? It is also interesting to see what conditions are necessary for extension on a larger scale. (e.g. Do we have to consider a credit system? How can the production of seeds for fodder crops be organised? How could specific equipment be built up and maintained locally?)

In this kind of experiment, the farmers take the risks.

1.3.4 Dissemination through extension of approved techniques to the entire zone of intervention

The essence of the work is the dissemination of techniques on a larger scale. It therefore relies on the local operators (extension officers) who should effectively assume responsibility for this task. This is called demonstration.

This classification (experimentation or trial, test and demonstration) is a convention for this chapter in order to gain a better understanding. These types of experimentation are absolutely complementary to one another, as well as mandatory. Each has its own requirements.

In summary, the management and the complexity of on-farm experimentation will differ according to what was decided at the beginning. Knowing exactly what is being investigated is another way of clarifying the division of tasks among individuals during experimentation (Who is responsible for processing? Who is in charge of the cost of inputs? Who takes the risk of the trials? etc.)

1.4 All techniques are not of the same nature ... and are not of equal interest to the farmers

If one focuses on the current problems which the farmers mention with ease, one should lead the discussion towards problems of a long-term nature. Moreover, technical changes are complex in the sense that they may result in important induced effects (see Chapter VIII: Extension services and the farm management advice).

> Problems such as maintaining soil fertility or preventing erosion are complex in nature and agronomists tend to show far more concern than the farmers do. These problems are, in many cases, addressed collectively and on a long-term basis, which is often frustrating for the farmers, as they are more concerned with the immediate constraints. As long as land is not a scarce resource, farmers tend to consider all problems related to it as issues without solutions.

To facilitate dialogue between farmers, researchers and local operators (technicians, extension officers) during on-farm experimentation, the following is advised:
- Do not try to tackle problems directly, but select situations or themes that are easy for the agronomist, in order for him/her to demonstrate that he/she knows something. At the same time, he/she will acquire some skills. For instance, variety screening trials are a good introductory theme as they are easy and interest many farmers.
- One should demonstrate one's ability to promote factors of change that perform well, if not more so than before, at least as well as previously.
- Do not try to do too many things at once. It is better to carry out a few trials that are well conducted, instructive and well monitored by technicians and farmers.
- Later, one will be well equipped to face difficult situations, this time with the support of the farmers. By then, one will be prepared to deal with trial failures. As long as there is collective responsibility and the farmers understand the reasons for failure, it will not really matter.
- The trials should be visited frequently with as many farmers (from the community and the neighbouring communities), technicians and other researchers as possible participating.
- It is advisable not to start an experiment if it is impossible to devote the necessary time to it.

2. Methods and tools

The propositions below are organised around five steps:
- Conception and development of the experimentation
- Crop development and harvest monitoring
- Some practical tips about the organisation
- Data analysis
- Valuation of the results

2.1 Setting up a trial

2.1.1 Define the problem accurately and set the objectives to be achieved

The theme for research must emanate from the diagnosis agreed upon, discussed and formalised by the farmers. The choice of themes must be a result of negotiations between the different partners involved in the experimentation and a compromise between the researchers' and the technicians' viewpoints on the one hand and the farmers' concerns on the other. (See Chapter I: Local planning, and Chapter II: Diagnosis)

2.1.2 Choosing procedures

The different role-players will have identified several possible solutions during the diagnosis. These solutions provide important elements in the choice of experimental procedures. The support of experts, researchers specialised in a particular crop or scientific field (i.e. crop protection, crop varieties, weed science, etc.), is essential at this stage. It is crucial to avoid complicating the analysis by factors that are not part of the experimentation. The choice of procedures must be decided on according to the expectations of the trial and the hypothesis being investigated. It is imperative to clarify all aspects of the trials from the beginning.

> An example: the intention is to verify the effects of herbicides on weed control.
>
> Seeds are treated, even if this practice is not common among farmers, since the objective of the trial is not to investigate the destructive effects of soil insects on germination, but rather the effects of herbicides.
>
> Likewise, one may treat the crops against an unexpected caterpillar attack (for example), even if this was not part of the initial research protocol.
>
> One can fertilise the trial, in order to have better initial weed growth and thus evaluate the effects of herbicides better.

2.1.3 Choosing control treatments

The choice of treatment that will serve as a basis for comparison in order to measure the effects of a given technique will be made according to the objective/s of the trial and the mode of analysis to be performed. Here are some alternatives:
- Absolute zero (no manure, no fertilisers) and optimal agronomic control of the dose of fertilisers, for example. This allows one to measure how the crop responds to increasing levels of inputs.
- Practices that are recommended by development institutions (e.g. dose of fertilisers prescribed by cotton companies).
- Average farmers' practice – this can change according to the type of farm targeted.
- Individual farming practices at the farm where the trial takes place. This is often convenient for the test itself but more difficult for the statistical interpretation. The large variation among practices requires that an average be established.

2.1.4 Choosing the number of treatments

There must be a compromise between
- a realistic number of treatments (up to six) to facilitate the set-up and the observations, without tramping all over the farmer's plot, but still keeping the exercise comprehensible to the farmers and visitors (which means that they must understand the trial so that they can discuss the results),
- and, on the other hand, enough treatments to widen the range of responses and provide answers to the questions investigated.

References at research stations can allow the number of treatments to be tested on-farm to be limited.

In Mali, the Office-du-Niger (a development agency) wanted to test different phosphate fertilisers on paddy rice. Only three or four treatments were tested:

a) zero phosphate control (zero P)

b) treatment with an imported phosphate chemical fertiliser

c) treatment with a local natural ground phosphate

d) the farmer's current practice, if it differs from a, b or c

One single dose of phosphate was tested in b and c. Other trials at research stations or on-farm controlled experiments have provided references on the effects of doses according to soil types.

2.1.5 Designing the experimentation plan

Using a particular experimental design on the pretext that other researchers have used it successfully in other circumstances is not recommended. Deciding on whether or not to make blocks, how many replications to include, which size to adopt for the plots and any other questions relative to the experimental protocol depends on the specific problem occurring in a particular location. Relevant literature is listed at the end of this chapter. The reader is invited to refer to it for further information on experimental designs and sound statistical analysis.

The validity of the analysis and its statistical interpretation depends on respecting the following three principles:
- Define the location of the different treatments (toss-up, random draw) randomly to prevent the involuntary placing of one factor at the best location all the time.
- Repeat each treatment to avoid mistaken generalisation owing to a localised and peculiar plant growth accident.
- Control error by striving to minimise all causes of variations other than those whose effects are under investigation.

> In a trial that crosses variety and fertilisation, the objective of which is to investigate the interactions between varieties and doses of fertilisers, one must make sure that the homogeneity of sowing dates and of sowing densities on the elementary experimental plots is strictly controlled.

A. Investigating only one factor (e.g. a variety, a dose of fertiliser or a sowing density)

The team must decide on the experimentation design according to the heterogeneity of the plot.

a. If the plot is homogenous (which is seldom the case) a completely randomised design is used (random draw) (see Figure 1).

N pieces of paper corresponding to the number of plots are put in a bag (n = 24, when 6 treatments are used and duplicated 4 times). The name of the treatment is written on each piece of paper. There are as many pieces of papers with the same treatment name as there are numbers of replications of treatments. One piece of paper after another is drawn randomly without returning them to the bag. The drawing order corresponds to the number of the plot on the field (i.e. plot number one is the one drawn first, and so forth). The risks of error during the setting up are numerous, and the visits to the trial difficult.

b. If the plot is heterogeneous and of a single gradient that varies from one side of the plot to another (or soil type), clusters are used (blocks). The division of the field into blocks (blocking) must be based on the information about the heterogeneity that has been observed. Blocking is a means of controlling the influence of one gradient or soil type (i.e. a known source of variation) since homogenous sub-plots are set up (see Figure 2).

The treatments are randomly located inside

Figure 1: Example of a completely randomised design
Six treatments: A, B, C, D, E & F

				A	C	D		
		F	B	D	A	B	E	
	A	D	C	E	F	C	B	
E	F	C	A	D	B	F	E	

Figure 2: Example of a randomised block design
5 treatments, 4 replications

A	B	E	D	C	Block 1
E	A	C	B	C	Block 2
D	B	A	C	E	Block 3
C	D	E	B	A	Block 4

(The gradient effect is vertical, from top to bottom.)

every block. K pieces of papers are put into a bag (k being the number of treatments). The name of the treatment is written on each piece of paper and a draw is done for every block. These blocks are not necessarily close to one another, and may even be far apart. However, one should avoid anything that causes serious differences between blocks (soil types, history of cropping systems, etc.). Plots are not necessarily square.

This experimentation design (blocks) is more often encountered because it is easy to set up and to visit.

c. Lastly, when there are two perpendicular gradients one can utilise a Latin-square design to control the heterogeneity in the two directions. However, it should be used only in exceptional cases as this design is very demanding. There must be the same number of duplicates as there are treatments, because the plots are square. Studying ten treatments would require 100 plots to be set up.
The random plan is drawn in the following way:
• One starts by designing a Latin square in which the squares are filled with treatments in such a way that each treatment must appear only once in each column and only once in each line.
• Then, a random move is performed along the lines.
• Finally, another random move is performed along the columns (see Figure 3).

The Latin-square design is a crossed block-like design, along the rows, and along the columns.

This experimentation design is not accurate with less than four treatments. It is difficult to set up and to manage with more than seven treatments. However, it makes sense in an area with heterogeneous features.

B. Investigating two factors

a. If the plot is homogenous and the experiment without practical constraints a total random factorial plan will be used (see Figure 4). This design is difficult both to set up and to visit.

In the following example, two varieties of rice and three types of fungicides were studied

Figure 3: Example of a Latin-square design
Four treatments, four replications

D	B	A	C
A	C	D	B
C	A	B	D
B	D	C	A

Figure 4: Example of a total random factorial plan
Six treatments x four replications = 24 elementary plots randomly selected

		E	F	C	B	
	F	A	E	B	D	D
F	C	E	D	A	B	C
B	D	C	A	E	F	A

(6 treatments).

A – IKP + FONG X
B – IKP + FONG Y
C – IKP + FONG Z
D – Dj 12 + FONG X
E – Dj 12 + FONG Y
F – Dj 12 + FONG Z

b. When there is one gradient and no practical constraints the plan in blocks is utilised. Six treatments will be placed randomly in each block (the same principle of a random draw from a bag as for the plan with one factor). Setting up such experimentation may be tricky (see Figure 5).

c. When there is a practical constraint (e.g. a tillage technique that cannot be applied to small experimental plots) the split-plot plan is used in setting up one of the factors.

In this plan one starts by drawing the first factor inside each cluster (block) randomly and then sub-blocks are defined. Then, one draws the second factor inside each sub-block randomly. One gains a lot in terms of ease with which experimentation is set up, but some accuracy is lost on the first factor. This design is commonly used because it is easy to read in the field.

d. When there are practical constraints in setting up both factors investigated (e.g. tillage in terms of the lines and the application of herbicides in terms of columns) a criss-cross plan is designed (see Figure 6). In this type of design, it should be borne in mind that the study of simple factors is sacrificed, and only interactions are tested. This design is easy to set up and easily understood by visitors.

As opposed to station experimentation, the choice of the design must not be orientated towards the best statistical performance, but towards the ease with which it can be set up and understood by visitors. In on-farm experiments, priority is usually given to plans in blocks (1 factor per block), split-plot plans (2 factors per split plot) and criss-cross plans (2 factors criss-cross). In these designs there are always some blocks. These may be either in

Figure 5: Example of a factorial plan in blocks

B	A	D	C	F	E	Block 1
F	A	C	E	B	D	Block 2
E	F	A	B	C	D	Block 3
A	E	D	C	F	B	Block 4

(The gradient effects are vertical, from top to bottom.)

Figure 6: Example of a criss-cross factorial plan
Example with 3 types of tillage (A, B & C) and 2 herbicides (h & j), 4 replications.

Ah	Bh	Ch	Block 1
Aj	Bj	Cj	
Ch	Ah	Bh	Block 2
Cj	Aj	Bj	
Cj	Bj	Aj	Block 3
Ch	Bh	Ah	
Bh	Ah	Ch	Block 4
Bj	Aj	Cj	

the same farm or in many separate blocks located on many farms.

2.1.6 Size of the elementary plots

The size of the elementary plots depends on many criteria related to the nature of the trial, the crop or the plot, as follows:
- The nature of the treatments: a trial on herbicides or on ploughing implements requires bigger plots than a trial on varieties.
- The type of crop: the spacing recommended for a maize crop is larger than that for groundnuts. Therefore, bigger plots will be required for a sufficient density of seedlings. A trial in agroforestry is another such case.
- The homogeneity of the plants: if the seeds are of doubtful genetic material, it is wise to increase the size of the plots to compensate for failures in germination.
- The actual usable surface: this determines (within the farmer's plot) the size of the blocks and the size of the elementary plots. For example, the presence of a hedge or trees, former burns, and the like can lead one to reduce the size of the plot.
- The nature of the criteria to be measured: it is unrealistic and inconvenient to use small plots when investigating economic factors (e.g. work time, work productivity). For instance, fatigue when implementing certain tasks (like manual sowing) usually occurs after half an hour to an hour, which means after 500 to 1000 m^2. Also, investigations concerning tillage generally require large plots.
- The objective of the trial: for demonstration purposes bigger plots are required. Smaller plots can accommodate simple tests to check certain results first obtained at a research station.

2.1.7 Number of replications

For simple trials with one factor, the number of replications must not be fewer than four. Accuracy improves with the number of replications. However, these replications are expensive and difficult to set up and a plan with more than six is seldom affordable. When factorial analysis is done with two factors, one can decrease the number of replications (fewer than 4) because the real replications for the same level of factor are more important than the number of apparent replications.

Rather than carrying out many replications in the same plot, one can use many plots on

With 1 farmer: 6 blocks

With 2 farmers: 2 x 3 blocks   F1 =   F2 =

With 6 farmers: 6 replications F1 =   F2 =   F3 =   F4 =   F5 =   F6 =

several farms. This system is known as dispersed blocks design. It is usually acknowledged that 10 to 15 trials are sufficient for a good evaluation of the technical data in a homogenous zone (this is called adaptation-purpose trials). This number provides sufficient variations in yield data and then the following design, with one factor and six replications, may be used:

2.1.8 Choosing the location (villages or communities)

The elements guiding this choice are the following:
- It is necessary to cover the diversity of environmental situations, a diversity that is normally explored and agreed upon during the concerted diagnosis with all stakeholders (see Chapter II).
- Practical considerations: it does not make any sense to set up trials that will not be accessible during the rainy season, even if these trials are aimed at achieving diversity and maintaining peace of mind. It will not be possible to monitor these properly under such circumstances and operators will waste their time.
- Frequent visits must be made by all operators, including farmers (and sponsors/donors although they are often in a hurry).

2.1.9 Choosing the farmers

It is advisable to work with volunteer farmers who are highly motivated and are selected if possible with the agreement of the community (including local authorities, traditional leaders, etc.). This choice should have been made during information meetings and discussions held in the village. The most important thing is that villagers must understand the objectives of the trial, its development and the division of tasks.

When a farm typology exists (reflecting farmers' diverse needs and interests), one should rely on it to discuss with the various interest groups the need for sertain trails (see Chapter II: Diagnosis).

2.1.10 Choosing the plots

Plots must be selected according to their homogeneity and representativity and also with regard to the trial requirements.

2.1.11 Division of task: the rules of the game

When it comes to developing a technical innovation, the risk is carried by research, which pays for the inputs. The problems of compensation should also be taken into account so that the farmer will not lose. In other words, there should be an objective comparison with any possible loss to the farmer's usual yield.

In Brazil (Massaroca development project, north-east area), researchers never work with isolated farmers but with communities. The problems and the priority themes are discussed with communities, from which interest groups arise. These groups of farmers are motivated by a specific theme identified during the diagnosis. They form the actual planning entities for on-farm experimentation for given themes. Task teams are constituted from these interest groups, and they focus on one or several experiments to be carried out with contributions from researchers and local development operators.

The theme under investigation

If the aim of the trial is to test herbicides, a weed-infested plot will be chosen so that differences among treatments can be evaluated easily.

Homogeneity of the plots

If the trial is about fertilisation, it is not advisable to select a sloping plot, or a plot with a complex history of diverse crops (e.g. one part with maize and the other with legumes), or a plot with different soil types.

Representativity of the plots

A fertilisation trial may be carried out on different levels of a sloping plot (blocks on the upper, middle and lowest levels of the slope). In the same manner, a variety-screening trial may be carried out on different soil types.

When testing a new maize variety that gives a low yield, the farmer should be compensated for the loss in yield by comparing the lower yield to the normal productivity of the farmer's variety (compensation can be in grain or its cash equivalent).

All these remarks on setting up a trial should be completed with the following important observation: A failed trial may produce a wealth of information as long as one knows how to benefit from the mistakes made, with trials being conducted in close collaboration with the farmers and both sides being aware of the risks being taken. There are, however, certain conditions. In order for such a trial to be considered beneficial, its development and the reasons for failure must be consistently monitored.

2.2 Setting up a test

The primary focus of on-farm tests remains the reaction of farmers to a new technique. The criteria for such tests are the following:

- There must be simplicity of experimentation design and a sufficient number of replications on different farms. If a farm typology has been developed, the tests could be implemented on the different farm types.
- It is interesting to have farmers participate in establishing protocol, but this does not mean building up protocol to reflect unrealistic expectations. There must be a consensus with farmers regarding the treatments to be tested, the measures to be taken and the criteria for evaluating the results.
- Two or three techniques (e.g. herbicide application vs. hoeing) or a set of techniques (e.g. double tillage vs. single ploughing + herbicide) should be compared, with one

Examples

A variety-screening trial was set up too late because the ploughing equipment was not available in due time. Consequently, the crop suffered from drought during the final stages of its cycle. Or, perhaps a farmer did not control the weeds in the trial as planned (hoeing or herbicide application), owing to sickness. Such trials can still help in the selection of varieties that are resistant to harsh environmental conditions (weed overgrowth, drought).

replication per farmer, on large plots, and, above all, with many farmers being involved (at least 10).
- The farmers themselves should be involved in setting up the tests, with the assistance of the development operators. The responsibility of paying for the inputs depends on each situation. Generally, it is agreed that development operators pay for the inputs during the first year and farmers pay for the following rounds. The farmers' possible reluctance or even refusal to buy the required inputs usually highlights mistrust towards the technique tested, and/or probable economic hindrances on the farmers' side.

2.3 Setting up a demonstration plot

Techniques that have already been tested and confirmed in previous steps will be used in demonstration plots with three objectives:
- To propagate new techniques at regional level;
- To train technicians and extension officers who will acquire skills about the techniques and exchange ideas and experiences with one another and the farmers while visiting the plots;
- To evaluate the techniques in a larger number and diversity of situations. Demonstration plots also offer an opportunity for the experimentation team to gather feedback from other stakeholders during visits and informal field discussions.

The success of such demonstration plots depends on the simplicity of their design, and on the actual participation of numerous farmers and development operators, from the diagnosis phase to on-farm trials and then to demonstration plots (visits, field discussions).

Sound management of demonstration plots requires the following:
- Demonstration plots should be organised by the local extension officers, possibly with the farmers' support, and should be visited by other development operators (especially researchers) several times.
- Demonstration plots must be kept accessible all year round, since they will be monitored and visited frequently. If the likelihood of success is high, one can even select a plot along a road. The risk of theft must not be a concern, as long as the objective of display and disseminating information is achieved. Explanatory noticeboards, with text and drawings, are advisable to make visits more instructive.
- Demonstration plots may also be combined with the different tests mentioned above.

2.4 Some practical hints in organising on-farm experimentation

To obtain sound results, limitations in means and time must be seriously and objectively evaluated before starting a programme.

2.4.1 Selecting the task team
Team members must have a basic training about on-farm experimentation and should be carefully selected. They must participate in the development of protocols, data collection and analysis so that they understand the importance of data accuracy. Each team member must be assigned interesting jobs to motivate him/her (the idea is to avoid the usual top-down hierarchy of researcher / extension officer / field worker / farmer). Sometimes, it is better to have under-qualified but motivated team members who have a local rural background, which may help with community integration. Young farmers working in local farmers' organisations or interest groups can play this role.

2.4.2 Detailed preparation
Outside the controlled environment of a research station, the following should be considered:

- A preliminary visit of plots with farmers to ascertain heterogeneity and identify abnormal locations (e.g. livestock tracks, threshing floors, burns), previous cropping systems and to discuss with farmers the objectives and division of tasks and responsibilities;
- Practicalities: effective availability of inputs, transport facilities, implements, necessary equipment for measurements, and so on;
- Preparation of trial designs and protocols in wording understandable to farmers, so that they can participate and accept them;
- Preparation of monitoring data sheets that must accompany the protocol and be discussed in the same way with field officers and farmers.

2.5 Monitoring and harvesting

2.5.1 Observations worth noting
Three points should be noted:
- Monitoring can be carried out in different ways, that is comprehensively and accurately on plots where reliable technical references are expected and much less strictly on other plots.
- Numerical data and quantitative analysis are necessary but they must be combined with qualitative observations and on-field discussions with farmers.
- It does not make any sense to collect data simply because it is what is usually done and then not use it. The number of observations and amount of data collected can be huge. Drawing up and then filling out a monitoring sheet is advisable as a guideline for observations and data collection.

2.5.2 Harvesting
This is an important stage of any on-farm trial. Nothing is more frustrating than seeing farmers undertake the harvest of a trial alone, not as scheduled, and mixing up the harvests from different plots. The harvest must be comprehensively organised. The date must be established in consultation with the farmers and must be respected or rescheduled collectively. Some practicalities must be kept in mind. It is necessary to
- discard the borders, then harvest the actual plot area,
- record the date,
- put reliable labels inside and outside the bags,
- evaluate the grain moisture content at the harvest date (a sample must be taken per treatment).

Two harvesting methods are possible. The choice depends on the surface to be harvested, the available means, and the farmer's harvesting techniques:
- The whole actual surface (excluding the borders) may be harvested, generally with the assistance of the farmers. This method is easier in the sense that results can be discussed with the farmers. However, it is advisable to harvest the different elementary plots at the same time and to thresh the harvests separately, in order to avoid mixing them up. This is difficult when the plots are large.
- Alternatively, four to five small elementary sample plots may be harvested. This method is quicker, but it may not prevent the rest of the plot from being damaged. The farmer can then harvest the rest of the plot, which can be threshed according to treatments or not, depending on his/her interest in the quantification of observations made. For the farmer, more practical quantification can be made, for example number of bags harvested.

Whatever the harvesting strategy, sampling is compulsory for the analysis of yield components (number of cobs, etc.).

3. Analysis of results

Different types of analyses can be used. Statistical analysis is the first one. Analysis of variance (ANOVA) remains the basic one and may be carried out on most designs.

3.1 Statistical analysis

Should the experimental design allow it, sound statistical analysis is indispensable as a support in the process of decision making and evaluation of treatments. This analysis makes it possible to know whether one can rely on the results obtained.

Even though field operators may be reluctant to use statistics, it is advisable that they should at least be familiar with the main ideas. For the experimentation team, the objective is to highlight the best treatments. The issue there is to identify, on the one hand, which of the differences in treatments are the result of errors (e.g. plot heterogeneity, measurement errors, etc.), and on the other hand, which are as a result of the actual impact of treatments.

The error is the difference between observed yield and calculated yield. In all statistical calculations, a margin of error in the final conclusion must be acknowledged from the beginning. This error represents the first and second-category risk:
- The first-category risk concludes that two treatments are different, while they are actually identical.
- The second-category risk concludes that the two treatments are identical while they are actually different. Yet again, this means huge research efforts without any results.

Clearly, the statistical analysis is not easy to carry out. However, it is useful in that it ensures that results are reliable (95%) and that errors are avoided, which makes it worthwhile.

Nowadays, pocket calculators (hand-borne computers) are available and easy to use. On-farm experimentation designs are generally simple enough to be treated easily. Otherwise, a desktop computer may be used along with statistics software.

Should the reader require more details on statistical methods for analysing the different on-farm experiment designs, some key references are provided at the end of this chapter.

3.2 Agronomic analysis

(See Chapter II: Diagnosis)
Agronomic analysis makes it possible to explain the biological mechanisms that account for the way in which crops respond to treatments.

This analysis is based on the study of the existing relations between cropping techniques, environment status and the plant population that is cropped. The qualitative model that enables these relations to be explained is the yield build-up model (see Chapter II). To set it up requires a lot of control and observation. This analysis is complex and may lead the experimentation team astray amid a flow of data, especially if initial assumptions and plans were not properly set up. Too much data may eventually hide something that a simple observation in the field would have revealed at a glance. It is therefore necessary to have a good understanding of this approach before using it.

Experienced researchers should undertake such agronomic analysis. However, its principles can still be useful to extension officers and field operators. Especially if the base work on yield build-up for a given crop has been done at research stations (which is often the

> For a farmer, the choice of a new variety that is no better than the previous one is not very costly (cost of seeds) and has no serious consequence. For research, this mistake generates time and money loss (experimentation).

> **Example 1. Rice cropping**
>
> In rice cropping, although it is not necessary to monitor the rice throughout the cropping season, some specific observations and measurements should be made at harvest time. These elements will explain when differences in yield build-up took place:
> - The number of plants per m^2. This number is an indication of the farmer's techniques and the problems that occurred when the seedlings emerged.
> - The number of panicles per plant. This gives information on the evolution of tillering and of panicular initiation.
> - The humidity content of grains harvested. This enables the differences between crop fields harvested at different maturity levels to be corrected.
> - A mass of 1000 grains (200 grains can suffice if a good scale is available). This mass informs us about the grain filling conditions and can help in calculating the number of grains per panicle and therefore tells us more about the flowering process and the beginning of grain filling up.

case), then relevant criteria should be chosen for monitoring, measurement and observation during on-farm experiments.

3.3 Agro-economic analysis

This type of analysis must enable the economic implications of proposed innovations to be foreseen. The experimentation team may have existing reliable data on labour costs, time and organisation and on the cost of inputs (cropping budget). Otherwise, they must collect the data themselves, during on-farm experiments.

The calculations depend chiefly on the convention adopted in evaluating certain data such as family work, costs of inputs, and so on. These aspects sometimes require tricky choices. Estimations are indispensable when advice is to be formulated for farmers, that is when the stage of on-farm experimentation in small trial plots is already over.

Treatments (techniques) that are investigated during on-farm experiments must match the farmers' practices, meaning that they must be acceptable to and affordable by farmers, given their technical and socio-economic circumstances (e.g. amount of inputs). Also, more reliable agro-economic references are obtained from experimental plots of a size close to that of the farmers.

In the case of tests, or even of demonstrations, the economic balance sheet may be done during feedback sessions or farmers' visits. The beneficiaries themselves then define their own criteria for evaluating the innovation.

This analysis must not be normative and judgement cannot be formulated from one single figure, namely the monetary added value brought about by the innovation. Although it may be easier for the experimentation team to

> **Example 2. Rain-fed crops**
>
> It is necessary to record the rainfall patterns in the experimentation area, as well as the possible differences in sowing dates. This can help to explain why fertilisers have given better results in one village than in another. One can then discuss with farmers the differences in sowing dates and investigate the reasons for such differences (e.g. the workforce was focussed on irrigated plots while rainfall made it favourable to sow rain-fed crops. This generated a delay in sowing, which resulted in a poor response to fertilisers.). In this case, the solution is to be found in the irrigated system rather than in a better formula of fertiliser for rain-fed plot.

translate everything in terms of money, this does not always make sense to rural people.

3.3.1 Financial implications of an innovation

Farmers can, of course, be sensitive to the strict financial implications of innovation (extra cost of inputs to be used vs. extra marketable or consumable yield). In the case of staple food crops (mainly self-consumed), other activities must generate the necessary cash for the technique to be sustainably adopted.

> For instance, in the Senegal valley, the main problem is household food security. In order to increase irrigated rice cropping productivity, more fertiliser may be used. This does not mean more rice is sold, as very few farmers are self-sufficient. Remittances from emigrated members of the family will then be used.

3.3.2 Labour-related implications

Farmers are also sensitive to implications in terms of labour. Are the newly introduced techniques very demanding in labour (time, physical efforts)? What type of labour? When? Beyond work time and money, what should be evaluated is the place that this extra work or new labour organisation takes in the family schedule (working calendar).

> For example, it can be assumed that extra work will be accepted as long as it happens during the dry season. If this work is to be done during the rainy season by the female labour force, which is busy at that time of the year with the market garden crops, adoption is unlikely. If family members cannot undertake such work, but are still interested, then establishing a money value for this work makes sense, since a salaried labourer may be hired at this specific time of the year.

3.3.3 Equipment and capital-related implications

Economists often use norms to calculate the utilisation cost of equipment, and its depreciation (see Chapter X: Product management).

The farmer, who is supposed to purchase this equipment, is more concerned about the actual number of hours he/she will be effectively, not theoretically, using it during a year, before deciding if he/she should buy it or rather resort occasionally to a contractor.

> For instance, certain equipment is assumed to have a five-year usage life (obsolescence). One hour is required per operation. The utilisation cost of this equipment will be calculated by dividing its price by 5 years and 200 hours of utilisation per year.

3.3.4 Implications with other farm components

One must not rely too much or solely on financial calculations, especially when studying small-scale family agriculture. Also, the interactions between cropping systems and livestock systems must be investigated and kept in mind when analysing the implications of an innovation.

> For instance, drastic weed control using herbicides on cropping systems can cause problems to livestock systems by removing certain weeds that are a forage resource. Conversely, changes in the livestock management organisation may prevent easily available manure being used for crop plots.

For example, for a rice variety-screening trial, the question should not only be: "So, what do you think of this new rice variety?" as the answer might be inaccurate or subjective, for example, depending on whether the respondent likes the extension officer or not.

Questions should be more precise: Did you notice any differences during germination, the speed of emergence, seedlings' strength, or colour? Which variety had a better tillering? Which one had a better reaction to nitrogen application? Which one gave the highest number of panicles? The biggest panicles? The biggest grains? Which one had a better yield and why? Did you notice different weed infestation rates within the different varieties? What about attacks from insects and from birds? What did the merchants think of the different varieties harvested? Did the women express any problems during threshing, shelling, and grinding? What does the meal look like (colour of the grain, taste, size of the grains, sticking effect, etc.)?

It is only at this stage that one can usefully acquire a comprehensive point of view from the farmer who has begun to feel that he/she is talking to someone who is really interested in his/her opinion as a skilled farmer and as an expert consumer.

The last question may be: "What about next year, will you carry on alone with the new variety? Are you willing to undertake further experiments, test other varieties? Are you interested in other types of trials?"

3.4 Analysing the farmers' reactions

Experiments should not be conducted in rural areas without keeping comprehensive and exhaustive records on the farmer's opinions about the experimentation process itself and the results. Reactions collected during report-back meetings or visits with farmers may be biased owing to the influence of the group and the results that have just been presented. Before reaching this stage, it is necessary to conduct a quick survey among all farmers involved in the experiment. For this survey to be successful, a precise questionnaire must be prepared, in which all important points are addressed (farmers' interest, problems vis-à-vis experimentation, etc.) and which also leaves room for free comments.

It is only at this point that the farmers' real competencies can be used and monitoring during experimentation becomes much easier. Farmers do not do much counting (quantitative data collection and analysis still remain in the experimentation team's hands), but they do observe a lot. In general, they are very reliable about most qualitative and comparative aspects (e.g. insect attacks, impact of a fertiliser application, etc.).

4. Using and making the most of results

The duty of the experimentation team is not limited to writing an annual report on the campaign in which the computerised analysis and results of the trial are mentioned. For the results of on-farm experiments to be really useful, they must be discussed with farmers, researchers and extension officers. They should be used during training sessions with farmers and development operators. Their evaluation must serve to adjust development support programmes.

4.1 Meeting and discussion with the farmers

Report-back sessions and discussions with farmers should not take place only in the final stages, after harvest. If the trial allows it, visits during the cropping season are very interesting when undertaken at certain well-defined times (i.e. when differences between treatments are likely to be obvious: emergence, tillering, flowering time, right after certain operations, etc.).

For the visits to be instructive, it is necessary for the experimentation design to be easily understood by farmers. Thus, simple designs must be selected, even though they may not be the best performing ones in terms of statistics (1 factor with blocks, 2 factors with split-plot or criss-cross designs). The visit must be well prepared. Overusing noticeboards and detailed explanations are not advisable. One should first make sure that the farmers involved in the experimentation understand the trial and the location of different treatments. It is far better for him/her to present his/her trial projects, than for the development operators to do so. He/she must be confident about the trial and have things to say (the idea is to make the farmer comfortable). From then on, generally, discussions start easily without any need for external intervention.

After the harvest, it is interesting to organise a group discussion on the results obtained. Depending on the interest in the trials and on the objectives pursued, a discussion can be organised either with the farmers involved directly in the experimentation, or with other interested farmers. Their discussions should be accompanied by simple figures, graphs and pictures that will highlight the differences observed in the results. These results must be presented while the observations made by the different farmers (mentioned by name or not, depending on the mood of the session) should be particularly noted. These observations, more than the experts' records, will generate discussion, which is more important than the presentation itself. The time devoted to discussion must be longer than the presentation time. Researchers must remain discreet here, and leave the farmers and extension officers present to discuss the results.

It is advisable to compare the opinions of the farmers who did not participate in the experimentation with the opinions of those who did. This usually allows the whole community to confirm and strengthen its trust in the experimentation team (experts, technicians and farmers), and therefore to maintain motivation among all partners. This time is also a time for social acknowledgement of a job done and it is important that the initiators of the trials present all those who took part in the experimentation (i.e. farmers-experimenters, local extension officers, village people actively involved in the project, etc.). At the end of this meeting, the preparation of the next cropping campaign can be discussed with the group.

4.2 Meeting with other experts, development agents, and institutional partners

This kind of meeting can take place after completion and distribution of the experimentation report. This will help to avoid a long, boring and redundant presentation of results and enable the discussion to start immediately:

- Are the results obtained (observations and farmers' opinions) conducive to the reorientation of work from certain research stations?
- How can the results be utilised by extension officers (further training or extension programmes) and by farmers (adoption, transfer to other farmers)?
- What results deserve verification?
- Are there any aspects that affect farmers' credit needs or equipment subsidy policies (or, more broadly, the farmers' agribusiness environment)?
- What were the costs and benefits of the programme (compared with usual experimentation)?

These last questions are very important since many works, reportedly successful in rural areas, have later been criticised for the cost involved (in terms of personnel or transport requirements). The results obtained must therefore be appreciated by the other development operators and decision makers. Field visits and discussions with the farmers are, of course, highly appreciated but they must be carefully prepared since policy and decision makers are usually sensitive to time being wasted. Using field information boards has proved useful with this kind of audience, as it brings about an element of objectivity to the explanation provided.

4.3 Training technicians and farmers

At the end of an experimentation campaign, certain themes and results may appear particularly important. A technical file on these themes can be established with the technicians or farmers.

Besides its role in allowing presentations and the collection of opinions and observations, visiting should also be an opportunity for training farmers and extension officers.

> For instance, in order to explain to farmers and local officers the role and mode of action of nitrogen in plants, a trial on this theme would be an ideal opportunity for training. In contrast to a visit, where simple observations of differences between treatments are underlined, one would go further and explain why and how these differences come about.

> In the Alaotra Lake area, in Madagascar, two technical files were established as a result of on-farm experimentation (one on weed control, the other on rain-fed rice cropping systems). These files were used as memo boards (reminders).

4.4 Adjusting the process

It is essential to obtain the concerted analysis of the cropping season results as soon as possible so that the following questions may be ask:
- Have the objectives been achieved?
- What new problems have been identified?
- What difficulties emerged?
- Which adjustments need to be made?

These questions underline the importance of clarifying the objectives of the protocol from the very beginning, so that progress can be evaluated. A group far larger than the experimentation team itself must discuss these questions. The group will include researchers and experts, development institutions, farmers' representatives, sponsors, policy makers, and the like.

5. Conclusion

The frequent inadequacy of results obtained in stations when applied to rural settings has led many researchers to involve farmers and rural stakeholders in all levels of their research. This does not undermine the merit and the interest of thematic research conducted in experimental stations. On-farm experimentation is simply complementary to research in stations and has its own advantages, in particular:
- Acquiring information about actual farmers' needs through concerted diagnosis
- Establishing solutions and implementing improvements that will be technically viable, as well as socially and economically sustainable, thanks to the combination of results from research stations on the one hand and from farmers' practices and realities on the other
- Establishing connections between research issues and extension issues by implementing research approaches within rural settings and by having them evaluated collectively and locally.

Herein lie the choices for the researchers who intend to finalise their research work by obtaining concrete results, notably in the improvement of production systems in a rural setting. Such a process requires research tools and methods (i.e. the choice of experimental designs, the evaluation criteria for the results, the division of tasks and risks when setting up experiments, etc.) to be adapted. It also requires the farmers, as true rural operators, to become real partners, whose practical competence will be complementary to that of the experts.

As difficult as it seems, both from a methodological and practical point of view, involving rural people in research is the best guarantee of success, as extension and training are natural features of on-farm research.

6. Recommended literature

Ashby, J.A. (1987) The effects of different types of farmer participation on the management of on-farm trials. *Agriculture Administration & Extension* (24): 235–252.

Byerlee, D.B., Triomphe, B. & Sebillotte, M. (1991) Integrating agronomic and economic perspectives into the diagnostic phase of on-farm research. *Experimental Agriculture* (27): 95–114.

Gomez, K.A. & Gomez, A.A. (1984) *Statistical procedures for agricultural research* (2nd edition) John Wiley & Sons Publ., London, UK.

Reza Hoshmand, A. (1998) *Statistical methods for environmental and agricultural sciences* (2nd edition). CRC Press, Boca Raton, Florida, USA.

Mettrick, H. (1993) *Development-oriented research in agriculture. An ICRA textbook.* ICRA publ., Wageningen, The Netherlands.

Moock, J.L. (editor) (1986) *Understanding Africa's rural household and farming systems.* Westview Press, Boulder, Co., USA, pp. 71–90.

Mutsaers, H.J.W., Fisher, N.M., Vogel, W.O. & Palada, M.C. (1986) *A field guide for on-farm research with special reference to improvement of cropping systems and techniques in West and Central Africa.* Farming Systems Programme Report, IITA, Ibadan, Nigeria.

Steiner, K.G. (1987) *On-farm experimentation handbook for rural development projects: Guidelines for the development of ecological and socio-economic sound extension messages for small farmers.* GTZ publication num. 203, Eschborn, Germany.

Zandstra, H.G., Price, E.C., Litsinger, J.A. & Morris, R.A. (1981) *A methodology for on-farm cropping systems research.* IRRI, Los Banos, Philippines.

Chapter IV: Monitoring and evaluation
M.-C. Guéneau

The following chapters are connected to monitoring and evaluation:

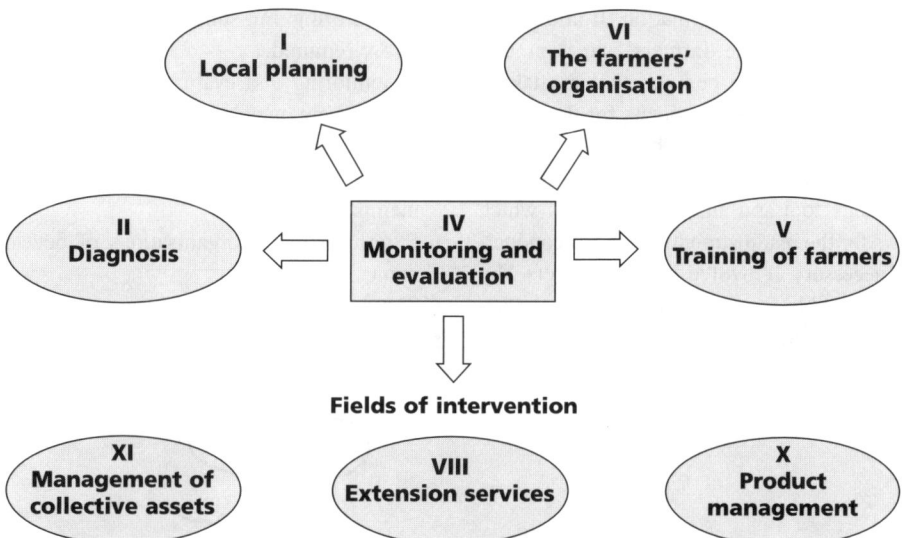

Since the 1970s, the concept of monitoring and evaluation has been very much in use among rural development projects, particularly in those having extended ambitions (i.e. integrated projects, covering many fields of activity). Setting up a system of monitoring and evaluation leads to questions that are easy to formulate but difficult to answer, such as: What is the impact of the development actions conducted? Does the outcome meet expectations? How should the actions conducted be adapted in order to fill the existing gap between initial targeted objectives and actual results?

Nowadays, development projects conceived from outside seem to be replaced by more flexible interventions involving more participation at a local level. This does not decrease the necessity of monitoring and evaluation. Whether support programmes are co-ordinated by development institutions (projects, NGOs, public services, etc.) or initiated by farmers' organisations to support their members, it is essential to set up monitoring and evaluation mechanisms. They must address several requirements:

- Measuring the viability of a project or an action before it starts (*ex-ante* evaluation);
- Following the progress of a project, on a long-term basis, the difficulties encountered and the conditions of its implementation;
- Evaluating the results during and at the end of the project (*ex-post* in order to learn from experience and to have accurate information that can be used to make adaptations to the project and to define new actions).

Evaluation at the outset, monitoring throughout the project, evaluation stages during and at the end of the project must all be seen as a coordinated and continuous process. Even though monitoring and evaluation are two different operations and can be presented separately, it is necessary to have a good understanding of how they complement each other.

Monitoring is a permanent process of collecting and processing information. It allows observations about the farmers' situation and their area and about changes that are taking place. Monitoring is the means for deepening the initial diagnosis (see Chapter II: Diagnosis) and needs of the project. Monitoring is a knowledge tool and an evaluation tool which allows for the readjustment of actions conducted if necessary. It involves all role-players who are responsible for or concerned with the actions conducted.

Evaluation is about questioning and analysing the information collected. It is a tool for reflection, which makes it possible to question the impact of the actions conducted and to reframe them into a larger context. Evaluation is also a continuous process with regular outcomes in relation to and to the benefit of all the different role-players concerned or responsible for actions. A significant part of the evaluation is derived from the information collected during monitoring, but sometimes additional surveys are required.

Monitoring and evaluation feed an *information* system, which can guide decision making by the different role-players:
- Leaders of farmers' organisations and their members
- External support organisations of development
- Sources of funds
- Policy makers

Monitoring and evaluation are therefore two distinctive steering tools which are closely linked and complement each other. They are presented separately in this chapter.

2. Monitoring: landmarks and hints

Throughout the duration of a development project, it is important to find out what is happening, to monitor development indicators and to make a situation assessment.

Monitoring is a continuous process of collecting and processing information. Basically, it is an internal activity by the role-players who are performing an action. Monitoring makes it possible to detect abnormalities during the execution and to bring about corrective measures in the management of the project.

2.1 Usefulness of monitoring

Monitoring is a tool for obtaining comprehensive and critical knowledge about the action under execution. It is a means of control. The data collected can be compared with
- initial targets or expected results,
- established technical norms.

> A collective maize thresher installed in a village is expected to produce 700 kg of cobs per day. Is the actual average performance different?
>
> A given tractor is theoretically capable of ploughing one hectare per day. Is this achieved? Its fuel consumption may not exceed ten litres per hour. Prices of agricultural inputs have been estimated during planning. Does the actual data correspond to reality?

Monitoring is a *management* tool. Any information collected should cause a reaction from the role-players. It enables corrections and adaptations to be made.

Monitoring enables the diagnosis to be completed. Information required for a better understanding of the environment must sometimes be collected over a long period, especially when long-term evolutions are being monitored (see Chapter II: Diagnosis).

> The local demand for maize is about 800 kg of threshed cobs, while the maize thresher is only capable of treating 500 kg of cobs. Should it be replaced by a high performance one? Should another one be bought?
>
> The tractor ploughs less than half a hectare per day. Is the ploughing service well organised? Is there sufficient demand for this service? In both cases, several practical measures can be taken.
>
> Agricultural inputs are more expensive than initially forecast. Is there an alternative source of supply? Is there another technological model which can be adapted?

2.2 The monitoring role-players

What is the use of monitoring? Who must implement it?

Monitoring is useful to all role-players who are responsible for the sound implementation of an action. Each level of intervention corresponds to a particular method of monitoring with the involvement of different stakeholders:
- At the action level, farmers and other direct beneficiaries of the action must be involved.
- At the support level, the different organisations which provide advice and training must be involved.
- At the financing level, the fund providers and sponsors must be involved.

The most important aspect of monitoring is the action itself. Local role-players who are directly responsible for actions in the field must

have a clear vision of what they are carrying out. They must make sure that
- actions take place according to forecasts set during planning, so corrections can be made or targets modified;
- the effects of actions are in accordance with what was expected (see Chapter I: Local planning).

Next is the monitoring of the work done by the support institutions. *Support institutions* strive to organise their modes of intervention, their schedule and their work methodology for the most effective monitoring. In particular, they strive to promote a flow of information downwards (to the decentralised institutions) and upwards (to the decision-making centres). Moreover, they must monitor the use of resources made available for implementing their work to ensure optimal use of these resources.

The *financing agencies* must also ensure that the projects they have financed are subject to a regular monitoring procedure. However, this does not mean that they should get involved in the decision-making process unless such decisions require some major reorientation. The monitoring of their own internal functioning should allow them to respond to their contractual obligations. Here the monitoring consists of ensuring sound connections between all different levels.

Monitoring is essentially an internal procedure at each level. Also, each level of role-players must ensure that other levels, to which they refer, have a procedure of monitoring as well.

2.3 Characteristics of an efficient monitoring procedure

It should be *light, well targeted* and *participatory*. It should combine *quantitative* and *qualitative* components.

The monitoring system must be *light*, meaning that it must not require a lot of time and money. Big multifunctional projects may include a specific team for monitoring and evaluation, but it should remain small (two or three agents and possibly a few part-time surveyors and interns) and must work in close co-operation with the different operational services.

Monitoring must be *well targeted*. After having clearly identified the information to be monitored, one will choose the relevant indicators and process the information. This selection process allows for a reduction in the number of indicators, rejection of the most complicated ones and avoidance of duplication.

The monitoring must be *participative*. Experience has shown that monitoring is often perceived as being imposed by the sponsoring organisations, and that it does not always lead to a sound reflection of the operations carried out by the local role-players. Too often locals are assigned a passive role in the monitoring process. Monitoring must be designed as a collective task for all the local stakeholders through regular data collection, periodical meetings, and workshops.

The information to be monitored will be chosen in collaboration with all parties involved. Only the information whose usefulness is understood and acknowledged by all will be retained. Participation is also required in the process of collecting and analysing data. Instead of designing a separate "monitoring and evaluation task force", it is advisable that the majority of role-players participate in the monitoring and understand it completely. Participation is also the best guarantee that recommendations from the monitoring are really applied.

Monitoring must combine quantitative and qualitative elements. Quantitative information is easier to collect than qualitative. Many projects accumulate quantitative indicators and observations and neglect qualitative information. Qualifying a fact, when you cannot quantify it, is indispensable and often constitutes reliable information.

3. Monitoring: procedures and tools

3.1 Indicators

The basis of monitoring is to collect information in order to establish and measure **indicators**. An indicator is a sign that can be easily observed (e.g. the existence of pastures, the absence of trees) or measured (e.g. depth of the water table, number of animals sold on a market). Indicators enable measurements and evaluations to be made over a period of time. (How an indicator changes from one month to another, from one year to another.) Indicators also serve to measure spatial changes. Similar indicators may be compared in different regions.

> For example, in one valley in Zimbabwe a development project aims to reclaim marshland by drainage and then to develop crop farming. One indicator could be the monthly quantity of drainage pipes installed. Another indicator, which could be more relevant and complementary to the first, could be the increasing surface under agricultural production. A third indicator could be measuring the evolving agricultural production in the reclaimed areas by recording the quantities of crops produced.
>
> In the cotton area of south Mali, a programme set up a savings bank and small-scale credit systems. It was noted that women, although important economic role-players, actually hardly participate as savers or credit beneficiaries. The participation indicator (3%) was compared with a participation indicator in a similar programme in the Dogon area of Mali, where 30% of the credit beneficiaries were women.

Indicators must be defined during the identification phase. Along with the development action, new indicators will continually arise, while others conceived earlier will become useless and therefore will need to be discarded.

3.1.1 Quantitative indicators

Each action generates its own indicators, therefore it is difficult to establish a standard list. The action undertaken should be examined from diverse angles. Below is a basic list of some of the more common indicators.

A. Technical indicators

These indicators make it possible to appreciate the development of the technical aspects of a project, for example, if it is a production programme, the following should be considered:
- The cultivated areas (surfaces), as per agricultural production
- Inputs used
- Labour used (manpower units per day, for instance)
- Production obtained (yields), and so on.

In the case of a programme on agricultural production, a similar approach is advisable.

> In the case of a press which produces karite butter in Burkina Faso:
> - The quality of the karite nuts supplied to the press
> - The quantity of butter produced
> - The inputs necessary to the process
> - The labour force required

If the action is concerned with a programme of commercialisation or the management of a supply storehouse:
- The quantities sold
- The unsold quantities
- The distances covered to reach the markets or the sale points, and so on.

If it is an agricultural extension programme, the following should be considered:
- Indicators of the realisation of the forecasted actions

- Impact indicators (adoption rate of the diffused techniques)
- Efficiency indicators (from the output and yields obtained)
- Indicators on the reaction of farmers
- Cost indicators, and so on.

(See Chapter VII: Extension services and farm management advice.)

B. *Economic indicators*

Economic indicators allow the economic impact and the profitability of an action to be identified. These are often prices and cost records (unit costs and cumulative costs which refer to a given period of time, or to a specific quantity):
- Input costs, labour costs and other variable costs (e.g. operating costs);
- Fixed costs (e.g. equipment);
- Total production costs, which determine the basic value of the products sold;
- Turnover, and so on.

For specific structures, such as savings and credit co-operatives, examples of indicators are provided in Chapter XII: Financing local development.

Examination of the economic and technical indicators during the course of a project often enables the discovery of serious abnormalities (see example on previous page).

C. *Organisational indicators*

(See Chapter VI: Farmers' organisations)

These indicators can help to develop a better understanding of the organisational problems associated with the operation of a project. A classical example relates to the functioning of co-operatives, associations and other local institutions which are set up to manage collective activities. The suggested indicators rely on
- the diverse degrees of participation in the economic activities, in the meetings, in decision making, and in the levels of information discussed and views expressed;
- the number of members on the management committee or the board of directors, the turnover rate and the quality of its members.

A project which promoted utilitarian handcraft in Mozambique aimed to help local carpenters improve their standard of living by increasing their production.

Technical indicators (quantities produced and sold) quickly demonstrated an increase in production. The chosen economic indicator showed a favourable increase in turnover. However, this economic indicator was incomplete, as the costs to be deducted from the turnover to establish the profit were not well controlled.

When an indicator of labour costs (its valuation) was introduced, it showed that the carpenters were earning less from a work day than those who were employed as unskilled labourers.

It was noticed, for example, in a community irrigation scheme in the high valley of the Senegal River in Mauritania that the board of directors was composed of old, influential people chosen more for statutory reasons than for their competence and that no regular procedure of re-election was in place.

The same indicators were applied to a neighbouring scheme and revealed a more balanced group, composed of young, dynamic farmers who were advising some of the influential people renowned for their wisdom all within the same board of directors.

Through comparisons of the two schemes a viable training programme on the modes of collective management in these communities was developed.

> The limited number of women found on the management committees of grinding mills is commmonplace. This weak female representation is out of line given the fact that women are the main users as well as the beneficiaries of the mills.

D. Indicators of social utility

They allow information to be collected on sociological realities and on the ensuing corrective measures that may need to be taken.

> For example, several relevant indicators were selected for a particular savings bank and credit facility in Mali:
> - The proportion of savers within each of the various social categories (rich farmers, middle-class farmers, poor farmers; merchants, craftsmen; women, men; elderly, youngsters, etc.)
> - The proportion of credit applicants within each of the various social categories
>
> The comparison of these two indicators allows a better understanding of the movements between the money saved (effort) and money borrowed (credit demand). These indicators allow for the analysis of the motivations of the two categories of beneficiaries and the necessary improvement of services within each category.
>
> Efficient monitoring of the indicators has enabled the identification of villages with low-performance records and has helped the board of directors to improve their work and develop the awareness of prospective beneficiaries.

3.1.2 Qualitative indicators

Some indicators are qualitative; they can neither be quantified nor can their value be accurately assessed. To describe facts, one can use a scale of values. These indicators are not summarised numerically, but are described in words.

> Example: Rate the farmers' participation in meetings: Good / average / bad

One can try to specify the indicator by using more than one word. Groups of words or sentences provide a clearer picture and provide comprehensive information.

> Examples of groups of words that could be used:
>
> The reception by men of female projects: not receptive / open and passive / open and active.
>
> The dynamics of local associations: Organisation of periodic meetings / Decision making without any delivery / Fund raising and implementation of projects
>
> The integration of different social groups in the decision-making process: The chief makes decisions alone / A few influential people and the chief make the decisions / Different social groups are consulted before decisions are made.

Diagrams and charts can be used to display qualitative as well as quantitative indicators (e.g. Venn diagrams).

3.2 Processing and analysing indicators

Collecting information relative to the indicators is obviously not sufficient. Information must be processed to make appropriate decisions and follow them through. Indicators are one of many tools.

3.2.1 Comparison between indicators

At the most elementary level of processing information is the comparison in time and space of the same category of information.

In a forestry project, the distribution of nursery plants (sales) went from 100 000 plants in one year up to 200 000 plants the following year. However, the same project had a success rate of 60% in the first year (60 000 plants developed into young trees), which dropped to 40% in the second year.

The role of monitoring and evaluation is to focus on the reasons for the expansion, the apparent success of the project and the conflicting reduction, which is cause for concern.

A village co-operative, specialising in agricultural input supplies, has been run with a working capital of US$ 2 000. During a given year, this working capital has been used several times and the shop has sold US$ 10 000 worth of inputs and small equipment. This is its turnover. The total goods sold were initially purchased for US$ 8 000 from retailers. The total costs of the shop's operation (transport, managers salary, warehouse rental, etc.) were US$ 1 800. The profit was therefore US$ 200 (= 10 000 – 8 000 – 1 800).

Turnover is an indicator. One can compare this indicator with the previous year's turnover, or to that of a neighbouring private shop. Profit is another indicator. (Is the business profitable? How can the community benefit from it?)

3.2.2 Tables showing comparisons between indicators

Another interesting way to compare the indicators is to group them in a table where they can be seen together, and to combine them in various ways and thus draw conclusions.

Operations Monitoring indicators	Collection of leaves	Processing	Marketing
Technical	Density of the tree leaf production, seasonal variations, etc. Transport of leaves: distance, means	Eucalyptol content of leaves Need for firewood and need for water distillation	
Economic	Valuation of labour (salaries) Transport costs Licence costs (the right to utilise the trees)	Firewood costs Still maintenance costs	Eucalyptol market price (development)
Organisational	Regularity in supply (delivery) of leaves Organisation of farmers with regard to the supply of leaves	Rate of still utilisation Organisation of the farmers concerning this usage	Regularity of eucalyptol supply Frequency of collection by the purchaser

> **Small-scale eucalyptol processing in Costa Rica**
> Eucalyptol is an essential oil extracted from eucalyptus leaves. Processing is performed by farmers in a mountainous valley where large areas have been planted with eucalyptus trees to control erosion on the steep slopes. In the fall, farmers collect the leaves and boil them in a basic still developed by the local university. A chemical company ensures the marketing of eucalyptol.
>
> An independent analysis of each indicator resulted in a variety of improvements in both the technical and organisational aspects (e.g. improvement of the still productivity through better use of both firewood and water, improvement in the organisation of work by those in charge of the still, etc.).
>
> However, an overall reading of the table of indicators highlighted relationships between very diverse indicators (e.g. the relationship between the work time required to collect a basket of leaves and transport it to the still, the eucalyptol productivity per basket of leaves after distillation, and the market price of the final product compared with the indicator on salaries, and whether the activity is profitable for those farmers who practise it).
>
> Another reading of the table of indicators allows one to see that individual processing and marketing (i.e. each farmer processes his/her own leaves harvested, and sells his/her product directly) were not economical procedures because lower prices were allocated to the small quantities marketed and the stills were used inefficiently.

3.3 Operation monitoring worksheets

In order to collect the necessary information and to measure the indicators, one has to make use of a large variety of monitoring worksheets. They allow information pertaining to needs and circumstances to be organised. Some of the most common types of monitoring sheets are detailed below. It is not advisable to duplicate the sheets or the programme could get lost in a bureaucratic avalanche of paper. Using a minimal number of well-chosen monitoring sheets constitutes a strong foundation.

3.3.1 Crop-monitoring worksheet (monitoring of a cropping season)

A crop-monitoring worksheet must display the target costs and benefits expected at the outset, the real costs and benefits achieved at the end of campaign and the difference between the two.

A simplified crop-monitoring worksheet is useful since it allows for an analysis of the gaps, which leads to an understanding of why the results expected (in a realistic way) were not achieved, and to the correction of the gap. It may happen that the gap is positive, in which case it is useful to know the reasons for success so they can be replicated.

The crop sheets allow not only the reasonable use of several indicators but also the analysis of their development.

3.3.2 Monitoring worksheet at the community level

This example of a monitoring worksheet allows a specific sector of activity to be understood. It does not enable one to follow the activities of a village as a whole.

It is, therefore, necessary to use a combined sheet to enable one to follow all the actions simultaneously and to have an overall perspective of the situation.

Example of cotton production: Crop-monitoring worksheet				
Product	**Planned / expected**	**Achieved**	**Differences**	**Explanation**
Cotton	Surface: 10 ha	8 ha	–2 ha	Insufficient land
	Yield: 1 ton /ha	0,8 ha	–0.2 ton/ha	Clearance
	Selling price: US$ 320 /ton	US$ 320/ton	No gap	Fertiliser application neglected
	Receipts: US$ 3200	US$ 2048	US$ – 1152	
Costs	**Planned / expected**	**Achieved**	**Differences**	**Explanation**
Seeds	10 ha x 100 kg/ha x US$ 0.4 /kg = US$ 400	US$ 500	US$ +100	Seeds had to be brought from far away (transportation cost)
Fertilisers	800 kg x US$ 0.8 = US$ 640	US$ 300	US$ – 340	There had been a fertiliser distribution at a subsidised price from the Ministry of Agriculture

This community-monitoring worksheet often turns out to be an excellent tool for encouraging the participation of local people (community members, stakeholders, and beneficiaries) and for monitoring (a graphic illustration of this point is presented in the section on evaluation).

3.4 Operational accounts

For all activities of an economic nature, the operational accounts method remains a very useful tool (see Chapter X: Product management).

Operational accounts can be easily visualised and provide reasons for profit or loss

Community-monitoring worksheet (example)
Name of the community
General information on the community (indicators)
Socio-economic information
Information on institutions within the community
Technical information
Operational information
Information by activity or sector obtained from activity monitoring worksheets:
Market gardening programme: all relevant indicators
Land reclamation and management programme: all relevant indicators
Grain sales programme: all relevant indicators
Vaccination programme: all relevant indicators
Borehole-drilling programme: all relevant indicators

Example of a local farmers' group managing a truck for small-scale local transport activity

Operational accounts (Year y)

Expenses			Receipts	
Fuel	US$	7 200	Receipts (truck rental to users)	US$ 30 000
Maintenance	US$	5 000		
Depreciation (over 5 years)	US$	8 800		
Salaries	US$	3 000		
Insurance	US$	1 000		
Overhead expenses	US$	1 000		
S/total	US$	26 000		
Profit	US$	4 000		
TOTAL	US$	30 000	TOTAL	US$ 30 000

and ways to increase receipts and reduce expenses. The operational accounts organise the information collected on a given activity. It is an accounting tool which can clarify the relationship between receipts and expenses during a given period (month, quarter, year, etc.).

Receipts and expenses are kept separate. The comparison between the sub-total of expenses and receipts will show a profit (or a loss) that will balance the account.

In conclusion, the collection and processing of the different relevant indicators make it possible to build up a real instrument panel, which can be sectorial or general, or by geographic zone, and which can be reflected on maps, tables or graphs.

Studying the differences observed (in time, in space, between what was forecast and what was achieved) allows a periodic diagnosis of the projects by the people concerned and in turn provides an opportunity to make the necessary corrections. Monitoring is therefore an important steering tool.

4. Evaluation: landmarks and hints

Monitoring, thanks to the information collected, is able to support evaluation. It constitutes the database of evaluation. Evaluation is the final stage of the analysis process. It can take place at a time scheduled at the outset; evaluation determines whether the project has achieved the assigned objectives and helps determine if the objectives are really useful, realistic and attainable. It is a time to stop and reflect on the progress made and redirect the project, if necessary. Monitoring is the movie and evaluation is the picture.

4.1 Evaluation: a permanent attitude of questioning

Evaluation is an ongoing process, which should begin with the identification of the project and establish whether the project is viable.

Evaluation continues throughout the life of the project as questions arise about the relevance of the various strategies chosen. It can be implemented by using the monitoring tools. At certain points a more in-depth evaluation should take place to coincide with the critical stages of the project (e.g. at the end of a growing season or a funding phase). At these times the entire programme should be assessed. Evaluation during the course of a project is closely linked to monitoring.

The final evaluation is done after a project has been completed, sometimes long after, so that one can appreciate the long-term impact on the conditions that one was expecting to change. When one examines the degree to which the long-term objectives have been achieved (improvement of conditions of life, social changes, changes in the economic environment, etc.), it is called *ex-post* evaluation.

4.2 Utility of evaluation

Evaluation, as a critical form of analysis, must be designed as a dynamic process. Evaluation leads to the revision and enrichment of the initial identification and planning phase of the project.

Unfortunately, local stakeholders, support institutions or financing organisations do not always perceive it this way. Instead, it is often perceived as a device for controlling, which may be followed by sanctions. Some are afraid of evaluation, but submit to it thus contributing to it in a negative manner rather than with a constructive attitude. For those who are in charge of evaluation it is sometimes perceived as an instrument through which they can demonstrate their administrative and financial authority and control. This can affect the success of a project and the types of decisions made.

Nevertheless, there has been a positive trend towards a more constructive attitude to evaluation. A clearer distinction has been made between the control function (finance and accounting control, audit and technical control) and the function of reflecting on the project. Evaluation is being seen as a management tool. Its educational dimension is also being recognised, since it makes all stakeholders involved in a project stop and reflect on the project's progress.

It is important to distinguish clearly between these two functions (both of which are necessary), because any ambiguities always tend to benefit the control function at the expense of the reflective, which is the most important function as far as the local stakeholders are concerned. In the rest of this chapter, the focus will be on this second function of evaluation: the promotion of collective reflection within the group involved.

4.3 Participants in evaluation: internal or external evaluation?

Evaluators who do not take part in the project perform external evaluation. They are chosen by the financing organisation or by the supporting institution. External evaluation can perform either one of the two functions mentioned above (internal reflection or control), but should not perform both at the same time.

Internal evaluation basically promotes and feeds the collective process of internal reflection. It is carried out in an autonomous manner, with the possibility of external methodological support.

A variant of internal evaluation is participatory evaluation or self-evaluation. It is used to evaluate projects in which success depends on the participation of the beneficiaries, or when projects are not scheduled in advance and make use of flexible funding systems (beneficiaries undertake projects when the need arises). Self-evaluation allows a large number of stakeholders to be informed and to participate in information analysis.

In practice, one observes that the two types of evaluation (external and self-evaluation) are combined in order to overcome the specific problems of each type.

4.4 Evaluation must be negotiated between the different stakeholders

The different participants of a project must be involved in the evaluation process during which time possible conflicting interests (first raised in negotiations), as well as solidarity (by accepting the inherent constraints of the process) can surface. It is not fair for only one participant (e.g. the main fund provider) to impose his/her views.

To avoid ambiguities, which can spoil the evaluation process and transform it from a valuable management tool into a worthless burden or a trial, the evaluation should be developed during the planning phase. At this time various elements can be determined:
- The type of evaluation: external evaluation, self-evaluation, mixed evaluation, functions to be performed, etc.
- The modality: Who should participate? At this point, it is very important to prepare the terms of reference carefully, determine the framework of the evaluation and present the themes worthy of being discussed.
- The schedule: What will be the duration of the evaluation process? When will it take place?
- Methodology: What information is necessary? How will it be collected? How will it be processed? This point is important as it determines how monitoring should be orientated to facilitate the evaluation.

5. Evaluation: procedures and tools

5.1 Criteria for evaluation

Since evaluation consists of assessing the performance of a project, it is advisable to specify from the outset what the criteria for evaluation are. A criterion is an angle of observation, a viewpoint from which evaluation and judgement will be determined. It specifically refers to an objective being evaluated. It allows one to make judgements based on this objective.

For example, if one wishes to assess the way a project is carried out and make a judgement on its mode of operation, criteria of *efficiency* will help to orientate the evaluation in this direction. (e.g. Are all scheduled stages properly addressed and accommodated in the project?) If one wants to assess the effects of the project or to check if its results correspond to the initial objectives, criteria of *impact* should be used.

The usual criteria or angles of observation frequently include:
- Effectiveness (maximum benefit achieved or objectives reached)
- Efficiency (optimal benefit or objective reached with minimum effort, means of achievement, etc.)
- Impact
- Viability/reproducibility
- Strategy for intervention
- Participation/satisfaction of the beneficiaries

These constitute a solid foundation for the evaluation of a development-support project. They can be used at the various levels of intervention (grass-roots groups, support organisations and funding organisations).

5.1.1 Effectiveness criterion
Measuring effectiveness consists of comparing objectives to results, and in assessing the gaps between what is achieved and what was planned.

This criterion is sometimes difficult to address because many projects define their objectives badly or too broadly, or they quantify their objectives incorrectly from the outset.

5.1.2 Efficiency criterion
Measuring efficiency consists of comparing the results with the means used to obtain them. The means taken into account must be extend-

In a water supply project in rural Senegal, plans were made to drill 110 boreholes per year. The measure of effectiveness consisted of observing the degree to which this goal was realised and of identifying any problems which could prevent the realisation of the objectives.

In this specific case, the objective was achieved for the first two years (and even exceeded during the second year). This form of analysis made it possible to conclude that the project was effective. It also highlighted the problems which could affect the continuation of the project: the speed with which the equipment wore out, and the difficulties involved in gathering and organising the labour force especially from certain communities.

In many agricultural projects, the objective assigned has been "to reach a level of self-subsistence and food security". If the indicators for such an objective and the stages which should be taken to reach it have not been clearly defined, it would be difficult to measure the effectiveness of the project.

Monitoring is like a movie, whereas evaluation is a picture.

> In the example of the borehole drilling, it is not sufficient to state that boreholes were drilled. One also needs to know whether they supplied water effectively all year round, at what flow and at what cost (water price per unit delivered).

ed to their broadest: financial and human as well as material. All sources should be considered, not just the aid provided by the external organisations but also the local resources used from the communities involved in the project.

Analysing efficiency requires a cost-benefit analysis where one evaluates both costs (relatively easy) and achievements (more difficult).

In a broader sense, it is also a method that is used to control the costs. It must end with questions as to what was achieved, and whether the project could have been done differently or at lower cost.

5.1.3 Impact criterion

In its broadest sense, analysing the impact of a project means evaluating all its effects on the environment (technical, economic, social, political and ecological). The impact of a project can also influence national policy.

5.1.4 Viability (or sustainability) criterion

To analyse viability one must assess the sustainability of a project in the long run, and,

---

> In Sao Tome & Principe, the success of marketing garden crops motivated smallholder farmers to set up pressure groups in order to obtain better land tenure security. As a result, the government has had to set up and implement land policy adaptations.
>
> In the Casamance area (Senegal), several villages organised the building and management of a tourist camp for foreign travellers.
>
> The economic impact seems to be favourable.
>
> Each village receives substantial annual benefits, which provide the finance for further collective equipment.
>
> Three to five jobs have been created in each village – a benefit to the local people.
>
> New activities have been started which encourage local craftwork and provide food (basket making and gardening, which generate individual profits for the camp).
>
> The shopkeepers' turnover in these villages has increased.
>
> Technical impact: The tourist lodges have been built using improved traditional techniques (traditional building materials with modern techniques), which have in turn been adopted to improve local buildings.
>
> Cultural and social impact seems to be positive: the project has generated pride, self-confidence and prestige among the people who participated. Contact between local people and tourists has generally been very positive, even though some behaviour on the part of the tourists demonstrating little respect of local customs has shocked the local people and generated some misunderstanding.
>
> Political impact: Different ethnic groups from the villages have learned to work together, which has established co-operation, exchanges and more peaceful relationships between them. The Senegalese government has been impressed by this experience to such an extent that it has included a rural tourism component in its tourism policy.
>
> Ecological impact has not been negative: constructing camps does not harm the environment or create water resource problems since the peak tourist season coincides with a period of water abundance.

more specifically, estimate its likelihood to survive without further external support. Viability is a multifaceted issue, which includes technical, organisational, economic, cultural and social aspects.

> A project supporting emigrants who return to Senegal after 20 years in France allows them to settle as farmers without any serious training in farming or any technical expertise.
>
> The technical viability is a concern, since they have decided to produce food crops, using mechanisation, and this is not profitable. The financial viability is difficult to ascertain: are income levels sufficient? The organisational viability is questionable as production work is organised collectively. Conflicts and complaints arise around the different working capacity and goodwill of each individual.

Reproducibility is one variant of the viability criterion. During an evaluation, one must determine whether a project can succeed if reproduced elsewhere, or if the means involved to ensure its success are too sophisticated so that the experience must be considered an isolated one and not replicable.

> In Honduras, a diversification project allowed groups of farmers to develop tropical fruit crops and export them to the USA. This project was very efficient and viable, but it could not be extended to further groups because the very small marketing network that was created was already saturated.

5.1.5 Strategy for intervention
The set up and development of a project must be thoroughly discussed. In the light of certain problems, were the solutions used suitable? Was the approach chosen the best?

> For example, in the project involving Senegalese migrants, the reason for each operation must be questioned: the collective organisation of the work, the use of mechanised equipment and the choice of crops. The entire foundation of the project should be reviewed.

These five criteria (effectiveness, efficiency, impact, viability and strategy) constitute the classic criteria used in an external evaluation. They make an overall view of the project possible. But there is a sixth criterion to be taken into account, which has a major influence on the others, particularly with regard to impact and viability.

5.1.6 Participation and satisfaction of the beneficiaries
This criterion is used to collect the opinions of the participants in the project. It makes room for the rural people's voice. The term "beneficiary" here must be taken in a broad sense:
- Those who actively participate in the project
- Those who participate on a small scale
- Those who refuse to participate or those who are excluded from the project

It is difficult to make an in-depth analysis of this criterion through external evaluations, which generally do not allocate enough time to collect the necessary information from the grass-roots stakeholders. To make an acceptable analysis this criterion must be integrated into an internal evaluation (or selfevaluation).

These six categories of criteria and their derivatives enable one to assess all the aspects of an action. However, it is not necessary that all these criteria be considered in all cases. Depending on the requirement of the end users of the evaluation and on the time available, some criteria will be more deeply evaluated than others, some might even not be used at all. Under such circumstances the evaluation, although partial, will still be useful.

5.2 Steps of evaluation

One can identify three stages in the process of an evaluation:
- Collection of information
- Processing the information
- Using the information

5.2.1 Collection of information

This first step may appear confusing or even disorganised as the information piles up with no one knowing how to make the best use of it. However, it is important not to rush to conclusions, but to let the information speak for itself and to let the contradictions express themselves.

A compromise should be found between the targeted information sought, as planned in the research schedule, and a broader collection of information, which may not directly link up with the theme under evaluation.

Some important tools for this step are discussed below.

A. Evaluation grid (or evaluation schedule)

The evaluation grid defines the general framework of an investigation by organising information. It is a board that combines all the questions that arise at the outset of an evaluation.

The grid is an internal tool for the evaluation team. Beneficiaries also add some relevant questions and participate in the development of the evaluation grid. The grid groups questions and answers according to the different criteria listed above. The grid should be developed at the beginning of the evaluation. Thus it may happen that during the survey or interview some questions may prove useless or may be asked in the wrong way.

The following example shows the evaluation grid of an inland valley rice development project in the south of Mali (the project is building up infrastructures to increase the flooding potential of rice-cropping areas). The evaluation team envisaged describing the project from the viewpoints of the people involved and those who could provide relevant information. For each theme, there is a specific category of people dealt with:

V = Village people (as a whole)
E = Extension officers (who facilitate meetings and train farmers)
F = Individual farmers
S = Support and service organisations

THEMES / QUESTIONS	STAKEHOLDERS / INTERVIEWEES
I. What is the situation?	V
1. How was the idea born?	
The previous situation	
* Land tenure	
* New vs. traditional organisations	
* Traditional crops? In different areas?	
* What were the expectations of the village?	
2. Objectives of the project	S-V-E
3. Technical description	
* Infrastructure build-up	E
* Role of each partner: topographical survey, bricklaying, provision of materials and training	
* Infrastructure management (management committee)	V

4. Organisation of production
a) The farmers: collective or individual work
   * Participation of women and other labour forces
   * Compatibility between the usual schedule of
     activities and the new activities
b) Village facilitators
   * Role – nature of their support
   * Criteria used to choose the facilitators
   * Perception by village people
c) The relationships between the project and the
   institutional environment
   * Collaboration with other projects
   * Relations with the traditional local authorities
   * Relations with the technical services
5. Prospects / beneficiaries' satisfaction
   * What is your view on the project?
   * What projects do you foresee in the future?
   * What are your future steps?
	V
	
	E
	
	S
	
	V

II. Results
1. Technical results
   * Infrastructure (dam): areas to be flooded for
     rice crops, indicators on the water table, etc.
   * cropping practices (impact of training)
Transplanting: yes or no
Other crops?
2. Physical results
   * Production / yields
   * Use of the production: self-consumption,
     sales, processing
   * Post-harvesting / processing issues,
     conservation, processing, use of by-products
3. Financial results
   * Income from crop production
   * Utilisation of the revenue: consumption,
     saving, investment
   * Cost of the project
   * Farmers' participation in the project's cost
4. Social results
   * Beneficiaries: number and profile
   * Excluded: number and profile
   * Conflicts linked to the beneficiaries
   * Conflicts linked to the uneven increase in revenues
     between certain categories of the population
5. Unexpected results
   * Indirect results: positive and negative
	E-S-F
	
	E-F
	
	E-F
	
	F
	
	S
	
	V-E-F
	
	F-S-V

III. General reflections and comments
   * Strategy of the intervention
   * Is it relevant or should it be different? What
     went wrong? What worked well?
   * Miscellaneous comments
	F-S-V-E

B. Documentation research

Preparing an evaluation requires the study of existing documentation on the area, of the project itself (whatever form it takes), of the monitoring sheets and records, of the accounting documents, and so on. Different recommendations can be made:
- Not everything in the documentation is useful to an evaluation. It is not a matter of doing exhaustive, encyclopaedic-style work. A *selection* must be made.
- The existing documentation may prove useful when cross-checking information to validate it. Tracking incoherencies and inconsistencies and searching for their causes can provide useful elements to the evaluation.

C. Interviews

Interviews involve discussions and the raising of questions with individuals. The critical synthesis of information provided by evaluation depends on the information collected during the course of the project. Interviews are very important, since they allow the direct and indirect collection of critical opinions and views of all the participants.

D. Meetings

Group meetings round off the individual interviews. They can be small, specific group meetings (interest groups) or village meetings.

E. Surveys

To obtain certain information, a survey may be necessary. It should remain light, that is based on limited but representative samples, and address a limited number of questions. Some evaluation teams take on the burden of surveying large samples using an impressive number of variables. They are seldom capable of processing the information and considerable energy is wasted. The results in this case tend to have limited use and impact.

5.2.2 Processing information

Once the information has been collected, it must be organised and analysed in order to make sense of it. This work is done by using tools such as comparison tables, analysis of operation accounts, studies of the indicators, and the like.

A. Table of roles

Such a table describes the roles played by the different participants. First of all, the different aspects of the project are listed and described: decision, realisation, financing and management. Thereafter, one identifies the participants involved at each level: village people, people's representatives, management committees, extension officers, NGO staff members, and so on.

The table of roles below was developed during the evaluation of a project which introduced a grain mill to a village farmers' association (VFA), which was a member of a regional farmers' association (RFA) in Senegal. It underlines that
- the decision to introduce the mill was taken by the RFA, which did not want to delegate this responsibility to the VFA members;
- the pricing at the mill, which should have been part of the VFA's responsibility was, in the first year, fixed by the RFA;
- during the first year, the maintenance of the mill was the responsibility of the VFA, but was taken over by a technician from the RFA during the second year because of technical mistakes;
- the management of the mill, which should have been taken into account by the VFA, was taken over by the RFA in the second year owing to mismanagement during the first year.

Roles	Scheduled	Realised year 1	Realised year 2
Who makes decisions?			
Introducing the grain mill	-	RFA	RFA
Contents of training	RFA	NGO	
Task partitioning	VFA	VFA	VFA
Pricing	VFA	RFA	VFA
Who does what?			
Technical training	RFA	NGO	-
Management training	NGO	NGO	-
Purchasing the mill	RFA	RFA	
Maintenance	VFA	VFA	RFA
Who pays what?			
Training	NGO	NGO	-
The mill	RFA/VFA	RFA	-
Operation costs	VFA	VFA	VFA
Who is managing what?			
The mill	VFA	VFA	RFA
Accounts	VFA	RFA	RFA

The table of roles is filled during the course of the project by indicating what was expected and what was realised. This enables one to understand the development of the project over time. This kind of table may serve to show possible conflicts of interest between different participants.

B. Table of partners' expectations of other partners

Such a table, created from information collected during surveys, allows one to visualise the views and expectations of the beneficiaries. It presents different viewpoints and highlights potential or existing conflicts of interest.

One column may display the beneficiaries' expectations and another the expectations of the support organisation, which are usually the terms of reference of the project document.

In most cases, a huge gap between the expectations of the different partners explains the poor results and achievements of a given project. The other partners' expectations could be used to complete the table.

Project	Beneficiaries' expectations	Supporting organisation's expectations
Health improvement	Benefit from free food handouts	Monitoring child health Decreased child mortality
Afforestation	Satisfying the NGO's expectations in order to establish a climate of confidence, then requesting a borehole development programme	Soil conservation
Vegetable crop production	Increasing income	Improving the diet balance

C. Comparitive tables
To facilitate the analysis of the indicators, they can be represented in comparitive tables.

For example, during *the evaluation of a rice-cropping programme in Mali* some accurate information was available from the following indicators:
- Yields per ha per farmer
- Work times per ha (day per worker)
- Input costs per ha
- Input prices
- Sale prices (per kg) on the market
- Average debts per ha and per farmer

Four categories of comparative tables were established to interpret these indicators. The three first tables allowed an in-depth analysis of the reasons for successful changes and the differences between locations.

It was observed that the yields increased because of the excellent organisation of production by the farmers and because of an increasing consumption of imported inputs. An alarming increase of input prices was observed, which was not counter-balanced by the sale price of the rice.

It was concluded that the yield decrease in 1989 could have resulted from the decrease of soil fertility resulting from the lack of fertilisation: farmers used all but a few inputs because of the higher prices.

Project development over time

	1985	1986	1987	1988	1989
Average yield/ha	2.5 tons	2.6 t	2.8 t	3.5 t	3 t
Average work time (work day per ton)	200 days/tons	220 d/t	220 d/t	230 d/t	240 d/t
Fertiliser price/kg	100 FCFA	100 FCFA	20 FCFA	130 FCFA	200 FCFA
Price of rice/kg	85 FCFA	85 FCFA	85 FCFA	110 FCFA	110 FCFA

Comparison between the activities scheduled and realised

Year 1989	Scheduled	Realised	Difference	Explanation
Area farmed	15 ha	12 ha	–3 ha	Land unavailable
Working time	200 days	220 days	+20 days (hoeing for weed control)	Underestimation
Yield/ha	3.7 t	3 t	–0.7 t	Less fertiliser used
Fertiliser/ha	150 kg	100 kg	50 kg	Increasing price
Benefit/ha	7000 FCFA	5000 FCFA	–2000 FCFA	Low yield and production

Comparison between different irrigation schemes

	Scheme 1	Scheme 2	Scheme 3
Mean yield	2,5 t	3,5 t	2,8 t
Average work time (work day per ton)	220 d/t	240 d/t	1120 d/t

The next table was developed in an attempt to understand the impact of successful changes and the differences in the standard of living of farmers.

In 1985, despite a mediocre yield, a farmer had to sell 1120 kg/ha of rice in order to cover his/her input costs. He/she received 1380 kg for self-consumption or further sale, for a work time valued at 1100 kg of rice (working day = 500 FCFA in 1985). But in 1988, when yields were higher (3,5 t), the increased use of fertilisers, and the increased input price forced the farmer to sell 2545 kg/ha, for a working time valued at 1260 kg.

One could therefore conclude that urban consumers benefited more from increased rice production (more rice provided) than rural farmers.

Comparative table of economic indicators (over time or per village)

	1985	1988
Rice price (F per Kg)	85 FCFA	110 FCFA
Average yield (t/ha – F/ha)	2.5 t – 212 000 FCFA	3.5 t – 385 000 FCFA
Working time valued in monetary terms as well as quantity of rice (Work day valued at 500 FCFA in 1985)	1100 kg — 100 000 FCFA	1260 kg – 138 000 FCFA
Cost of fertiliser valued in monetary terms as well as in quantities of rice	1120 kg — 95 000 FCFA	2 545 kg — 280 000 FCFA
Debt	940 kg – 80 000 FCFA	2090 kg – 230 000 FCFA

5.2.3 Using information

At this point one can attempt to draw some conclusions. A few representations should be used, which are not necessarily quantitative:

A. Qualitative assessment sheet

In concluding an evaluation, a table can be drawn which shows both the strong and weak points in order to synthesise the different elements.

Example: Programme for the development and management of an irrigation scheme in Mali

Strong Points	Weak Points
Increased cultivated areas	Inefficient water management committee
Increased production	Fragility of the whole organisation, which relies on the quality, willingness and skills of the extension officers
Increased food security	
Improved food quality and consumption through vegetable production	

Another example: A project supporting the organisation of an agricultural co-operative which was created by Sarakole emigrants who returned to Senegal.

Present strengths	Present weaknesses
Strong personality of the leader of the co-operative	Insufficient technical skills
Adapted technology (no sophisticated equipment, easy maintenance)	Lack of labour force (women do not work in the fields)
Potential strengths	**Possible risks**
Tenacity of members	Problems in finding profitable crops
Entrepreneurial spirit	Saturation of vegetable market
Possibilities for improved processing, allowing for better conservation and easier commercialisation	Remoteness/poor access to the area

B. Report back

It is important for the results to be reported back to the local authorities and to the local stakeholders in an appropriate and informative way, particularly when outsiders carry out the evaluation.

This provides closure and allows for corrections to be made (certain data only appear at this time). It also fuels a debate on the potential improvements of the current or the scheduled operations (see Chapter I: Local planning, and Chapter V: Training of farmers).

6. Conclusion

One must be aware that the main hindrances to monitoring and evaluation are not of a methodological nature, but rather of an institutional nature. In fact, monitoring and evaluation always present critical views, which challenge the method of interventions, project conceptions, most initial ideas, and practices at the farmers' level as well as at the local and national institutional level. They enable permanent adaptation of interventions to real problems facing farmers and rural people, and promote a renewed kind of relationship between rural people, authorities and financial institutions.

This is generally true in theory, but not always in practice.

A better integration of monitoring/evaluation into the process of development and the acceptance of their function requires a clear definition of their content, methods and modalities from the outset and in a manner that makes sense to all the participants.

7. Recommended literature

Anandajayasekeram, P. et al. (2000) *Agricultural project planning and analysis.* FARMESA / University of Pretoria, South Africa.

CEMAS (1987) *Basic control of assets: a manual on prevention of losses in small cooperatives*, CEMAS / International Co-operatives Alliance, Geneva, Switzerland.

Cusworth, J.W. & Franks, T.R. (editors) (1993) *Managing projects in developing countries.* Longman, Harlow, UK.

De Beer, F. & Swanepoel, H. (2000) *Introduction to development studies* (2nd edition). Oxford University Press, Cape Town, South Africa.

Douet, H. (1988) *L'auto-évaluation dans les actions de développement: pratique et réflexion.* UED, Geneva, Switzerland.

Esgcomb, E. & Buzzard, S. (1988) *Monitoring & evaluating small business projects: a step-by-step guide to private development organizations, and facilitator's manual.* PACT, New York, USA.

Feuerstein, M.T. (1986) *Partners in evaluation: evaluating development and community programmes with participants.* TALC, St. Albans, USA.

Geneva Group (1986) *How to run a small development project.* Intermediate Technology Publications, London, UK.

Gentil, D. & Dufumier, M. (1984) *Le suivi-évaluation dans les projets de développement rural: orientations méthodologiques.* AMIRA Paris, Working paper, num. 44, Paris, France.

Harper, M (1990). *Consultancy to small business: the concept/training the consultants.* Intermediate Technology Publications, London, UK.

Keehn, M. & Kniep, W. (1987) *So you want to evaluate? Building evaluation into programme planning for development education.* Inter-Action, Washington, USA.

Otero, M. (1989) *A question of impact: solidarity group programmes and their approach to evaluation.* ASEPADE & PACT, New York, USA.

Pfohl, J. (1989) *Participatory evaluation: a user's guide.* PACT, New York, USA.

Porter, D. & Clark, K. (1985) *Questioning practice: Non-government aid agencies and project evaluation.* Australian Council for Overseas Aid, Development Dossier num.16, Canberra, Australia.

Prio, J. & Lefilleul, M.F. (1988) *Guide pratique de l'evaluation.* Editions Ouvrières, Paris, France.

Rughs, J. (1986) *Self-evaluation: Ideas for participatory evaluation of rural community development projects.* World Neighbors, Oklahoma City, USA.

PART TWO:

THE TOOLS

V. Training of farmers
VI. The farmers' organisation
VII. Contracts between role-players

Chapter V: Training of farmers
J. Mercoiret & M.-R. Mercoiret

The following chapters relate to training of farmers.

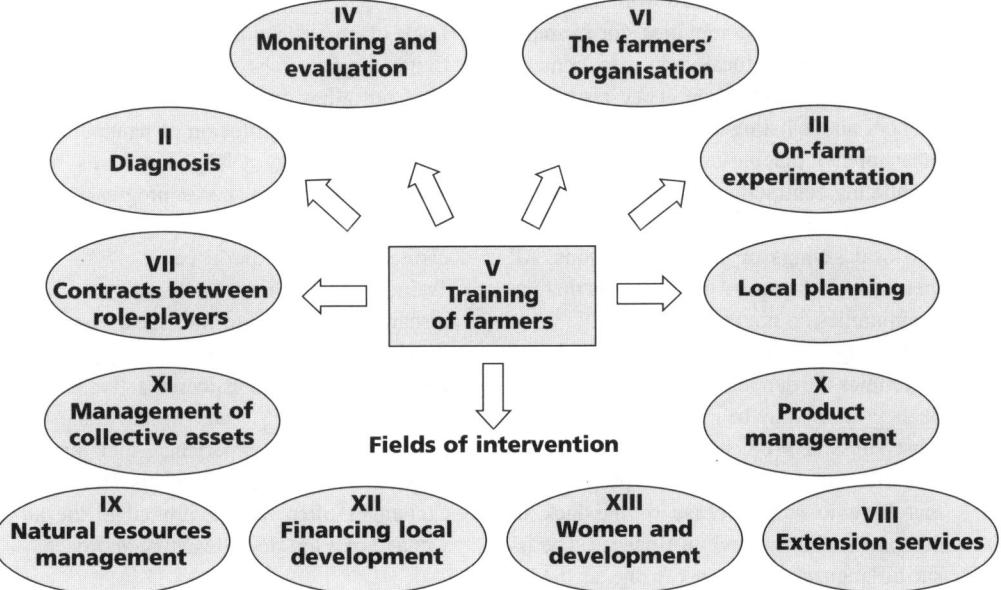

Over the past 30 years, very much, and at the same time, very little has been done about training small-scale developing farmers:
- *Very much* because several formulas have been defined and tested. They are often grouped under the term "informal education", as opposed to "formal education" of a school type. This category includes disseminated actions (local activation, mass education, human capital promotion), pre-professional and professional training (earmarked for youth, adults, couples), literacy programmes and various training techniques. For some years now, much energy has been mobilised to train people in management.
- *Little* has been done because the means set aside for training farmers are still insufficient in sub-Saharan Africa. The share allocated to training farmers within developmental projects is often minimal; when it is significant, it more often concerns extension than training in real terms.

A huge demand for training

Development operators find there is a huge and increasing demand for training nowadays. The farmers themselves and the leaders of local organisations feel and express the same need. This includes the following:
- Training of expert farmers, which is often seen as a priority (e.g. tractor drivers,

smiths, borehole makers, etc.) but also of farmer advisers, who are able, in turn, to train other farmers thus reproducing the training they have received (e.g. seed treatments, crop protection programmes, etc.).
- Management training programmes, which become crucial once the community has overcome the stage of small-scale, irregular or limited production or processing. The programmes are concerned with the management of sectorial activities, for example market gardening, local shop management, cereal storage, and so on. They also teach farmers about listing needs (for seeds and other inputs), placing orders, debt recovery, negotiating contracts for commercialisation, winning or keeping certain markets, managing local savings or credit institutions. All these activities require competences that the communities do not always have.
- Depending on the country, the need and sometimes the demand for basic literacy (alphabetisation) may be important. One of the most stressful problems for farmers is the lack of continuity in certain programmes that have to stop because of the lack of equipment and shortage in finances. The issue of language itself is a key one, as the local tongue may not always be appropriate for negotiation and trading with external partners. Farmers may then be trained to speak English, French or Portuguese (depending on their needs and the country).

Multiple responses, diversity in quality and focus

Support institutions are in favour of training, but there is often a gap between the activities conducted and the real needs. The following major issues are encountered:
- The absence of real training skills among trainers: a number of development operators act as trainers, relying upon their own experience as former pupils.
- The lack of thorough identification of the objectives and modalities of training (priorities, duration, location, period) by both trainees and trainers.
- Lack of background educational media (documentation) which, combined with a shortage of equipment, often reduces training sessions to more or less an elusive talk show.
- Lack of co-ordination between training activities organised by different services. This does not allow for economies of scale (see for example the profusion of handbooks on market gardening, syllabus hand-outs, management files, etc.), nor a progressive improvement of contents and methods, nor capitalisation on experience.
- Training is sometimes badly integrated into research activities and into local economic development issues.
- Farmers are rarely considered active players in training, which leads to the under-exploitation of their experience, their knowledge and capacity to train others. Horizontal exchanges often remain limited to the advantage of a top-down teaching approach.

However, there have been some rewarding and original experiences, which are confirmed by the results obtained (farmer's interest, operational efficiency of the beneficiaries). Some are well known, while others do not go beyond the local level where they are taking place.

1. Points of agreement ... and controversy

1.1 Objectives of training small-scale developing farmers

Globalisation, the state withdrawal process, decentralisation and liberalisation trends increase the need for training at local level everywhere. In most countries, there is a trend to transfer responsibility to local rural organisations. A transfer of competency must accompany this transfer of responsibility, if it is to be effective, efficient and sustainable. In other words, rural people and farmers must acquire the knowledge and the necessary skills to exercise their new responsibilities. However, this consensus hides some important differences:

- Most development operators acknowledge that farmers should have increasing responsibilities in the technical field (e.g. maintenance of equipment, adoption of new technologies) and in the economic field (e.g. inputs supply, credit management, commercialisation and products processing). This can be realised through the implementation of training programmes that allow the farmers to acquire new technical skills.
- Other operators assign more importance to the transfer of responsibilities. They consider that farmers need to regain their decision-making power (e.g. regarding the management of their local territory, irrigation schemes, choice of cropping systems and productions) and to acquire real negotiating power with their partners (public and parastatal services, private sector). Some specific skills are then required of the rural people: they must analyse their reality in order to define priorities, identify and test adapted solutions, define performing local organisation, acquire equipment and methods for planning, monitoring and evaluation, and the like. (see Chapter I: Local planning and Chapter VII: Contracts between role-players).

1.2 Modalities for training and issues raised

1.2.1 Short or long training sessions?

Rural people are often busy with their social lives and livelihood systems (on-farm and off-farm activities, household management, community-related commitments, etc.). This encourages external operators to favour short training sessions (lasting between a few hours and a few days), focussed on one or two themes. As operators are concerned with adapting training to the farmers' socio-economic reality, access to structured knowledge can be delayed owing to the prioritisation of information awareness or extension aspects rather than providing real skills and knowledge. Thus, longer training cycles are sometimes proposed (lasting between a few weeks and several months). The younger the beneficiaries, the longer the training duration and the more varied and harsh the criticisms are regarding these different formulas (e.g. disconnection from the reality and the environment leading to an inability to readapt at a later stage, etc.).

This debate actually masks three others.

1.2.2 Training in or outside the local environment?

In western Africa during the 1960s, training centres were emphasised, resulting in the multiplication of such centres that often offered accommodation because of their distance from the beneficiaries' homes. Owing to their disappointing results, there was a tendency to get closer to the communities and to develop on-farm training. Both locations for training may be relevant. Certain training sessions must take place in the field in order to take into account local specificity: others need the trainees to leave their usual environment behind. This allows one to avoid the local constraints, to acquire a more objective viewpoint and to widen the experience. Training sessions, in which the community and the training centre alternatively accommodate the session, have shown their worth, as have studies abroad (regional and international exchanges between farmers).

1.2.3 Mass training or expert training?

Today it seems indispensable for certain farmers to acquire particular skills in order for them to fulfil certain tasks (e.g. tractor driving, milling and food processing, management, etc.). However, to avoid forming a small group of experts keen to use their skills to the detriment of the grass-roots members, it is essential for the base community to exert a control over them, especially regarding management. This control can only be effective if the community has the means of exerting it. This means that the base has the information and training required to control and, if necessary, enforce (see Chapter X: Product management).

Besides, training farmers should not be reduced to training a few experts, which would seem to be the current tendency. If these experts have an important role to play in collective services (e.g. downward and upward production levels, health, etc.), their efficiency will be reduced if farmer practices and behaviour do not change. What is the use of a local health centre if there is no improvement in the hygiene of the village?

1.2.4 Training the youth or adults?
In the past, the trainee population was often divided. The youth was perceived as receptive and innovative. Thus, long-term training outside the community was often planned for the youth, while on-farm training activities where designed for adult farmers that took into account their time availability and their gradual progress (demonstration, extension).

The training system for rural youths has elicited considerable interest and has been the subject of much literature. These systems were often based on the hypothesis that a young farmer, who is well trained, well equipped and practises a modern agriculture will subsequently be able to make a good income and influence the rest of his/her farming environment (as a sort of a model). Although there have been some promising results, in general, the overall outcome has been fairly disappointing. The training package was often of good quality, but the youths trained were unable or did not want to get involved in the production system, owing to land-related problems, conflicts between the youth and adults, and so on.

Often support services favour the youth for specialised tasks. This practice can be legitimated for several reasons:
- The youth have a lot of free time available.
- Their capacity to acquire new knowledge is high.
- Having them stay in their village is linked to the acquisition of a social status.
- By investing in the youth, of course, one is preparing for the future.

However, one has to avoid going too far (or too fast) in this sense. The availability of the youth is also the cause of their instability. This search for social status may lead them to external opportunities and, in any case, farmers trust adults more than young people for certain tasks.

1.2.5 Homogenous attendance or training based on gender differentiation?
(See also Chapter VIII: Women and development.)

As far as technical training is concerned, women have long been assigned duties related to the domestic economy. This includes several productive activities such as market gardening, small-stock production, small-scale food processing, and the like, whereas new technologies concerning the main production process were earmarked for adult male training.

The progressive recognition of the role of women in different stages of the production process has led to some, more or less successful, revisions. A desire to be more open and to treat men and women equally, as far as training is concerned, has resulted in researching mixed gender audiences. Depending on the rural society, this co-educational system has been more or less efficient.

If dialogue can be instituted, what is the use of insisting that women speak at village meetings when they do not wish to talk, as they consider that it is neither the right place nor the right time for them to do so? Conversely, why should nutritional education or hygiene basics be reserved for women only?

Taking into account the specifics of men and women as is essential in training as in any other aspect of development. It does not automatically lead to co-education. This depends on the needs identified by both men and women and on the specific roles they play in the local society. Finally, it is important to point out that there may be a gap between what is *educationally sound* and what is, at a given moment, *socially possible*.

1.2.6 Indigenous vs. external knowledge

In the past, many training programmes tended to minimise the importance of the farmers' knowledge during training sessions. This knowledge has long been ignored by the trainers, and treated in some cases with disdain ("your way is wrong"). Farmers' knowledge was not really taken into account and seldom appreciated. Although ludicrous, situations where "those who do not know" teach "those who do know" have existed and still exist.

In reaction to these top-down approaches, a common thought has been developed concerning indigenous knowledge and has lead to pedagogic practices that aim to rehabilitate and integrate this knowledge into the training processes.

Without reporting comprehensively on these approaches, one can make three remarks:
- Rural people possess a great deal of complex and structured knowledge even though some of their agricultural practices are no longer adequate in the new ecological and socio-economic environment.
- Farmers do not have answers to all their problems. It is therefore justifiable to help them with solutions, innovations and technologies, which are sometimes simply essential to their survival. The transfer of skills and technology is then necessary.
- The search for a solution to farmers' problems cannot be limited solely to the transfer of knowledge (technical, economic, managerial or social) from the outside. For each theme and every problem, be it general or specific, knowledge should be collected from every available source and favourable conditions created so that farmers may also express themselves and so that their knowledge can be analysed, compared and tested and therefore truly taken into consideration.

1.2.7 Trainers

A common assumption is that any extension officer or development operator, who has a more or less sound technical knowledge, is de facto a good trainer. This shows little respect for farmers and clearly disregards the farmers' own knowledge. For many years, there has been a struggle, the aim of which is to get the trainers well prepared for their tasks and to prevent them from showing themselves up through unprepared talks and misplaced recommendations to adult audiences.

Moreover, the role of the trainer must not be a privilege reserved for external operators only: it is important for the farmers to also assume this role. Several examples demonstrate that it is possible and fruitful
- to train the youth, by taking advantage of the adult farmers' experience (and not only about their local history);
- to take advantage of a multiplier effect (i.e. a trained farmer becoming a trainer to his own peers).

This needs proper preparation, adequate content and sound pedagogic support.

2. Methodological orientations

Farmers need to acquire new skills for them to exercise the technical and economical responsibilities that should be transferred to them and also for them to negotiate with their external and local partners the objectives and modalities of their development programmes.

Farmers cannot acquire all the necessary expertise from a few quick meetings or training sessions. This means that long, mid and short-term objectives should be set up and targeted from the beginning of an intervention and be kept in mind at all times, while the adequate means to reach these objectives will be mobilised at each and every step of the intervention.

A training strategy is not just an accumulation of training days or sessions. It is, first of all, a mindset that characterises those who are in charge of supporting the farmers and their development (leaders of farmers' organisation, extension officers). It results from the combination of three types of actions:
- Encouraging farmers to have increased access to information;
- Evoking and supporting a collective review by the farmers about all the steps of the development process;
- Organising and carrying out genuine training actions, covering technical aspects, literacy programmes and management training.

2.1 Farmers' information

2.1.1 Elements of diagnosis
The higher one goes in the development-support hierarchy, the more available and accessible one realises technical, economic, social and cultural information is. In getting closer to rural communities, one realises how scarce the information is.

As information is selected by the development apparatus and reformulated by specialised services, the final result is that farmers are left with little information at their disposal. Many factors explain this situation: linguistic problems, adaptation of contents to local realities, weakness of distribution mechanisms, and so on.

Moreover, farmers are not equal when it comes to information. Certain farmers' organisations are well structured, experienced and have well-trained facilitators. Hence, they are capable, firstly, of looking for the juridical, technical and economic information; secondly, of treating it according to their objectives and own projects, and thirdly, of disseminating it to their members. Conversely, many communities are under-informed. They receive only little or very selective information, which is seldom conducive to making informed choices about diverse solutions or proposals.

Messages such as "get into groups", or "the pump X is the most suitable", and the like are simple straightforward messages that are believed to be appropriate to the beneficiaries, but which, in fact, do not stimulate the farmers' thinking.

And yet, access to widespread and diversified information is a stimulus for critical reflection; it favours autonomy and fuels both creativity and innovation.

> A few years ago, the federation of Senegalese farmers' organisations (FONGS) took upon itself the task of training extension officers capable of providing available and relevant information in diverse sectors to its grassroots members.

2.1.2 Which information should be distributed?
Small-scale developing farmers need diversified information that is presented in an accessible and attractive manner and linked to their daily preoccupations.

A. Juridical and institutional information
This should focus on the following:
- The local, regional and national administrative and institutional framework and the possibilities that it offers to small-scale farmers (e.g. local public authorities and their role).

- The different legal forms of organisations, that is farmers' groups, co-operatives, associations, and so on. It is interesting to focus on the access conditions for each form of organisation (social share or not, procedure), the cost (if there is a registration fee) and their advantages and limits.
- The means of getting access to credit (Who can apply? What are the proposed rates?) and the diversity of existing credit sources (rural or agricultural bank credit, savings and credit co-operatives, credit programmes supported by NGOs, etc.).

B. Technical information

(See Chapter VIII: Extension and farm management advice.)

Adapted to each region, this information should revolve around the constraints and problems facing farmers, according to their priorities, for example water (boreholes, drilling, pumping techniques), fencing (a problem raised by all communities undertaking market gardening in sub-Saharan Africa), access to high yielding seeds, alternative crop management sequences, small stock production, feedlot systems, and so on.

C. Economic information

This should include different sectors:
- Existing development programmes and projects at local or regional level
- Public and parastatal technical services (location, new role, modalities of collaboration)
- Service supply structures (private or public)
- Existing local possibilities for seeds and inputs supply
- Existing outlets for the different products, the prices practised and the main commercialisation problems that arise in local areas
- Possibilities for diversifying rural activities in agriculture or not

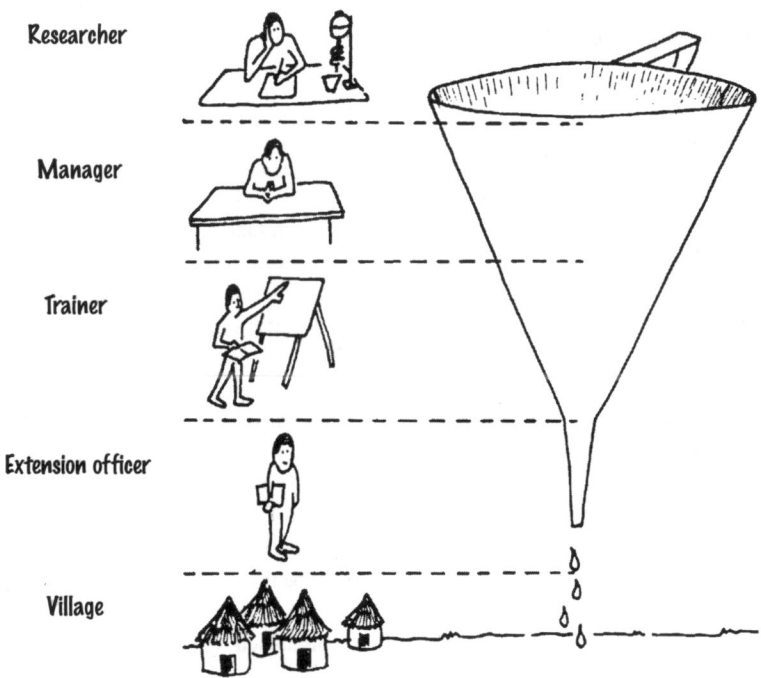

Scarce information trickles down to rural people

- Organisation of the large commodity sectors (i.e. maize, sugarcane, wheat, tobacco, groundnuts, rice, cotton, coffee, etc., depending on the location) and their economic results (with information on the situation and constraints of the global market as well)

D. Socio-cultural information

For each region, this may concern the following fields:
- Existing training sessions organised by different public and professional structures (objectives, content, and modalities of access)
- Existing educational documents produced in the languages used in the area
- Meetings, rural workshops, cultural week organised in the area

2.1.3 How can information be distributed?

A. Three principles
- To be of real interest to farmers and rural people, information must combine both general information and local and regional information.
- Information should not come from one source only (e.g. project, public services, etc.); it may also integrate local sources, from farmers (case studies, farmers' experiences, questions) to other role-players (public services, projects, NGOs, private sector). Messages coming from diverse sources stimulate the farmers' review and their search for solutions.
- As far as possible, information must follow different channels and be adapted to the different types of beneficiaries.

B. Concrete modalities

These depend on the zone and the means available:
- Rural radio has long been used successfully, provided it is not used as a simple medium for government propaganda and also that it gives the farmers a voice.
- Rural newspapers are very successful, especially when they provide space for the farmers to express themselves and when they combine information on the general and local economy with technical information.
- Posters, technical sheets and booklets have been efficient in a number of programmes. Moreover, like the rural press, they are an excellent medium for post-literacy support and can stimulate further literacy demands.
- In some programmes, showing movies, videos and slides has proved efficient in as far as it has allowed the information to reach a larger public audience. One can also imagine itinerant video programmes or portable information boards which can be set up in any local meeting place (at weekly markets for example).
- Opening information and innovation centres is a more ambitious formula which enables documentation and exhibition equipment to be shown in a given place (photos, drawings, machines, objects, etc.). It also allows one to organise open days, debates, demonstrations, and the like periodically.
- Exchanges between farmers, in a given area, or from different regions, are a good way to spread and share information. They have the advantage of getting visiting farmers away from their daily reality, and therefore making them take a step back, open their minds, compare and think. These exchanges require financial support and thorough preparation concerning practicalities (trip, accommodation, food, reception) and educational aspects (organisation of visits, collective review and discussion sessions). Exchanges between commercial and small-scale developing farmers (from different areas, or even countries) can be equally interesting. For such exchanges to be really profitable they must be thoroughly prepared from both sides and an efficient educational structure must be organised by people who know both realities.

2.2 Support to collective reflection

2.2.1 Full participation from farmers

Improvement in farmers' knowledge and skills is not only a result of their participation in training sessions; it is also the result of effective co-operation with farmers that involves them at all levels of an intervention or development programme.

Therefore, the most efficient way to increase the farmers' analytical capabilities is to involve them in the preliminary diagnosis of the intervention and the ensuing process (see Chapter II: Diagnosis). In the same way, farmers will progressively acquire skills by participating, firstly, in the planning and programming of the action which concerns them (see Chapter I: Local planning), and secondly, by evaluating the results obtained (see Chapter IV: Monitoring and evaluation).

The objective is therefore to shift from an external diagnosis for programming, monitoring and evaluation, managed only by the external operators (or only by the leaders of a farmers' organisation) to a more consensual diagnosis, programming, monitoring and evaluation procedure.

This approach has two advantages:
- It creates favourable conditions for the ever-increasing involvement of farmers in actions.
- It develops the farmers' ability to diagnose, analyse, monitor and evaluate the activities that are conducted in an autonomous manner.

2.2.2 Methodological elements

This support to farmers' group works requires methodology:

A. Importance of reporting back the external diagnosis

In order to stimulate analysis and prospective research, it is necessary to fuel the farmers' view with themes and contents of direct interest to them:

- Reporting back the external diagnosis is an excellent means of mobilising the farmers' analysing ability. It allows them to confront the external reality (that of the external operators) with the internal vision (of those who live in it). It can be an opportunity of improving the diagnosis, completing and sometimes of correcting it, provided that, firstly, reporting back is not a formal exercise; secondly, the farmers are invited to criticise what is presented to them, and thirdly, the ambience created by the external operators favours exchanges and the expression of all concerned (see Chapter I: Local planning, where an example is provided).
- The solutions envisaged to solve an identified problem should also be the subject of exchanges and of collective review. Therefore, instead of choosing a technique for pumping water, a standardised crop management sequence, a form of organisation, and so on, for the farmers, it is preferable to present them with a range of available, imaginable solutions, the advantages and disadvantages of which will be discussed together.
- In the same manner, reporting back the result of an external evaluation to the farmers can allow a better understanding of the constraints, the search of solutions, and sometimes also create awareness among farmers of the importance of respecting their commitments (see Chapter VII: Contracts between role-players).

B. Importance of content and the reporting back procedure

For the farmers' review to be productive, certain rules should be respected:
- The information given to them must be understandable, accessible and attractive. Drawing board presentations and visits are effective if those who utilise them are well prepared and have mastered their use.
- Reviewing in large groups is effective if it is prepared by preliminary reviews in small,

homogenous groups (gender, age groups), otherwise the opinion of certain groups may not be voiced. The facilitator must, at all times, remain calm and be open-minded to all kinds of remarks and propositions. He/she does not have to decide on what is good or bad, or on what is relevant or not, but must rather make sure that
- farmers express their thoughts and their viewpoints fully;
- they listen to one another;
- they define orientations or make decisions based on real analysis. This does not prevent the facilitator from giving his/her viewpoint and making some proposals.

> When a training session has objectives and content that do not match farmers' priorities and needs, it usually results in the supposed beneficiaries showing little interest in it (absenteeism, passivity, poor motivation, etc.), and in inefficiency (farmers not practising what they have been taught). For instance, a number of educational programmes focus on the prevention of disease, whereas the farmers are more concerned about treatments. Experience has shown that their receptivity to health education increases when it is done in a programme that includes a treatment component, otherwise health education is seen as a diversion, not adapted to their daily problems.

2.3 Training the farmers

Training activities should allow the farmers to acquire directly utilisable skills, and also simultaneously to develop their abilities to analyse, act, initiate and negotiate. The training should allow the farmers to improve their economic and technical results and increase their autonomy vis-à-vis external intervention.

2.3.1 Principles

Past experience shows that the best results are obtained when the following six principles are respected:
- *Training is more effective when it is embedded in the farmers' reality.* It begins with the problem felt and analysed by farmers and contributes to providing the means of solving it. This supposes that training activities should be linked to concrete situations and realisations and that the farmers should be associated with the definition of the objectives targeted by the training programme.
- *Training must take into account the farmers' knowledge.* It must allow for the expression of this knowledge and its critical analysis. It is both a socio-cultural requirement (appreciation and capitalisation of the farmers' knowledge) and an operational requirement

about a break or a shift away from local knowledge, but rather to establish continuity in the knowledge process. In practice, this creates conditions for truly adopting what has been learned.
- *Training must include both practical and theoretical aspects.* Farmers often need practical training that will allow them to undertake concrete actions and quickly improve their production systems and livelihood. Responding to this demand does not mean that one should refrain from the review process underlying the practical training. On the contrary, training should explain current and proposed practices.

Besides, training should also help to integrate the new skills and sectorial knowledge acquired into the broader context of farming systems.

> Training on how to use agricultural equipment can progressively lead to training on the conditions for using animal traction profitably (fertilisation, crop maintenance, etc.) and also on land management, credit, and the like.

Numerous examples in western Africa testify to the interest of farmers in the so-called theoretical approaches (e.g. water cycle, organic fertiliser cycle, nitrogen cycle, variety screening principles), provided these approaches illustrate some locally observed phenomenon, farmers' practices or an alternative proposal. Providing information that explains why a new technique has been proposed certainly does not hamper its adoption, it simply leads to a better understanding.

- Diverse experiences (GRAAP in Burkina Faso, FONGS in Senegal) show *the interest of training sessions based on successive modules* (or with options). Each module has an operational meaning in itself and prepares future acquisitions. This build up of training in modules makes it possible
  - to take into account farmers' time constraints (each module lasts only a short time);
  - to satisfy farmers' demands for a targeted and useful training programme;
  - to build a learning programme over a length of time, during which the farmers may progress in acquiring more structured and complex knowledge gradually.

- The efficiency of training activities is directly linked to *the establishment of a real dialogue*, of a mutual learning relationship between trainers and beneficiaries. A training programme that aims to increase farmers' responsibility must not lead to top-down, directive, authoritarian practices (treating farmers like children).

- The training must also be subject to *monitoring*. This aspect is often neglected. In fact, it is rare for trainers to visit farmers in their areas once they have completed their training in order to see if it has resulted in a change in farmers' practices and to discuss any difficulties encountered during their implementation. And yet, it is a useful step, different from retraining courses, which can adapt training to requirements.

Example: Improving the result of a crop gardening scheme
In many cases this requires the acquisition of
- technical skills (related to production, water supply, and maintenance of equipment);
- management skills (data recording, bookkeeping, basic accounting skills, reflection on the possible improvements, making choices, etc.);
- information about the environment (supply, credit, commercialisation).

Farmers cannot acquire all these skills in a few training sessions. However, from a general objective defined with the farmers (improving the economic and technical results of market gardening), it is possible to define short-time training modules, spread over two or three years, in order to avoid repetition and to allow the beneficiaries to progress.

All the modules will not necessarily concern all the farmers in the scheme. Some specific modules may be organised, e.g. for pump operators, water bailiffs, managers, the managerial committee, and so on, even if modules are open to all those who wish to attend them.

Each and every module must target a concrete result (seedbed making, transplanting, bookkeeping, basic accounting skills, choice of the crops, etc.). Each new module must explicitly show its link, both with the previous module and with the general objective.

2.3.2 Training small-scale developing farmers: different components

There are three components: technical training, management training and literacy training. Even though each and every component has its own specific purpose, they need nevertheless to be set up in an integrated manner.

A. Technical training

In many cases, technical training remains the priority. Different technical training with different characteristics and pedagogic requirements can be envisaged: mass training, training of specialists and training of local trainers.

- *Mass training* concerns all the users of a technique in a given area.

> The adjustment of a sowing drill or of a plough, spreading of fertiliser, sowing density, water management at plot level, improving the nutrition of draught oxen, and so on.

This training must be on-farm training (in the field, next to the animals, taking into account crop farmers' and stock farmers' calendars, etc.). Practical demonstrations and farmers' days organised around precise themes are adapted means, as long as the farmers attending can participate actively (express themselves, interact, make criticisms, practise or train). Training officers must refrain from trying to convince at all cost. It is important to justify the proposed technique and to its constraints.

- *The training of community specialists*
This is meant for the farmers who are co-opted by their community to carry out a technical function, the beneficiaries of which will be the grass-roots members.

The objective is to train specialists who are often already competent (capable of using and

> The pump attendant in an irrigation scheme, the animal health care attendant, the miller, the smith, the tractor driver, and the like.

maintaining the petrol pump, the tractor, the milling machine, etc.), and to enable them to benefit from as much autonomy as possible (to be capable of facing unexpected incidents). They should, however, use their skills for the common interest, and not as a means to exert power over their fellow citizens.

This supposes that particular attention should be given to
- the collective definition of the role and tasks that the specialist must fulfil;
- the modalities of his/her choice, according to criteria defined in a unanimous manner with the future beneficiaries of these services;
- the compensation he/she will receive in recognition of the services provided (there are many limits to volunteering and compensation in work has sometimes proved to be unsatisfactory);
- the technical quality of services that he/she will provide, which supposes adequate theoretical and practical training that can be divided into modules over one or two years followed thereafter by monitoring and possibly a refresher course or further training;
- the practicalities of his/her services (period, timetables, if it is not a full-time job);
- the social behaviour that the community expects from him/her; this requires monitoring which is not exclusively technical;
- the definition of the modalities for regular evaluation, which will include both technical and behavioural criteria and which should be followed by sanctions or incentives.

- *Training of farmer trainers*
This deals with agents who are generally in

charge of a given sector (agriculture, livestock farming, human health care) and have to ensure a multiplier effect of the initial technical training and monitor the farmers' practices. This is increasingly important for the future because of state withdrawal and the consolidation of farmers' organisations with their own personnel.

It must include both extensive technical training (if one wants to avoid the transmitted contents being simplified and impoverished) and pedagogic training (so that farmers show interest in the contents). Particular attention should be given (as for village specialists) to the choice of farmer trainers and to the conditions under which they perform their functions.

B. Management training

(See Chapter X: Product management, and Chapter XI: Management of collective assets.)

Management training today is felt to be a priority by a number of local groups or associations, external operators, leaders of farmers' organisations and sometimes by grass-roots members as well.

It is first necessary to define management at a small-scale developing farming systems level:
- *Managing a productive activity*, individual or sectorial (market gardening, food processing unit, contractor enterprise in mechanisation, etc.) means being capable of measuring and evaluating its results in economic and financial terms: What are its production costs? What revenues are generated from the activity (in kind, in cash)? Is the activity profitable or not?

This supposes that
- farmers have reliable data at their disposal on the expenses and revenues of the activity and that they are as strict as possible with their bookkeeping;
- these accounts are organised (balance sheet, farming accounts) so that farmers know whether they are making a profit or a loss;
- farmers have to think about a given activity, its advantages and profitability. The accounts must indeed clearly identify the factors that limit the profitability of an activity and facilitate decision making in order to increase this profitability (solutions may be increasing production and productivity through technical innovations, reorganising the workforce, minimising production costs, seeking more profitable outlets, organising shorter marketing channels and integrating upstream and post-production activities, etc.). The provisional farming accounts can enable decisions to be made as to whether one can start an activity or not, or whether one can continue or stop an activity (see Chapter X: Product management, and Chapter IV: Monitoring and evaluation).

- *Managing* means being capable of *comparing different productive activities*. In fact, a productive activity may be profitable by itself, but it can for example compete with other productive activities (see Chapter I: Local planning).

It is wise for the farmers to think about
  - the possible competition between activities: this competition may relate to land (crop farming vs. livestock farming); it can also be about labour scheduling, and so forth;
  - the search for complementarity between activities (market gardening/animal fattening, agricultural activities/handcraft activities, etc.);
  - the search for the diversification of activities in certain circumstances (risk limitation).

- *Managing* also means being capable of *evaluating activities with regard to the protection and regeneration of natural resources, potentialities and capital*. Some activities, although productive and profitable in the short term, may result in the destruction of those resources.

> Profitable cropping systems can quickly lead to neglecting actions that are not directly profitable but that determine long-term sustainability and the future of agricultural activities: e.g. erosion control, soil fertility maintenance, afforestation, etc. (see Chapter IX: Natural resources management).

- Finally, managing means being capable *of evaluating productive activities with regard to non-productive ones* (e.g. health care, equipment maintenance, leisure, socio-cultural activities, etc.). Transfers must sometimes be performed in such a way that the first may contribute to support the second, without, however, jeopardising productive activities by excessive transfers.

Experience has shown that four main types of management training must be envisaged, owing to the diversity of situations and the needs derived from them.

a. Farm management
(See Chapter VIII: Extension services and farm management advice.)

In Latin America, Europe and Africa, diverse experiences have demonstrated that the farmers' bookkeeping of their production units is an important support in making decisions concerning the choice of production, investment, labour organisation, and the like.

These experiences have also highlighted the difficulties encountered. The farm, as a management entity, is often a complex reality, where several centres of decision may coexist with several production units and livelihood systems; sometimes they overlap or sometimes they differ depending on diverse modalities. However, the necessary adaptations must not lead to the issues of management and technical advisory systems being neglected at this level.

b. Management in farmers' groups or associations
(See Chapter X: Product management.)
Two levels are identified: on the one hand, the training of the managers (who are technically responsible for the bookkeeping), the board members or the managerial committee, and, on the other hand, the training of the members, so that the result may be known by all and lead to a prospective review.

- Training the managers
This supposes that the content and limits of their responsibilities should be precisely defined, as should the tools (books of accounts,

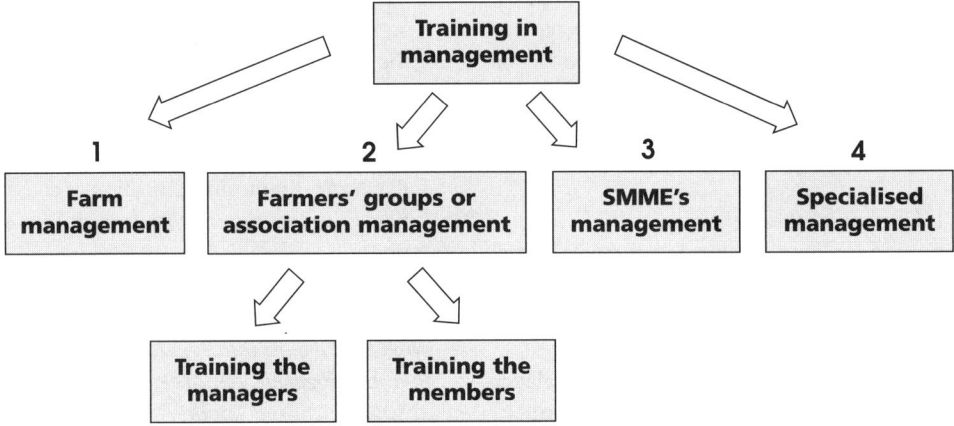

etc.) that they will have to use.
Experience shows that it is useful to
- avoid giving managers the responsibility of decision making, because this privilege should be given to groups (board, steering committee). Nevertheless, this does still sometimes happen and then the manager may believe that he/she has been vested with the power of refusing or of accepting an expense, for example;
- choose the manager unanimously with the members by defining the criteria of choice beforehand;
- develop tools, in language which is understood, that are simple and correspond to the different operations that are to be carried out: the parties involved must understand a management-support document; it is not a question of knowing how to fill it out, but of understanding its utility. Also, the number of management-support documents must be limited;
- define the modalities for reporting back results to the grass-roots members, so that collective control may be carried out: often the collective control is insufficient, in which case one has to think of bringing in an external control (see Chapter X: Product management);
- make it clear that the managerial position should not be confused with that of the cashier.

Training in management techniques is better assimilated when conducted in several steps or modules. The following steps may be followed:
- Highlighting the diverse operations is often a necessary precondition. Its aim, in particular, is to identify, through simulations, operations linked to the activity which must be taken into account by management and also problems which may arise in the course of each of them (in Chapter X details are provided on the tools that can be used).
- The transition from these concrete operations to the headings of record/account books deserves attention; the manager must acquaint himself with the headings, since a number of difficulties which arise in the course of recording come from an insufficient mastering of the headings. In this regard, it is useful to define or to adapt the headings (names, spelling) with the help of the managers themselves.
- Training in bookkeeping and data recording can be carried out during intensive training sessions, during which case studies and simulations should be dealt with.

Special attention should be given to the following:
- The importance of monitoring bookkeeping and data recording and the necessity of organising retraining and improvement courses that focus on managers' practices
- The wisdom of equipping the managers with tools that facilitate their work (calculators, for example)

- Training the members

The accounts of a group or of an association must be transparent and open, so that any member can have access to them and be as well informed about them as possible.
It is therefore useful to
- publicly explain the objectives of management and the modalities of its implementation;
- plan meetings to make a presentation and report back on the results to all members, at the end of the cropping campaign for instance; this will allow information to be widely dispersed and will stimulate review about the improvement of activities.

In line with the above, particular attention must be paid to the report-back activities which must make use of tools that make the accounts accessible to illiterate members. The manager must be trained in these report-back activities, which form an important aspect of his/her work (see Chapter X: Product management).

> If the farmers' association performs more than one activity, each and every activity must be recorded and managed separately before an overview is given to members.

c. Management of farmers' enterprises and SMMEs
In a context of state withdrawal, liberalisation and local empowerment, farmers' enterprises of various sizes are emerging and reappearing.

They may be *individual private small enterprises* managed by small groups (often families), for example village shops, local food processing units or trading businesses, small workshops, transport companies, and so on. They usually start with their own funds (savings, subscriptions), or are subsidised (NGO), loans are granted by banks, credit funds are set up by a NGO, and the like. Monitoring this type of enterprise, especially when it benefits from external financial support, should enable one to find appropriate management tools that are conducive to highlighting the condition of its economic viability.

Local enterprises may also be managed by an *individual or a small group which acts as a local service provider* (mechanisation, small equipment trading or maintenance, fencing, etc.) and which benefits from the equipment acquired collectively (donations, subsidies). This phenomenon is recent but significant in western Africa. In fact, some farmers' associations have found it more efficient to have their equipment managed privately instead of managing it collectively (see Chapter XI: Management of collective assets).

This system may appear efficient but it requires the modalities of management to be clearly defined and the different parties to agree on the sharing of charges and benefits. It is the duty of training programmes to highlight the need for some basic principles (separation of the enterprise's accounts from the association's accounts, setting up contracts) and concrete tools (pricing of service provision, regularity of payments, bookkeeping) that will ensure fluid operations and sound relationships.

d. Specialised management
Certain rural activities require specialised managerial training, for example local savings/credit banks (see Chapter XII: Financing local development).

C. Literacy and post-literacy
Even if one strives to limit its effects, illiteracy among rural people definitely limits their access to information. It also makes bookkeeping and accounting impossible, limits the control of managers by grass-roots members, and hinders the capacity of farmers to negotiate contracts with external operators. The principle of mass literacy courses has been adopted in western Africa since the 1960s. Its implementation has encountered many difficulties, but has also achieved successes (in southern Mali, Burkina Faso) that should undoubtedly serve as examples.

a. *Long-term* village literacy courses or *intensive* sessions?
A literacy programme can be carried out in the villages at a rate of a few hours per week, over a period of one or two years. It requires facilities (a classroom, some basic equipment and stationery), a network of local, trained literacy instructors, and supervisors/co-ordinators (themselves trained) that control the instructors. It is not easy to meet all these conditions, especially during the initial years. Moreover, a number of other difficulties have been encountered:
- An increasing level of learner absenteeism with time: socio-economic constraints and discouragement owing to the slow pace of literacy acquisition are often pointed out.
- Absenteeism is sometimes even increased by the literacy instructors' lack of compe-

tence and sometimes by their lack of motivation (the question of remuneration is always raised).
- Classrooms are often not appropriate or functional: uncomfortable furniture, deficient light, insufficient manuals, lack of stationery, and the like.
- There is an erosion of acquired knowledge, owing to a lack of practice, shown more or less rapidly after the literacy programme.

A number of programmes, in Mali for instance, have tested an alternative system consisting of a 45-day session, which requires the full-time participation of candidates in the literacy programme. This procedure has shown some advantages: absenteeism remains limited, learners are fully available and focussed during the sessions and it speeds up the pace of acquisition, which limits discouragement. Moreover, more qualified trainers can be called upon.

There are some disadvantages: it is difficult for certain farmers to free themselves from their household and family responsibilities (this is particularly true for women) and the costs increase because of the need for accommodation.

This formula remains, however, more efficient than a literacy programme extended over a lengthy period. Also, it is still the only possibility when the concerned beneficiaries are geographically scattered all over the area (e.g. elected members of local authorities, board members of a co-operative covering a large geographic area, leaders of federative farmers' organisations, etc.).

b. *Functional aspects* of literacy programmes
Farmers are motivated to learn to read and write almost everywhere. The beneficiaries must clearly perceive the objectives and advantages of literacy in order to commit themselves on a long-term basis as a literacy programme demands a great deal of commitment.

This functionality is not always easy to demonstrate. A number of experiments have proved that management is a more efficient support than most productive activities. While literacy is not indispensable to producing, it is indispensable to managing (a cash box, a market gardening scheme, a processing workshop, trading and commercialisation operations, etc.), and to a proper control of management and economic operations.

In practice, experience has shown that
- it is not realistic to try to teach everybody to read at the same time. One should rather begin with those who perform tasks that require literacy (managers, secretaries to committees or boards). Their motivation is stronger because they can see the use of literacy more obviously and can immediately put it into practice.
- youngsters are often the first beneficiaries of literacy programmes, which then enables them to occupy technical positions usually occupied by older leaders. This results in their gaining social status and they can be at the origin of strong innovative dynamics.

> For instance, the cotton and food syndicate SYCOV of southern Mali was born from a process initiated by the secretaries of village associations meeting with one another.

c. *Co-ordination* between literacy programmes and other training activities

Such co-ordination significantly reduces the issue of post-literacy, since technical training or management training may include aspects of monitoring and recycling. Instructors must promote the use of various types of written media to stimulate and motivate the neo-literates (booklets, files, information pamphlets, etc.).

d. *Literacy programmes in local languages* and learning another national official language (English, French, Portuguese, etc. depending on the country).

In western Africa, the most successful literacy programmes have used the most common languages at local level. This choice is based on cultural reasons (the national culture being highly valued, of which language is an essential component) and on educational reasons. It is much easier to learn to write one's mother tongue than to try to learn to speak, read and write a new language at the same time.

The major questions raised while translating national languages (difficulties as in Senegal in the 1970s) are now solved (e.g. invention of new words, or resorting to circumlocutions for the translation of certain specialised terms).

However, using local languages does not exclude learning another major national official language in the future (English, French, Portuguese), especially when this language is the major medium for information, exchanges, and the like, across the country. It is then important that part of the population should master it at least partially (leaders of farmers' organisations and all those, who, on behalf of a group or a community, have contacts outside, negotiate contracts, study propositions, etc.). The pretext of authenticity (resulting in the exclusive use of their local language) may confine farmers and rural people to their local zone, condemn them to remain sub-citizens, dependant on translators, and finally, hamper their ability to understand their environment and defend their interests.

2.3.3 A tool for organising farmer training: the training plan

Training farmers must be seen as a permanent process that adapts itself to their needs and their constraints. Two stumbling blocks are to be avoided:
- Developing a rigid, scholastic type of programme which cannot accommodate unexpected needs or social constraints

- The mere absence of planning, leading to session-based training (or even to improvisation), which is not conducive to systematic progression nor to the efficient use of available resources

When a medium-term development plan and a short-term programme of activities exist (see Chapter I: Local planning), it is possible to link together and define a medium-term training plan and a programme (on a yearly or a semester basis, etc.) of training activities.

To set up the training plan, it is advisable to answer the following questions: Who will be trained? On what? Why? By whom? When? Where? With whom? How? (What about the educational material and financial means involved?) This should relate to the development actions planned for the area.

It is important that this training plan should be established in conjunction with the farmers, who must be involved in the different steps of its development, its use and its evaluation.

A. Identifying training needs

Identifying training needs often remains the responsibility of outside training operators. It is therefore characterised by their analysis of the situation (often external) and by the objectives that they are pursuing (often sectorial, almost always determined by their institutional requirements). Thus it is decided, from the outside, what is good for the farmers, what is necessary, and later, the farmers' lack of enthusiasm to follow the sessions organised for them is deplored.

For the farmers to participate effectively in organised sessions, the training process should begin with the phase of identifying needs.

Various methods exist that enable farmers to express themselves and their needs:
- Farmers have certain training needs, which they may express explicitly (women may ask to be trained in certain craftwork techniques or in market gardening; youngsters may want to be trained in modern agricultural

techniques, or in small-scale service entrepreneurship, etc.). These needs are to be taken into account and have to be analysed with the community, notably by highlighting the conditions to be fulfilled so that the skills acquired through training can be used. It is not enough, for example, to learn a new production technique, without questioning the availability of equipment (how to get it?) and market (where to sell the product?).
- Other needs are not directly expressed, but come to the fore through the farmers' identifying of certain issues (e.g. the cost of contracting for tillage is considered excessive, yields are very low, credit rates for accessing inputs are too high, etc.). Training requirements are implicit in these remarks though not expressed directly. The role of the trainers is to reveal these needs and express them in an explicit manner. These needs are often heterogeneous (they appear during discussions with rural people), hardly prioritised and often sectorial (rural people do not necessarily establish a connection between the different problems, they are not aware of the upstream and downstream implications of a specific request). Finally, these needs must be analysed in terms of a training programme, which may contribute to finding solutions.
- The external operators, the trainers may also perceive some needs. They must also be discussed and analysed with the farmers.
- All the training needs perceived, expressed and analysed may not be satisfied. It is necessary to define priorities and to classify them. This classification should be a consequence of the development plan and of the programming of actions which is decided upon with the agreement of all parties (see Chapter I: Local planning) and which puts training terms into the operational decisions taken. The role of the trainers, at this stage, consists of highlighting this relationship between local planning and the training programme.
- The training needs are diverse and may lead to the design of diverse programmes. It is a good idea to distinguish them according to their objectives (mass or specialist training), or beneficiaries (men, women, youngsters, etc.).

B. Defining the syllabus for training

The contents should be defined from an accurate identification of the profile targeted: What task must be performed by the manager of an association or by a grass-roots member of a market gardening scheme? Which skills must he/she have? Which training topics are indispensable? This procedure has the advantage of revealing the differences between what is indispensable, necessary, useful and more or less useless. It also helps to establish the progression of the syllabus.

C. Identifying educational resources for training

There are external and internal resources:
- Internal resources are often numerous. They can be human resources (e.g. experienced people, young scholars, farmers who were exposed to training in the past, people with traditional knowledge about land, health, resources, etc.). Taking stock thoroughly of this local know-how and skills is a necessary prerequisite. It allows connections to be established between external knowledge and the search for complementarity and contributes to the continuity of knowledge and action. Internal educational resources also relate to local social structures (demographic patterns, neighbourhood network, solidarity network, etc.), and to infrastructures (existing clubs, groups, and the places where they meet, village hall, etc.). Cultural resources may also be resorted to (e.g. traditional institutions for discussion, control, decision, but also local proverbs and tales, etc.).
- External resources are indispensable but they have to go beyond the resources made

available by the training organisation: underemployed local people, idle extension officers, and the like, may also get involved.

D. Developing the training plan

The development of an education/training plan derives from the above: analysing the training needs, defining the contents, as well as mobilising educational resources.

The duration of the process is similar to the one proposed for local development planning (see Chapter I: Local planning). The different steps of the programme should take account of:

- The groups of beneficiaries
- The targeted educational and operational objectives
- The envisaged progression
- The concrete modalities of training

If the plan is reported back to farmers, it may become a contract which fixes the reciprocal commitments of the different partners concerned.

E. Basic educational scheme

The development of a training programme can help put the above principles into practice. It

---

Example: Improving fertilisation techniques

1. Problem statement:

Farmers are aware and concerned about decreasing soil fertility, which is revealed by certain indicators they observe (e.g. decreasing yields, changes in soil status–moisture content, increasing weeds and soil-borne diseases, soil structure degradation, etc.).

2. Expression and analysis of the farmers' knowledge:

Farmers know some traditional techniques for maintaining soil fertility (long-term fallow, manure application, etc.). These practices are limited owing to the reduction of available surfaces to be cropped, the decrease of livestock, or livestock migration, grazing problems as herds are no longer looked after, etc.

3. Information supply, training:

Modern fertilisation techniques are presented and explained to farmers (e.g. chemical fertilisers, rotational cropping, compost application, etc.). For each technique, the expected effects and the technical modalities of application are specified.

4. Experimentation:

- Setting up experimental or demonstration plots, visiting experimental plots already in place (see Chapter III: On-farm experimentation)
- Highlighting the experimentation conditions (isolation of a factor, duplications)

5. Analysis of the technical results:

The experimental results are explained (the nutritional needs of crops, nutrient exports because of cropping, the effects of inputs, etc.).

6. Evaluation:

This concerns the costs linked to the different techniques, their profitability, the optimal conditions of use and the conditions for access to the inputs. This kind of evaluation can be used as a basis for calculation exercises and post-literacy purposes.

This type of scheme can be implemented for a simple theme in a couple of hours, or for a more complex issue, over several months, with regular sessions. It uses the different steps of scientific approach: identification of a problem or of a question, formulation of a hypothesis, experimentation, evaluation of results, verification and formulation of a response.

should be guided by the following general educational framework, which remains open and may include infinite variations in modules and modalities:
- Precise *identification of the problem* that the training aims to solve (this is a reminder, since the training preparation already highlighted it).
- Expression and analysis of the farmers' skills (*indigenous knowledge*) in order to highlight their practices, know-how, justifications for these, their advantages and limitations and the factors which explain these limitations.
- *Supply of information* (training itself), aimed at transmitting technology, knowledge or skills capable of helping to solve problems or open up new opportunities.
- *Experimentation*: the propositions formulated during training constitute hypotheses, since their validity has not yet been proved locally with the beneficiaries of training. They must then be verified by comparing their results with those obtained by the farmers.
- *Analysis of the technical results*: this consists of comparing the results obtained and explaining the differences observed (see Chapter III: On-farm experimentation).
- *Evaluation of the socio-economic relevance of technical propositions (are they affordable? can the farmers adapt them?)*: economic profitability, access to the necessary inputs; definition of the conditions for adoption, and so on.

2.3.4 Implementing training: practicalities and tips

A. An educational and practical preparation

Training, be it for a long-term programme or a half-day session, must be carefully organised and prepared:
- Timetables must be respected (training sessions, breaks, meals, etc.); a transport system must be set up to enable the farmers to attend; premises should be in a good condition; educational equipment and media must be collected and checked ahead of time to create favourable conditions and to avoid wasting time.
- In the same manner, trainers must not improvise their training session but prepare it carefully; this does not mean that they have to talk all the time, refusing to let the farmers speak. This means that they must structure the content in advance, including examples, proper media, free discussion times and so forth.

B. Necessary ongoing dialogue

Dialogue is necessary so that the farmers may express their skills and knowledge and accept an analysis of these. In doing so they can also express their doubts and reservations about the propositions that are made to them. Dialogue is impossible if the trainer wants to convince at all costs, using arguments that he/she believes without exception, or if he/she accepts the farmers' views only when they support his/her message.

C. Diverse, meaningful and appealing training features generate success

Boredom arises from monotony. Keeping farmers locked up in a room for long hours is not very educational or efficient; the same applies to keeping farmers in the sun for a long demonstration. Media and training venues must be varied and attractive and always chosen according to the objectives pursued and to the theme under discussion.

3. Conclusion

Training farmers is necessary for them to master the technical, economic, social and cultural changes that they call for or that their environment imposes on them. Training must not be seen as a gift from those who know (the development operators, technicians, trainers, etc.) to the farmers who are considered as ignorant. Training will be effective only if its beneficiaries wish to acquire expertise, skills and knowledge, in order to understand and react in accordance with their interests. The role of training operators is, first, to mediate between the farmers' skills and external knowledge and secondly, to facilitate the access of farmers to new knowledge that will fuel their own creativity and help them to acquire increased or new expertise.

The efficiency of training activities is also linked to several other factors:
- The diverse activities conducted must be organised and be part of a coherent strategy which allows the proper evaluation of each specific action.
- Training is closely linked to action (to the extent that some teams now call their activities *action-training*). This co-ordination must explicitly link local planning, the implementation of development programmes and training activities.

An efficient training programme is demanding on human, educational, practical and financial means (even though these means are far less than the cost of formal education). Thinking that farmers will support their own training financially is a myth. The small contribution that they could possibly make would never match the financial requirements of training, because they have many other priorities that often determine their own survival.

In a democracy, training, capacity building and the promotion of human capital are public mandates, even when the beneficiaries contribute to their conception and management. This should be taken into account during the current processes of the transferral of responsibilities and of charges to rural communities. This means that governments and their diverse fund providers should
- support training programmes defined by local farmers' organisations (through transfers of personal subsidies, etc.);
- monitor and evaluate the results;
- promote the adaptation of programmes (objectives, contents and modalities of management) to meet farmers' expectations.

It is under these conditions that training can help farmers to become more autonomous and more efficient in technical and economic terms, while maintaining and respecting their cultural and social values.

4. Recommended literature

De Beer, F. & Swanepoel, H. (2000) *Introduction to development studies* (2nd edition). Oxford University Press, Cape Town, South Africa.

Easton, P. (1984) *L'éducation des adultes en Afrique noire. Manuel d'auto-évaluation assistée.* Tome 1: Théorie. Tome 2: Pratique. Karthala, Paris, France.

Fraser, C. (1994). *How decision-makers see communication for development.* Development Communication Roundtable, UNICEF, WHO, Agrisystems, London, UK.

Mettrick, H. (1993). *Development-oriented research in agriculture. An ICRA textbook.* ICRA publ., Wageningen, The Netherlands.

Moock, J.L. (editor) (1986) *Understanding Africa's rural household and farming systems.* Westview Press, Boulder, Co., USA.

Ramirez, R. (1998) Communication: a meeting ground for sustainable development. In: *The first mile of connectivity*, FAO, Rome, Italy.

SADC (1998). *Participatory rural communication appraisal handbook.* SADC, Centre for Communication, Gaborone, Botswana.

Selener, D. (1997) *Participatory action research and social change.* Cornell University, Ithaca, NY, USA.

Soveri, R. & Zunguze, M. (1999) *FARMESA/Masimanyane community video pilot project. A chronological account of the community video process as performed by the marginalized community of Woodville.* FARMESA, Bulawayo, Zimbabwe.

Whiteside, M. (1998) *Living farms. Encouraging sustainable smallholders in southern Africa.* Earthscan Publications, London.

Chapter VI: The farmers' organisation
M.-R. Mercoiret, J. Berthomé

For reference purposes, Chapter VI may be connected to the following chapters:

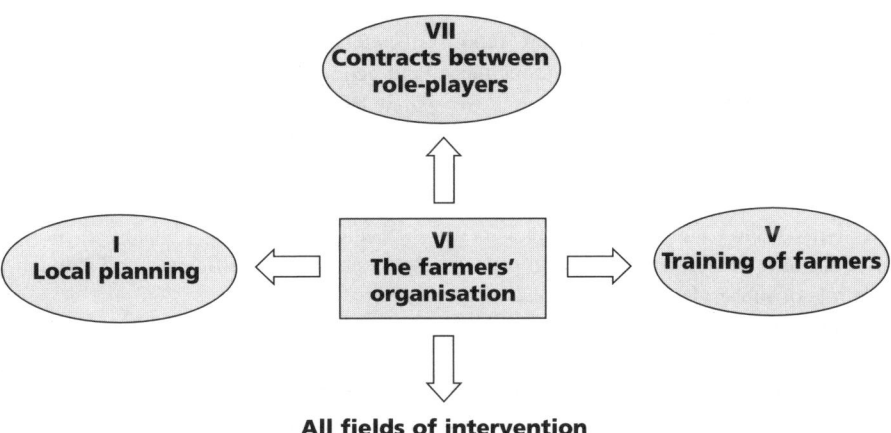

1. Introduction

The role that rural people and farmers can play in defining the objectives and management of agricultural development programmes will depend on how well they can organise themselves
- in an autonomous manner, with objectives and programmes in accordance with their priorities and interests;
- within efficient organisations, capable of bringing specific answers to the farmers' problems and expectations;
- at different geographic levels (group, village, local, regional and national), by participating in the different types of decisions that concern them.

This organisational process is sometimes called the professionalisation of agriculture. It can be defined as being the emergence of an organised agricultural profession that
- has the means and ability to take responsibility for support functions for farmers;
- has the required representativity to defend the farmers' interests while negotiating with other agricultural production stakeholders (state and public services, industries, traders, etc).

In sub-Saharan Africa, the evolution of the economic and political context opens new prospects for farmers' organisations:
- Political openness and democracy have been achieved (or are underway) in most countries. This guarantees the right to form associations, in which members define objectives and planning themselves.
- The liberalisation of the economy, by reducing the role of the state, theoretically creates room for local economic initiatives coming from farmers.

Moreover, most states and external operators wish to discard pre-established organisational models formally imposed on farmers in a more or less direct manner.

Some ambiguities still exist. In certain cases, the state has transferred responsibilities, functions and charges (notably financial) to the farmers, as it can no longer meet or perform them. This transfer is often decided upon in a unilateral manner, by the state and fund providers. It is then carried out without the farmers' participating or being consulted as to the modalities or time schedule. It often concerns more or less profitable functions and establishes limits to the farmers' responsibilities.

Well-established and strong farmers' organisations have reacted and refused to play a simple role of subservience to the state:
- They have grouped themselves into federations at regional or national level (e.g. creation of the Cotton and Food Crops Farmers' Union (SYCOV) in Mali; reinforcement of the federation of NGOs in Senegal (FONGS) and of the federation of Naam groups (FUGN) in Burkina Faso).
- These organisations represent the farmers' interests and negotiate the conditions under which farmers should be involved in society as a whole, at different levels. Thus, for example, they have demanded a role in the management of certain product chains, and of credit (in Mali, Senegal and Guinea).
- They create support services for their members (services in production, agricultural advice, training, etc.).

This development has had a number of consequences:

In many places, extension workers are in a redeployment/retraining phase. They have had to give up their former role that gave them enormous power over farmers' organisations (top-down approach) and become advisers. It is probable that, in the short run, some will be employed by farmers' organisations (some examples already exist in Burkina Faso – Naam groups, in Senegal – FONGS, Koupentoum group, etc.).

New development operators have appeared: rural leaders or members of local organisations have put their skills at the disposal of their organisation and created specific structures to support their members, outside of any formal development structure.

All those concerned are faced with many questions, such as: What are the actual social and economic circumstances of the organisation? How should it be analysed? How can one adequately define objectives, with members' participation, according to their needs and forces? How can one support the farmers' organisations so that they reach their objectives and increase their efficiency and autonomy?

2. A diversity of situations

In most cases, there is still a long way to go from the current situation of farmers' organisations to the objective of increased autonomy and professionalisation. Most of the organisations are still fragile.

Proper evaluation of the background situation is indispensable to define services and support that are adapted to a farmers' organisation.

In the following pages, two key issues are discussed:
- What are the different types of organisations that one may find in sub-Saharan Africa?
- What analytical schedule (grid) should be used in order to provide adequate support?

2.1 Different types of farmers' organisations

Political history, the successive options for rural development, as well as diverse initiatives that have come from NGOs and sometimes from the farmers themselves have generated many diverse organisations, which coexist in the same country, the same region, and sometimes in the same village.

These new and modern organisations are linked to traditional forms of social organisations, sometimes through co-operative relations and sometimes with conflict.

> Traditional authorities often interact with local farmers' organisations.
>
> In West Africa, in many cases, influential local people or traditional leaders have used their power over farmers' co-operatives, directly or by appointing some of their close relatives. Traditional authority has thus invaded modern organisations. Some individuals or groups marginalised by traditional power also benefited from the new organisation and acquired power, thus joining the circle of influential local people. Often political parties or farmers' unions also strive to keep control of local initiatives in rural areas. These interactions are not negative as long as they do not generate conflict and the organisation remains focussed on its objectives and serves its members.

Farmers' organisations are not new in rural areas. Organisations often existed before present interventions. Any support must take into account the reality of these organisations, their past experiences and the knowledge that farmers gained from these.

The following brief analysis only takes into account farmers' organisations that perform technico-economic activities, and are based at least theoretically on free adherence, a democratic way of operating and relationships with external development partners.

It excludes traditional organisations and local solidarity networks, which essentially operate on the basis of mutual services, and which operate much better without external support (external operators trying to rationalise them or to divert them from their initial aim).

2.1.1 A typology of farmers' organisations

Farmers' organisations (FOs) that meet in the field can be classified according to their origins. In western Africa, three categories can be roughly distinguished:
- FOs created in the context of large development programmes (mainly during the 1960s)
- FOs tied up with local external interventions (NGOs for example) (mainly from the 1970s)
- and FOs resulting from local initiatives (from the 1980s onwards)

A. FOs created in the context of large development programmes

These result either from direct state intervention, through its administrative and technical services, or from projects (or development institutions).

In the sixties, agricultural policies focussed on structuring rural areas so that farmers could become more autonomous, which, in a number of cases, led to the setting up of local co-operatives. The proposed (imposed) model was quite uniform and inspired by European agricultural principles and the realities of European co-operatives. This type of co-operative provided real services (equipment, credit and marketing), but farmers were rarely involved in them. In many countries of West Africa, farmers complained about state seizure, the lack of clarity in accounts, of specialising in export crops only, and so on.

Some years later, large regional projects, managed by parastatal development companies, created other forms of local organisations (village associations, farmers' groups) based on key economic functions (primary marketing of products, credit, input supply) or technical functions (water management, cropping systems, etc.).

Frequently situated outside the co-operative movement, but considered often as pre-co-operative groups (Guinea, Benin), these farmers' organisations have shown promising results (associations and groups in cotton crop-

ping areas, for example in Mali (CMDT), or Senegal (SODEFITEX)). These organisations often have the following characteristics:
- The organisation modalities are defined by outside operators and are rarely negotiable.
- Farmers are obliged to adhere to these groups in order to access inputs, agricultural equipment and credit and to sell their products, since these organisations often have a monopoly on marketing.
- Development companies have exerted control and management over these farmers' organisations. Dependency, and sometimes conflict, resulted.

The current withdrawal of the state has led to an intensification of support to farmers' organisations, as a necessary condition for the transfer of responsibilities. In many cases, however, such an organisation continues "to organise the farmers" according to pre-established and non-negotiable models. Thus, some regulations are still stipulated, defining new modalities according to which the farmers should be organised, with little room for local initiative (Water Users' Associations in Madagascar, Village Committees for Land Management in Burkina Faso, and many others).

B. FOs tied up with local external interventions

Besides large development operations, farmers' organisations have been developed around local projects, supported by diverse external operators.

Women's and youth groups have been created by administrative services supported by international donors (e.g. FENUE, UNICEF) to diversify activities (market gardening, fruit production, fishing, food processing, etc.).

From the seventies, NGOs started to operate and to accompany an increasing number of farmers' groups. Their diversity is extreme, not only in size, but also in their functions and the results they obtain. Some organisations aim at local integrated development, while others focus on a specific activity or function. They concern all village people or certain categories (women, youths, interest groups, etc.). NGOs focus on a local, participatory approach and do not necessarily collaborate with national technical services.

All these initiatives have often been too isolated, scattered, sectorial and brief to be really efficient and durable. However, one can note the following points:
- They have often solved the most urgent problems (health, water supply, alternative crops, etc.).
- They have sometimes started from pre-established but negotiable organisational models, allowing farmers to adapt them.
- Some of these groups have sustained themselves after the projects' completion, and developed further (federations between villages, widening of activities and objectives, etc.).
- Others have just kept going but have remained very weak from an economic point of view and have eventually collapsed.
- In many cases, farmers have discovered the utility of the organisation for solving key problems and diversifying productive activities.

These organisations therefore constitute a considerable potential for the future, although very few have developed so far.

C. FOs resulting from local initiatives (endogenous origins linked to the associative movement)

Since the eighties, or before for the older ones, farmers' organisations have stemmed from local or autonomous (no state intervention) initiatives and have been linked to the so-called "associative movement". This is undoubtedly a most interesting process, which has accelerated in recent years, but which is far from being general. It is most common in Senegal and Burkina Faso, yet is also found in other countries.

They are often inter-village organisations that federate farmers' groups, associations or local forums. They define their own general objectives (e.g. self-sufficiency and food security, improvement of standards of living, etc.), which they develop into a multi-sectorial action plan (collective equipment, technical support to farmers, organisation of collective support services to production, management training, literacy, etc.).

Some organisations have shown very interesting results in several sectors (market gardening, health, literacy, village water management, etc.) and are responsible for very promising local dynamics.

> In western Africa, some pioneering organisations are famous (Walo Farmers' Association, ASESCAW, Bamba-Thialène group, Kabiline group in Senegal, Naam groups in Burkina Faso, etc.).
>
> There are now hundreds of local organisations in Senegal, Mali, Togo, Madagascar, etc.

These organisations have the following common characteristics:
- Even though they sometimes have very strong leaders (out of the ordinary), and even if they benefit from external support (NGOs), they are the result of a local initiative. Farmers organise themselves instead of "being organised". This accounts for their diversity in names, inner regulations, activities, and so on.
- They show the will of the farmers to take over their own development, even if very different levels of awareness coexist inside the organisation.

This requires
- inter-sectorial planning at local level, the negotiation of contracts with external partners, and initiatives in all the sectors;
- setting up exchanges between organisations and creating national and regional federations which represent the farmers while negotiating with other stakeholders and taking charge of various economic and technical functions.

> **Two examples of federative organisations linked to the associative movement**
>
> In Burkina Faso, the local Naam groups (of at least 50 members) form 63 regional federations throughout the country and adhere to a national federation (FUGN). The local groups benefit from rotary loans, which are negotiated between the federations and a NGO (Six-S).
>
> In Senegal, many farmers' organisations are linked to the associative movement and adhere to the federation of NGOs in Senegal (FONGS). This federation performs several functions: training, setting up of an inter-regional programme for food security, guaranteeing short-term loans by agricultural banks, etc.

State withdrawal and the decentralisation process are favourable to local initiatives and organisations. The latter are the natural partners of development agencies, against whom they fought in the past. They demonstrate the emergence of farmers' movements, even if, in practice, their efficiency is still variable.

2.1.2 Rapid developments

This incomplete and quick overview emphasises the diversity of farmers' organisations: diversity of origins, of activities, of size and area of influence, diversity of results, diversity also of autonomy levels as far as external operators are concerned. The picture is even more complex owing to the rapid development of these organisations.

> **An example: The delta area of the Senegal River**
>
> Two types of organisations have long coexisted, sometimes with conflict. On the one hand, there are village sections of co-operatives and farmers' groups, supervised by the development company (SAED), and, on the other hand, there are local youth clubs, federated in the Walo Farmers' Association (now called ASESCAW) which is autonomous.
>
> The state withdrawal has led to
>
> - the creation of three other autonomous federations (UGIED, AFEGIED, UGEN) between 1989 and 1990, which merged in 1991 into one co-ordinated federation (ASSOPAF);
> - the creation of federative organisations inside the SAED irrigation schemes, responsible for managing the transfer of infrastructures. Some of these organisations have joined ASESCAW (Pont Gendarme), whereas others have remained autonomous (GIE in Thiagar).
>
> Moreover, the creation of a union is currently being discussed.

There is a correlation between the general political situation of a country and the types of farmers' organisations. The more advanced the economic and political liberalisation, the more the socio-professional context is diversified. Forms of organisation inherited from the past coexist with emerging ones, some collapse or develop and others federate or merge with organisations of different origins. Opportunistic initiatives then coexist with firmly rooted movements.

This proliferation sometimes gives the impression of anarchy and certain administrations would like to organise it. However, this disorder is undoubtedly inescapable in the present transition period, which bears the marks of the democratisation of public life, of economic liberalisation and of decentralisation in most African countries. It most probably precedes further gatherings, mergers, fusions and new co-ordinations, which will occur together with the awareness of the actual external constraints and stakes (e.g. contribution to national agricultural policies).

2.2 An analytical schedule (grid)

The above presentation is general. However, situations do differ according to countries or even regions. In the field, when it comes to supporting a farmers' organisation, it is advisable to take stock of the existing situation. This consists of identifying the existing organisations and analysing their activities in order to understand their nature, their operational modes, to appreciate the level of autonomy, and efficiency and identify the needs for support.

For each organisation studied, the following criteria can be used:

2.2.1 Organisation's history

- The initiators: Who created the organisation?
- Establishment: When did it start? Why?
- Evolution: Geographic extension? Widening of activities? Why?
- Did the FO experience internal conflict? Which? Why? How was it solved?
- Did the FO experience external conflict? With whom? Why? Consequences? How was it solved?

2.2.2 Organisation's activities: inventory of activities and functions performed

- Economic: Supply, tillage, credit, production, marketing, food security, and so forth.
- Social: Health, local water management, information, training, and so on.

- Nature of union: Representation of farmers' interest, negotiations, and the like.
- Is the FO economically balanced in each sector (each activity)? As a whole? Are there activities in deficit?
- Does the FO have a solid economic base? Does it have real economic weight at local, regional and national level?
- Does the FO have political weight at local, regional, national level? Why?

2.2.3 FO's inner organisation
- Characteristics of the members and non-members. What are the modalities of adherence?
- Legal status of the FO.
- Internal structure of the FO, the different levels of organisation (basic members, intermediate levels, leadership and management, etc.).
- The leaders: Origin, socio-economic status, renewal, renewal methods, remuneration, and so on.
- Salaried employees: Who are they? Who do they depend on?
- Mechanisms of decision and control: Do all different levels participate in the decision-making process? How are they involved? What is the level of decentralisation and delegation of responsibility?
- Financial resources of the organisation: Donations, subsidies (origin), membership fees; transfer of added value between activities.
- Financial management: Who does it? How? Is there transparency in the management?
- Does the FO have an ideology? What are the general objectives, perceptions of the past, present and future? Is there any reference to a doctrine? What are the key words?

2.2.4 Relationships with external institutions
- Relationships with public and parastatal institutions: What is the nature of the relationships? What are they based on? Characteristics of the relationships (control, conflict, co-management, minimal relations, etc.).
- Relationships with external institutions: Which ones? Nature of the support received? Respective share of external financial support and self-financing.
- Relationships with the private sector (business, trade, industries, credit banks, etc.): Nature of the relationship and results.
- Relationships with other organisations: Other FOs, unions, political parties, training centres, churches, etc.).
- What are the FO's expectations vis-à-vis the external operators?

2.2.5 FO and the state withdrawal process
- Is there any envisaged or current transfer of responsibilities towards the FO? If so, what is it about? Who has decided on its contents and time frame? Is there any debate regarding the transfer modalities?
- Are there any accompanying measures, for example training, transfer of resources, and so on?
- What does the FO think of that transfer (leaders and grass-root members)?

3. Methodological propositions

(See Chapter X; Product management.)
The diversity of farmers' organisations precludes any uniform approach. There are no recipes regarding support to farmers' organisations. The right attitude is to take the farmers' organisations as they are, with their strengths and weaknesses and to support them according to their needs by adapting support when necessary developments occur. This requires

- good assessment at the outset (see evaluation criteria above). This initial analysis should be discussed with the organisation to form a basis for further real collaboration.
- definition of realistic objectives, agreed up-

on with the local people and their representatives: Which way forward? What are the needs?
- collective definition of concrete modalities for collaboration: What kind of support is necessary? How will it be done? The nature of the objectives and the support modalities must also be negotiated with the farmers' organisations and be subject to contracts (see Chapter VII: Contracts between role-players).
- definition of modalities for collective monitoring and evaluation in order to appreciate how efficient the support is and to predict possible modifications.

3.1 Several principles
Past experience has enabled three working principles regarding support to FOs to be identified:

3.1.1 What are the real stakes?
For an organisation to set itself up, *there must be real stakes*: there must be problems to be solved collectively and considered important by farmers. These can be technical stakes (e.g. acquiring collective equipment, erosion control), economic stakes (e.g. access to inputs, to credit, to market networks), social stakes or broader issues (e.g. health, women or youth promotion).

Failure of land and resource management by farmers' organisation

Sahelian farmers are aware of the degradation of their ecological environment. They can provide accurate descriptions of its features, causes and consequences. They know that maintaining and regenerating natural resources requires various co-ordinated activities that cannot be set up without sound local organisations.

However, in practical terms, many organisations assigned to land and resource management are far from efficient. This gap is probably due to having their different priorities in as far as problems are concerned and therefore in the resulting actions of farmers and external operators.

The latter consider that managing natural resources is a priority in the Sahel because present trends are jeopardising the near future of agriculture.

The farmers agree with the above statement, but they are forced to find short-term solutions to the immediate problems they face. Their priorities are more often food security, monetary income, and so on. It is only after having found short-term solutions to these problems that they can engage in activities with medium and long-term effects.

Certain groups were initially motivated by a number of technical, economic and social issues (stakes). However, if the organisation is not efficient, meaning if the expected objectives are not reached, it progressively collapses.

A number of Sahelians involved in women's market gardening operations have learned the hard way (facing the problem of unsold products year after year has resulted in their becoming discouraged). The organisations have faded away since they were not capable of solving the problems they encountered or reaching the desired objectives.

These stakes are likely to mobilise the farmers only if they correspond to their strategies and priorities. Although external operators can objectively see a problem as important and worth solving by a local organisation, this does not, however, guarantee the farmers' involvement and commitment. The organisation may soon become an empty shell if the stakes, which are the force behind the organisation, do not interest the farmers, or if they do not correspond to their priorities. All efforts to form an organisation will be seen as a formal exercise and a burden.

3.1.2 Functions to be performed define the organisation

Organisations are more likely to be efficient if they are defined, in a concerted manner, from an accurate analysis of the functions they should perform. An organisation should be derived from a function. The following questions can structure the review process with local people:
- What are the farmers' objectives? The problems to be solved will define the operational objective.
- What are the tasks that need to be performed to reach the target objective? A detailed and exhaustive inventory is necessary.
- What are the tasks that should be performed by the local people? By other stakeholders? (see Chapter VII: Contracts between role-players)
- How will the local people be organised to accomplish their duties? Who will do what? And when?
- Which resources are necessary to realise the tasks planned? These may be material or financial resources but also human resources (expertise or skills to be acquired through training).

This approach, based on common sense, is hardly compatible with a standard model of organisation as defined by external operators and more or less imposed on farmers. Such pre-established models still exist. The farmers can only adopt them if they are negotiable: that is if they establish a clear relationship between the functions to be performed and the proposed organisation, give real room for adaptation, and so on.

Lastly, let us take note that each function often corresponds to a specific type of organisation. One cannot manage a community clinic, a collective borehole, agricultural equipment, seed supply or a credit system in the same way, for instance. (See Chapter X: Product management; Chapter XI: Management of collective assets, and Chapter XII: Financing local development).

> In some instances, the legislation or national regulations provide an organisational framework. If in practice, this framework can be adapted to local circumstances, the farmers will adopt it and are more likely to commit themselves to the development.
>
> For example, in the eighties, the GAO project in Mali was forced to develop within the context of existing livestock co-operatives, all of which were but empty shells at the outset. An in-depth review with the members allowed them to identify their problems, define action programmes and adapt operational modes. The livestock farmers therefore gave meaning and purpose to the co-operatives and took them over.

3.1.3 Farmers' organisation as an ongoing process

Farmers' organisation is a process. It should include the concept of progress and the right to make mistakes.

Experiences in developed countries, where the agricultural profession is strong and well organised, show that farmers' organisations were built over a period of time and became progressively more complex, integrating new responsibilities, new functions and defining new levels and new forms of organisation. This progression has been associated with some hesitation, mistakes, failures and changes. The organisational patterns that seem to have been the most adapted to a target objective, at any given time, have usually resulted from the analysis of successes and failures and are a criterion for efficiency.

Two observations can be made at this point:
- A number of farmers' organisations tend to become multifunctional and take charge of all support to rural development (technical, economic, and functions of general interest). They also try to perform all of these simultaneously with the same structure and sometimes using the same persons.
- As these organisations faced numerous problems, one can currently observe their progressive specialisation in a federative organisation. Base structures remain co-ordinated, but they perform several specific functions. This development allows for increased autonomy and operational efficiency, especially for economic activities, without compromising the unity of the whole organisation (federation).

Why should it be otherwise in Africa? The implementation of this principle should lead to an experimental approach to organisation, its shape being progressively modified in accordance with the results of monitoring and evaluation processes.

3.2 Methodological orientations adapted to the diversity of situations

In the following pages, three different situations will be discussed in order to highlight the diversity of the needs for support and the different forms it can take.
- Organisations that link groups or associations at local (inter-village farmers' organisations), district, regional and national level.
- Farmers' organisations linked to large development projects (farmers' groups in cotton production areas, irrigation schemes, etc). One can consider farmers' organisations emerging according to new legislative frameworks and requirements (e.g. Water Users' Associations in South Africa) as of that kind.
- Isolated groups that have a fragile economic base.

3.2.1 Supporting federative organisations
A. Specific needs for support
Whatever their original background (external or local initiative) and whatever their scale, local organisations generally have
- a multi-sectorial project for the future, which implies economic, social, cultural and socio-political objectives;
- an inner structure where many levels interact with one another, for example from the grass roots up to the federal office;
- leaders, often characterised by strong personalities, who are well informed and skilled, aware of their power and prerogatives and understandably enjoy their autonomy.

From an external viewpoint, these organisations often need to become more professional, more competent and more efficient at operational level. They are faced with the following inevitable questions:
- How can one translate general objectives into a concrete work programme that increases the internal credibility of the organisation (while avoiding any discouragement from the grass-roots members) and reinforces this credibility vis-à-vis external partners? This question refers to issues such as local planning, programming, technical advice, accounting and management, setting up economic functions, monitoring and evaluation and relationships with financial sources.
- How can one carry out the activities efficiently? This question refers to issues such as information, technical and management training, control and collaboration with ex-

ternal operators (research, technical services, private sector, banks, etc.).
- How can one prevent the risk of a break between base and top. This question refers to communication between the different levels of the organisation, and to setting up internal mechanisms of both control and decentralisation.

B. Concrete modalities of support
It is imperative that support to a federative organisation
- respect the autonomy of the organisation in terms of orientation and operational choices;
- establish with it a contractual relationship on the basis of a real partnership (see Chapter VII: Contracts between role-players).

In practice, this means
- *identifying the needs for support together:* The organisation can make demands or react to external propositions. This is possible as soon as the leaders of federative organisations have been trained. The needs for support are often various. They may be technical and management training (community leaders, village experts), educational facilities to intensify internal communication, increased access to the results of agronomic research, or financial means (working capital, credit funds managed by the organisation, etc.). In some cases, there is a demand for qualified personnel (e.g. secondment of technicians to the organisation). The opinions of organisations sometimes differ on this matter. Some fear that technicians in an organisation may try to take over, or destabilise their own agents (extension officers, advisers). Most are greatly in need of qualified personnel and assume that the best solution is to have these personnel paid by the organisation. The state should be paying a subsidy to the organisation for that purpose.
- *defining the modalities of collaboration together:* It is necessary to formalise this in a written document (agreement protocol, memorandum of understanding, contract, convention, etc.). It must state precisely the respective engagements of the different parties, the monitoring, control and evaluation mechanisms and the possible sanctions.
- *setting up a contract strictly by both sides:* Outside partners must provide services (technical, economic or organisational) and adapt them to the needs of the members and also take care of their distribution on a larger scale.

Respecting the autonomy of an organisation does not exclude control, as long as its modalities are based on a common understanding. It does not exclude external propositions either, as these propositions remain the core of the support provided. It requires new attitudes that promote teamwork, advice and discussion. A partnership should be built up on a day-to-day basis and offhand or dictatorial attitudes may result in defiance.

3.2.2 Supporting farmers' organisations linked to development projects
A. Needs for specific support
In some instances, farmers' groups or associations have been set up by development projects focussing on commodity crops (cotton, irrigated rice, coffee, cocoa, palm tree, etc.). The idea behind this was that local institutions would support extension activities and perform economic tasks, as defined by the umbrella project.

Two cases may arise: either the groups are technically and economically efficient or they have a fragile economic base. Only the first situation is discussed here, the second will be discussed later (see point 3.2.3).

Those technically and economically efficient groups that focus on certain specific functions are usually faced with the following key questions:
- How can one increase their autonomy vis-à-vis the umbrella organisation or project, which is often in the process of withdrawing

from its functions and responsibilities? This question refers to information needed by the members to support collective review, technical training, accounting and management training and to support the establishment of direct relationships with external partners (agricultural credit, businessmen, transport services, industries, etc.).
- How can one increase the functions assumed by the groups in order to move progressively from a sectorial perspective (partial management of one or two crops, upstream and downstream) to the general management of different interconnected sectors of activity, without which no local development dynamics can take place. This leads to support for collective review, local planning, multi-sectorial programming and support for the emergence of new forms of organisation.
- How can one transform the organisational structure that links each group separately to the supervision organism into a horizontal structure which federates the base groups and increases their negotiation power with external partners. This question refers to exchanges between the groups that enable a progressive restructuring at higher level and that avoid a top-down approach which usually favours only bureaucracy and centralisation.

> This should prevent decisions being taken at national level, for instance, quickly to create federations (female groups, livestock farmers, artisans, etc.) without the base groups expressing any real need for them.

B. Concrete modalities of support

The objective should be to increase the autonomy of existing groups, expand their functions and field of intervention and strengthen them without decreasing their technical and economic efficiency.

An in-depth review is required prior to implementation. This can be carried out within each base group and can lead to exchanges between neighbouring communities (according to location, type of activity, or socio-ethnic composition, for instance), or between groups sharing similar interests (e.g. irrigation scheme). In some instances, this review can start straight away with exchanges between several groups. This preliminary work, which can last several weeks and requires various meetings, may refer to the following issues:
- Taking stock of the existing activities, the results obtained so far and the investments made from resources produced by the principal activity. This balance sheet activity should be carried out mainly by official members (family or household heads), but also by those who are indirectly concerned (youths, women, dependant relatives). If it involves different social categories, this assessment may raise an interesting debate while being reported back at community level.
- Gathering and sharing information regarding the concrete modalities of state withdrawal and its consequences for the organisation. This may lead to a review of the autonomisation process and of the resources that are necessary to avoid decreasing economic performance. Specific needs may then be pointed out (training, contracts with private sector, etc.).
- Taking into consideration the problems facing local people with regard to the current activities of the farmers' organisation. This may provide some rationale and background for an expansion of activities. It can also highlight new agendas and objectives of members, who could not express themselves openly until then. Most tools and local planning and programming methods can then be used (see Chapter I: Local planning), either by a whole community or by some sector.

The supplying of various information, exchanges between groups and visits to federative organisations that exist in the region or in the country can often boost the collective review process.

Carrying out specific programmes is often easier with this type of organisation, since it often has acquired efficiency reflexes. However, it is advisable to be careful about the following aspects:
- Avoid the accumulation of functions by certain individuals that result from giving new responsibilities to persons already holding various responsibilities.
- Open the membership to various social categories of people, who should be directly concerned by the new activities (women, youth, etc.).
- Identify the expertise and skills required by new functions or activities planned (see Chapter V: Training of farmers).
- Establish co-ordination, monitoring and control mechanisms to ensure that actions are integrated and adapted to the needs of local people.

In southern Mali: From village associations (AVs) to a Union of Cotton and Food Crops Growers (SYCOV).

Village associations were created by the Malian Company of Textiles (CMDT). This was generally considered a success story in terms of cotton companies structuring the rural environment.

AVs first had a limited mandate, essentially focussed on the primary commercialisation of cotton. They progressively enlarged their economic activities (inputs management, credits). With the resources accumulated (discounts on collective input supply, refunds), they made various investments of collective interest (health care, community houses, flour mills, cereal storage facilities, etc.), but also of private interest (loans). They also carried out activities related to the dissemination of new techniques.

The efficiency of AVs was, of course, variable but some proved competent. The specific effort put into training the associations' secretaries and bookkeepers undoubtedly contributed greatly to these results (literacy, numeracy).

In the late 1980s, diverse evaluations highlighted the dependence of the AVs vis-à-vis the CMDT, which remained the supervising and central body. This dependence was aggravated by a star-like mode of organisation, characterised by a lack of relations between the AVs.

From 1989 onwards, diverse spontaneous meetings between several AVs' representatives paved the way for a decisive mutation, which eventually occurred in 1991:
- The farmers created a federation of AVs, and compiled a list of grievances, made up of 12 points, to be discussed with the Ministry of Rural Development.
- Some months later, the Cotton and Food Crops Growers' Union (SYCOV) was created.

This significant change was indeed partly linked to external factors, such as political changes at state level (pulling out), wage increases for civil servants and CMDT's employees, increasing cotton production costs, while cotton prices remained desperately steady, and so on. However, this development can further be justified by deeper local circumstances:
- On the one hand, farmers and rural communities faced increasing problems: decline in their income, and standard of living, and of their natural environment (soil fertility decrease).

- On the other hand, rural societies and their leaders showed increasing awareness, autonomy, and willingness to break from their past ways of full dependence on public services.

In its objectives and orientations, the movement now focuses on various types of actions that directly concern the daily reality of farmers (e.g. animal food supply and its cost, technical advice systems). However, their actions also concern involving the farmers' representatives in decision-making processes at regional level (with local authorities) and at national level (market prices, commodity sector management, renegotiations on the contract between CMDT and the government, etc.). Many factors favour this movement and its improvement in power and efficiency:

- It has an important economic base and therefore real negotiating power, especially with regard to cotton, which is a key product for Mali.
- It defines itself as a socio-professional and rural movement, which overcomes the risk of it being classified as political, or strictly agricultural. It is also a dynamic movement which is based on past and present technical, economic and social changes.
- Its leaders are strong and realistic farmers, fully aware of the problems and open to discussion with CMDT and public services. They are also aware of the need to strengthen the movement internally, but are determined to carry on step by step and feel confident that they will succeed.

3.2.3 Supporting isolated groups with a weak economic base

A. Needs for specific support

There are probably thousands of these fragile groups throughout sub-Saharan Africa. They result either from local initiatives (farmers are eager to do something) or from sporadic interventions of administrative services or NGOs. Their social base is often very important, especially within women's groups, but their economic base is often weak. Some struggle, without realistic economic objectives and without adapted technical content. They often limit themselves to sectorial activities such as market gardening, fruit production, dyeing, and so on, the market outlets of which are sometimes uncertain.

These organisations obviously need support. They are unfortunately often neglected by most development support services. The need for the support of these communities is considerable, yet these needs are often not really analysed by the members, who rather formulate grievances. The main issues they face are the following:

- *How can the technical and economic results of their activities be improved?* This question refers to adapted technical choices, to technical and management training, and also often to access to input/output networks.
- *What are the objectives? What are the priorities?* The groups often need a sound diagnosis of their problems, the resources at their disposal and their constraints in order realistically to define a framework that will stir local people into action. This refers to local planning, be it integrated or sectorial, in the short or the long term.
- *How can the members' horizon be expanded?* How can their negotiating capacity be increased? The groups first need to come out of their isolation. Improving their access to information and connecting with local/regional structured organisations may help.

B. Concrete modalities of support

For each community, the support should target three operational objectives.

First, it is important to strengthen, even slightly, the activities conducted, so that the farmers' efforts are not in vain. It is a matter of generating self-confidence through the results obtained. Provided that the activities do have market outlets, support can consist of organising the commercialisation network for raw or processed products, organising input supply or improving the technical results by short and efficient training sessions.

> Certain activities may have poor prospects, for example when there is no market outlet for products (e.g. certain home-made handwork) or when there is little access to the required basic inputs. Before attempting to intensify an activity, one has to make sure it can be carried out steadily and that its products and/or by-products can be sold.

Second, one should take advantage of the enthusiasm raised by quick results to undertake an in-depth review (diagnosis, definition of a framework, action planning). Information, visits to neighbouring groups and exchanges must feed this review.

Third, it is necessary to define the needs, then the modalities for an action programme. This often implies linking up with other groups or federations for sectorial or general purposes.

One of the constraints in supporting these groups is that they are numerous and geographically scattered, sometimes in the same village or ward. One of the possible solutions is to assign a development agent at local level for a given period (two years for example). This agent must be capable of providing the best and closest support possible. He/she may be a government agent or a development worker linked to a federative organisation. His/her action should be clearly defined and possibly correspond to an administrative area. His/her role should consist of

- supporting farmers' groups and communities who are improving their current activities and defining their programmes;
- facilitating and promoting links between neighbouring groups (same ward, village or rural area) in order to foster the emergence of federations even if they are modest in their objectives.

3.3 Supporting a local organisation: rules of the game

Whatever the circumstances of any given organisation, four major concerns must guide the agents supporting them:

3.3.1 Proper location and levels for the different tasks and functions

The roles and tasks to be performed must be located at relevant levels. They should be defined in a concerted manner, taking into consideration the objectives of efficiency and decentralisation. As soon as an organisation becomes established (federated or not), there is an increased risk of a concentration of tasks at the top. This results, on the one hand, in extra work for the leaders, and on the other hand, in

---

Example of seed supply management at CADEF (Senegal)

A farmers' organisation in Senegal (CADEF) has structured seed supply management into three geographical levels: farmers' base group (located in a community or a village), sector (regrouping several communities) and region (the highest level of organisation).

In order to get good quality groundnut seeds, in sufficient quantities and at the right time, a specific organisation has been set up. The following table summarises the tasks and their partitioning.

Tasks	Group level	Sector level	Region level
Assessing the needs in terms of seeds	✓		
Grouping the demand		✓	
Purchasing seeds			✓
Transport to the sectors		✓	
Transport to the groups	✓		
Seeds distribution to members	✓		
Getting back seeds from production and cost recovering	✓		
Transport back to sector		✓	
Seeds treatment and storage		✓	
Technical and management monitoring			✓

condemning the grass-roots members to a passive role. This may eventually lead to disappointment, disagreements and even conflict. Generally, the *subsidiarity* principle should be applied, meaning that the partitioning of tasks between the different levels should be done from the base and that only the tasks that cannot be carried out satisfactorily at lower levels should be performed at a higher one.

3.3.2 Taking on responsibilities requires skills

Small-scale farmers are seldom prepared efficiently and sustainably to take on responsibilities in local organisations. This often means that they must first acquire the necessary expertise and skills. Specific means, time and resources should consequently be mobilised for informing and training the farmers, and for supporting their collective reflection (see Chapter V: Training of farmers).

3.3.3 Need for farmers' organisations to function democratically

In the past, many local organisations became feudalistic in the way they operated, decision making and operational modes being over-centralised, opaque and in the hands of several local heads. Democratic modes of operation have long proved much more efficient and sustainable. They, however, require the following:

- That the grass-roots members are able to express themselves and be heard and that evaluation structures are set up so as to make collective assessment of the operations.
- Transparency in the choices made and in the financial management. That supposes that each and every choice should be justified (e.g. purchases, location of equipment, etc.). There must be a good and well-organised accounting system. The accounts should be reported back periodically to the base members.

3.3.4 Towards autonomy

The progressive autonomy of the ventures and services generated by farmers' organisations seems to be a factor favourable to their economic profitability, and therefore, to their sustainability. FOs are requested to create conditions that enable local enterprises to thrive (e.g. through encouraging initiatives, loans, working capital availability, entrepreneurial training, etc.). However, they must, in turn, allow the different economic units to become autonomous, even though the FO may perform some management control and monitoring, and recover the starting funds. This autonomy removes the burden of numerous managerial committees (see also Chapter X: Product management).

Extracts from report-back meetings at CADEF (a federation of farmers' associations in Senegal):

CADEF has a threefold structure: group, sector and region (see previous table).

A. CADEF members complain about certain working procedures at CADEF:

Certain groups would rather join another sector to which they feel closer and with which they would like to share resources and have good relations. Some groups complain about certain leaders and officers who do not deliver enough at local level, who spend too much time in town, who do not respect schedules and appointments, and who do not take account of their demands ... Other groups complain about the unequal attention paid to the different groups. Some receive a lot while others are neglected. Some groups say that their project, although very important, is not considered by CADEF and does not receive adequate support.

CADEF leaders and officers have accepted this criticism, although it is harsh.

B. CADEF also expresses certain grievances against the farmers' groups:

Certain groups are not organised enough, show little dynamism and remain passive, wait for support ... Some of them do not even have any local facilitating officer. Other groups start certain actions but stop before completion. Some groups do not respect their contracts and agreements with CADEF: they benefit from their rights, but hardly respect their duties within CADEF. This free-riding attitude does not benefit the organisation as a whole.

C. Proposals

It is useful to remind everyone that the farmers' groups form the backbone of CADEF. Therefore, groups have to operate well for the whole organisation to thrive. This implies that

- every group should have a clear framework, which must come from the members concerns and priorities and which is not just a reaction to external proposals;
- each and every group must have a local facilitating officer, who is well trained and motivated;
- each group should receive adequate support from the relevant levels of CADEF, according to the topics.

On a practical basis and according to the different grievances, the following is proposed:

- Even though they do not feel very close and have different interests, groups from the same village should co-ordinate with one another for the community's sake. Therefore, local committees should continue.
- However, groups from different sectors, who are used to working together and have a good relationship, may form a sector as such and organise it (manager, local facilitators, location of the sector's office, etc.).
- A small building is required to accommodate the sector's headquarters. The groups will contribute financially to the building and the acquisition of basic furniture. The building will host the sector's meetings and certain training sessions.
- All communications between groups and the regional level should go through the sector.

4. Conclusion

Current liberalisation and decentralisation trends urge small-scale developing farmers to establish more balanced and more favourable relationships with the state and other development stakeholders. Their self-organisation, which is autonomous and efficient, is a powerful tool for this purpose. This can happen if there are formal locations and structures for discussion between the different role-players (see Chapter VII: Contracts between role-players).

Past and current experiences show how efficient farmers' organisations can be when considered as true partners: They can
- take on or co-manage economic functions, as well as upstream and downstream production;
- perform functions of general interest in the field of education (rural newspapers and radios, information and training centres, see Chapter V: Training of farmers), and in the field of technical and managerial advice (see Chapter VIII: Extension services and farm management advice).

Other examples also highlight the following:
- Local management of natural resources is far more efficient and sustainable when carried out through a farmers' organisation or the co-ordination of users (see Chapter IX: Natural resources management).
- Organised farmers can actively and positively take part in negotiations and decision making, along with other stakeholders, regarding regional development, commodity chain management (e.g. cotton), and the like. Other examples show that they can also progressively contribute to the definition of sectorial national policies (credit, extension, research, supply, etc.).

However, farmers' organisations cannot (and must not) do everything:
- The state is responsible for creating a political, institutional, legal and economic context, favourable to the emergence and the reinforcement of farmers' organisations; it must set up rules between role-players, so that the emerging organisation is not exposed to an unregulated jungle.
- Professional farmers' organisations should co-operate with local administrative and political authorities, which are, undoubtedly, in a better position to foster sound relations between the state and civil society. FOs might then contribute to local planning (see Chapter I). This obviously makes sense only if real decentralisation is operational (i.e. setting up local power, decisions and budgets).

There are many possibilities for farmers' organisations to be considered as partners in the development process at local, regional and even national levels. It must be a process and not the result of short-term, unilateral decisions. It should first be a capacity-building process, during which farmers and local managers acquire the necessary skills.

Deviations from this approach are already well known:
- A transfer of responsibilities to the farmers that happens too quickly is likely to fail, which may, in turn, discredit this approach for many years.
- Conversely, government or developmental agents can also draw out the process excessively, on the assumption that farmers are not ready, that it is too risky, and so on.
- A symbolic participation from farmers can serve as an alibi, with everything remaining centrally controlled.
- A transfer of responsibilities can also result in the creation of a limited local elite of influential people who are totally disconnected from the members and from the base organisations. This elite group is often self co-opted and re-elected, allying itself with some government or project officers to the detriment of the farmers.

All these situations can be observed in Africa, as well as in developed countries.

A major issue for the coming years is undoubtedly the creation of a new balance and new relationship between the state, the private sector and the associative sector. The reinforcement of the latter is a guarantee for integrated rural development in which farmers' needs and responsibilities are considered.

This is not likely to be entirely spontaneous; adapted support systems are required. However, in the long run, it proves less expensive, more efficient and more sustainable. Such emergence and reinforcement is hardly imaginable without real social democracy. A democratic environment can allow the farmers legally to take initiatives, without any direct or indirect risk to their security. A sound and open economic context will also guarantee minimum profitability for their enterprises.

5. Recommended literature

Bebbington, A., Kopp, A. & Rubinoff, D. (1997) *From chaos to strength: social capital, rural people's organisations and sustainable rural development.* World Bank, Washington, USA.

Berthome, J. (1990) *Les associations villageoises de developpement en Afrique de l'Ouest.* Revue Economie et Humanisme, 314.

Bosc, P.M. et al. (2001) *The role of rural producers organisations (RPOs) in the World Bank rural development strategy.* CIRAD-DFID-ODI, Montpellier, France.

Chambers, R.J. (1977) Creating and expanding organisations for rural development. In: *Cliffe, L. (ed.) Government and rural development in East Africa. Essays on political penetration.* Nijhoff, The Hague, Netherlands.

Eponou, T. (1999) *Planning linkages between research, technology transfer, and farmers' organisations: results of an action-oriented project in Mali, Senegal, Tanzania and Zimbabwe.* ISNAR, The Hague, The Netherlands.

Esman, M.J. & Uphoff, N.T. (1982) *Local organisation and rural development: the state of the art.* Ithaca, Cornell University, USA.

Flora, B.C. & Flora, J.L. (1993) Entrepreneurial social structure: a necessary ingredient. *Annals of the American Academy,* Sept. 1993, 48–58.

Francis, P. et al. (1996) *State, community and local development in Nigeria.* World Bank Technical Paper #336, 55p.

Mercoiret, M.R. (1995) Peasant organizations in sub-Saharan Africa: some reflections on progress to date. *Rural Extension Bulletin,* 7 (1995):35–42.

Whiteside, M. (1998). *Living farms. Encouraging sustainable smallholders in Southern Africa.* Earthscan Publications, London, UK, 217p.

Chapter VII: Contracts between role-players
M.-R. Mercoiret

The following chapters relate to the establishment of contracts between role-players:

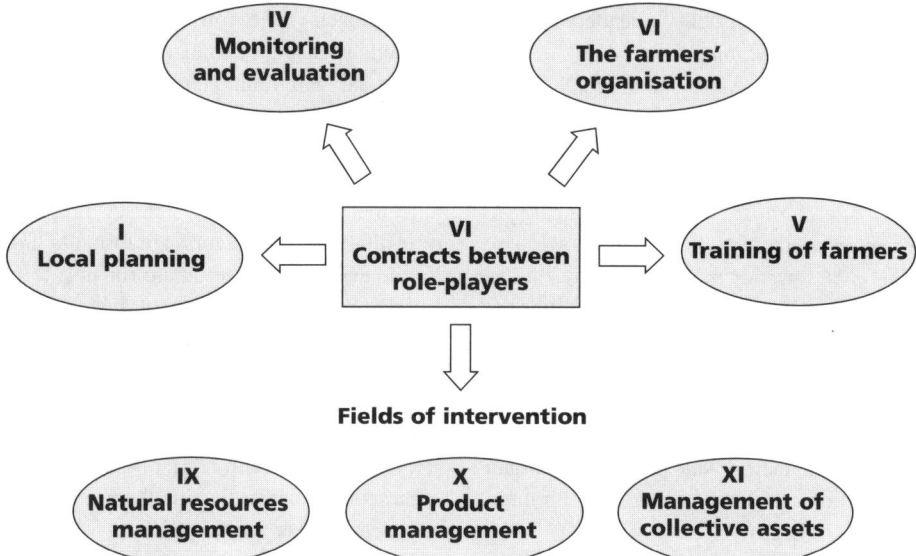

1. Introduction

"Contractual development" and "participation" were the mottos of many rural development policies in western African countries during the sixties. Those old tunes are back again, but this time rejuvenated with new concepts: contracts between role-players, partnerships, negotiated relations, new distribution of tasks between the state and civil society, and so on. Will this new speech remain incantatory, as it was so often in the past? Will there be effective changes in the relations between role-players and other parties involved in the development process? Is it possible to draw up true contracts between farmers, the state and its diverse sectors (technical services, development companies, administration, etc.), private role-players, and sponsors, while taking into account the strategies and interests of each?

It is necessary first to define some terms (these definitions are intended to simply refine the most common sense of the words and their meanings in the strict context of this handbook):

A *role-player* plays a specific role and then influences his/her close environment. Conversely, a *spectator* remains inactive.

An *economic and/or social role-player* is a person or institution that performs voluntary actions in an economic and social context, and who consciously establishes a relationship with other role-players. He/she may also be referred to as a *stakeholder*.

> In the context of decentralisation policies, local authorities correspond to a local administrative and political district or ward, managed by locally elected people. Their prerogative is generally to manage local affairs (infrastructures and services, land, development planning, education, etc.) within the strict confines of the general orientations defined at higher levels. To achieve this, they usually have specific means (local taxes and subsidies). They may be communes in Rwanda and rural communities in Senegal or municipalities in South Africa, and so on.

The following are usually considered as the main *economic role-players*:
- *The government (state) and its different sectors*: national and provincial administration, technical services and parastatals (development companies)
- *Local decentralised authorities* (local government structures)
- *The civil society's role-players*: by this we mean all private persons and institutions that act on behalf of citizens and not on behalf of the government. Among the role-players of civil society, one should not forget
  - *farmers*, livestock keepers, craftsmen/women, their private professional organisations, and the grass-roots communities (villages);
  - *private economic role-players*, traders, salespersons, rural entrepreneurs, local industries and the banks;
  - *private support structures*, such as NGOs, consultants, and the like.
- *External sources of funds* are important players in rural development in Africa, owing to the influence they can exert on official choices and policies and owing to the support they can provide (or not) to initiatives or to projects. One can distinguish between public sponsors / donors (co- and operation, bilateral and multilateral aid) and private sponsors (NGOs, firms, etc.).

Legislation usually defines a *contract* as an agreement according to which several persons or institutions (contractors) are formally linked to one another and are bound (or not) to perform something. This contract, which is often written, is intended to be enforceable by law, as specified by certain terms of the agreement. The term also refers to the document recording this agreement.

A *bilateral contract* is one in which the contractors reciprocally bind themselves. A *social contract* (as formulated by J.J. Rousseau, French philosopher of the seventeenth century) refers to a convention between the government and its citizens or between the members of a society.

A *partner* is a person who shares or takes part with another or others in a game or a sport, or in a joint business or common interest project, with shared risks and profits.

> Rural organisations are economic role-players that are classified in the private sector (as opposed to the public sector). However, as the interests of the strict private sector often differ (or are sometimes contradictory) from those of the farmers' organisations, it has been deemed necessary to discuss them separately in this context. Rural organisations can be considered as part of the associative sector, while rural industries, commercial operators, and so on form the private sector.

2. Past practices have not been very convincing ...

2.1 Gap between discourses and practices

There is often a gap between the officially proclaimed willingness to promote a contractual approach to development and the practices observable in rural areas. Even though this depends on the location and period, this gap is considerable. In the past, many pseudo-contracts were set up between unequal partners, some of whom were unpredictable and even irresponsible. Yet some innovative experiments took place in the early sixties.

> **Development staff members' seminars in Senegal in the sixties**
>
> Periodic seminars were organised by development services in the villages, where local and regional technical staff (agriculture and livestock officers, officers in forestry and water affairs and also local teachers, health officers, water services' officers) met rural people. Through dialogue with the local people, their problems were discussed, priorities were identified and work programmes were established. These programmes defined the actions to be conducted and the distribution of tasks between the community members and the development-support staff.

However, shifts and drifts often occur quickly in the contractual approach. These deviations are officially justified by an emergency or by certain changes ("we cannot afford to follow what was agreed upon"). In fact, they illustrate the government's desire to control the farmers and sometimes to obtain something from them. The services, which are supposed to promote the contractual process, sometimes become a mere cog in the political wheel (a special party for example) and subsequently lose their credibility. The idea of negotiated programmes is then replaced by top-down awareness-generating programmes, which are often seen as a means of making farmers agree to external decisions.

> **Groundnut Productivity** launched in Senegal in 1964 was inspired by post-colonial macro-economic considerations (end of French government subsidies on groundnut prices). It was highly technocratic. The farmers were neither involved in the situation analysis nor in the project development. The operation led to the creation of an extension service which was supposed to disseminate technologies. This service, centrally managed, was supplied with ever-increasing means. Secondly, it competed with participatory and contractual approaches. However, the latter showed sporadic but significant development. Eventually, from the early eighties onwards, the extension services lost credibility and became marginalised owing to the progressive emergence of NGOs.
>
> In some countries of central Africa, for instance, the initial contractual attempt did not even take place, or had a short lifespan ("popular education" initiative in Chad). Moreover, some awareness-generating programmes even became coercive and dictatorial (e.g. growing cotton became compulsory in certain areas).

During the seventies, contracts were more or less imposed on the farmers by parastatal development companies. These took the form of co-management projects, actually defined unilaterally by external operators. However, some promising results were achieved by such arrangements, especially in the context of cotton cropping. There were indeed converging interests among the different partners, even though such interests were not clearly formulated at the outset. The cotton company redu-

Development parastatals and public services often withdraw without any consultation with rural people.

ced its operating costs by transferring some functions to the farmers, who eventually benefited from this move (mastering cotton weighing as they did not really trust the company, benefiting from discounts on input supply, etc.).

Usually, when farmers did not find much benefit in those imposed contracts, they by-passed them. This has happened in many countries and has played havoc with agricultural credit passing through local co-operatives, the payment of tillage operations and water fees in irrigation schemes, and the like.

The new agricultural policies, which were defined during the eighties, have not yet succeeded in making fundamental changes in practice and real partnerships are still rare.

At present development systems are withdrawing their support without always agreeing with the farmers on the content, pace and modalities of transferring both charges and responsibility (smallholder irrigation schemes in Madagascar, Mali, Senegal, Zimbabwe, South Africa, for example). Economic decisions are taken without preliminary consultation with the farmers (such as on prices of agricultural products, cancellation of subsidies on inputs). Extension programmes are launched without consulting the farmers on the kind of support they need. Besides, many farmers are more concerned with securing upstream and downstream production linkages than with technical advice.

One should note that there are a limited number of exceptions that are mostly due to specific strategies implemented by local role-players or local innovative institutions, for example, the GAO project in Mali, community planning in Rwanda, the Rural Promotion project SAED/FAO in Senegal, and so on.

2.2 Reasons for such shortcomings

The difficulty of promoting real contracts between two or more development players lies, undoubtedly, in the respective situation of each of the potential partners. It is also highly dependant on the legal, political and economic circumstances.

2.2.1 Situation of local partners
A. Farmers

Farmers are not prepared to stand as equal partners in a development programme. Their capacity to defend their interests, to liaise with other stakeholders and to negotiate contracts depends on their local professional organisation. The latter unfortunately often remains poor, although exceptions do exist (see Chapter VI: The farmers' organisation).

Farmers often face tremendous ecological (drought, flooding, desertification) and economic difficulties (declining commodity prices, cancellation of subsidies on inputs). They are overwhelmed by numerous external, successive or simultaneous development initiatives, which are sometimes contradictory, and often ill co-ordinated. These initiatives have contributed to weakening the local mechanisms regulating the rural environment.

As a result, farmers have developed *short-term, individual and family strategies* to deal with uncertainty and risks.

Moreover, they generally no longer trust development institutions, which have promised much and continue to achieve little. Farmers address their grievances to external operators, but with a "win-some-loose-some" attitude. ("It may work this time, who knows?")

The memory of past unfulfilled promises jeopardises the farmers' own commitment to future contracts and partnerships. This vicious circle of mistrust has encouraged free-riding behaviour. Farmers have sometimes also developed bad habits and can act irresponsibly with their current external partners, which they would never allow themselves to do inside their own social groups.

B. Extension officers

Technicians, trainers and agricultural advisers have occasionally been discredited in the farmers' opinion, because of the past mediocre performance of development programmes, even though, more often than not, extension officers have been nothing but implementers in such programmes. They are often at the lowest level of the hierarchy in development organisations that pay little attention to their opinions and viewpoints.

Extension officers appear to be the scapegoats for poor achievement in many development programmes. Moreover, they compete with new forms of intervention from which they are excluded (NGOs' direct support to farmers' groups, for example) and are deprived of the proper means to carry out their work.

As a result, they are defensive and seek both legitimacy and identity, which does not help create new relationships with farmers. All over Africa, there is a great need for redeployment and retraining. However changes are not easy to implement among those who benefited from the past situation and the material and social privileges associated with that. The image of many officers has been tarnished by certain individuals' bad practices, be it embezzlement or the confusion of political power with administrative authority and public technical services.

In some places, establishing contractual relations has been made very difficult because of the past. Ideological discussions have sometimes been a cover-up for power games, top-down approaches or paternalism, thus making it difficult to discuss matters openly. Farmers have not forgotten the irresponsible behaviour of certain officers who belong to powerful institutions.

C. Private economic role-players

For different reasons (depending on the country's history), commercial agents, industries and bankers remained quite discreet in rural areas between the sixties and the late eighties. They are now making an appearance, but are generally involved in rural sectors in which they can expect a quick return on investment and profits. This is clear from their recent interest in upstream linkages to production (input supply) in high potential agricultural areas (irrigation schemes, for example).

They have played an important role in downstream production, in the commercialisation of some products for instance (such as animals, fruits and vegetables) and have managed to establish a power balance in their favour. Although preoccupied with short-term profitability, they are capable of high adaptability. Hence, their openness to negotiating contracts favouring small-scale farmers depends on competition or the monopolistic context. The presence of farmers' economic groups developing transport, processing and/or marketing activities is likely to generate a more balanced relationship. (See below and Chapter X: Product management).

It is important to note that farmers have a clearer (if not more favourable) relationship with private economic role-players than they have with extension officers. It is a more businesslike relationship based on a clear power balance, on fair-is-fair principles, and on commercial values, which are more explicit and more familiar to the farmers than the development officers' references to some "inter-ministerial commission's decisions, regulations and structural adjustment".

D. Public local authorities

Local government institutions should be the ideal place for negotiations between the national government and civil society. In certain countries (e.g. Rwanda, Senegal) they are, at present, slowly emerging, after long-standing uncertainties, prevarications and procrastinations caused either by centralisation reflex or by conflicts or a lack of capacity at local level.

At present, public local authorities do not yet exist in a number of countries or enjoy enough autonomy and the necessary mechanisms to enable them to play a significant role as players in development. Recent changes in South Africa, for instance, and the new structures and regulations in local government raise hope and expectations in this regard.

E. Local administration

Local administration is also often in a transitional phase and adapts to the changes resulting from the state's withdrawal relatively quickly. Local, regional or provincial administrations (decentralised branches of line ministries) have for a long time been co-ordinating all development activities that have taken place on their territory, especially cash flow. Certain staff members are still reluctant to give up this function and the power associated with it. They sometimes express the need for intermediation, for the homogenisation and systematisation of procedures, which eventually impair the farmers' initiatives aimed at creating direct relationships with new partners (including NGOs, commercial agents, transporters, etc.), especially when public services are not associated with them.

F. Foreign sponsors

Foreign sponsors and international development institutions generally favour contractual approaches to development; some have favoured them for a long time (NGOs, in many cases), while for others this has been more recent. Although it is difficult to generalise, one must recognise that their capacity to approve or disapprove the financing of a programme gives them considerable negotiating power with the other partners who are in dire need of external funds. Donors and sponsors may legitimately express their viewpoint on projects and programmes submitted to them for financing, as they cannot back ill-prepared projects. However, they sometimes set certain conditions (be it about national political choices or technical choices at local level), which are put forward in a very exclusive and constraining manner.

2.2.2 A diversely favourable context

Two factors hamper the establishment of negotiated economic relationships between role-players:

The *economic context* is uncertain and the economic role-players are not prone to taking risks. This context favours cautiousness and short-term strategies.

> **Two examples: input supply and credit**
>
> The cancellation of subsidies on input supply explains, in certain areas, the cautiousness of private economic role-players who are reluctant to sign contracts with farmers. Uncertain demand adds to this mistrust and as it is reinforced by the fall in product price, this reduces the farmers' solvency.
>
> Agricultural credit banks are usually doubtful about the smallholders' capacity to reimburse loans. Furthermore, current land management features in certain areas (commonage, lack of individual title deeds) prevent farmers from having collateral to back their loan application. Consequently, banks set up conditions of credit for certain categories of farmers that are not adapted and are non-negotiable.
>
> The farmers are reluctant to set up contracts with commercial operators and with agro-industry because they are afraid of not being in a position to fulfil their commitments.

The *political and administrative context* is equally unfavourable, depending on the country, even if some positive changes are quickly taking place. Ill-adapted legislation and slow procedures still exist in places, as well as certain controlling attitudes inherited from the past which contradict the proclaimed economic liberalism.

Lack of adapted and flexible regulations on contracts between economic operators, and the lack of proper organisations to enforce them, have resulted in a climate of laxness. Such conditions often generate frustration and disappointment.

In practice, farmers' failures are sometimes penalised, whereas this is seldom the case

> For instance, a farmers' group has no recourse, in the event of its harvest being ruined owing to a faulty water supply, when the contractor hired to maintain its water pump does not respect the maintenance contract.
>
> What can the same group do if the truck contracted to collect their harvested tomatoes misses the appointed pick-up time?

when it comes to failures by administration officers, who always have excuses (urgent meetings, lack of appropriate means, etc.). This is unacceptable in terms of contracts between role-players, and expensive and inefficient in the long run.

The general context is not conducive to entrepreneurial spirit and initiatives. The different partners are seldom ready to define negotiated contracts beyond short-term prospects. The search for short-term opportunities and bargains often overshadows the development of sustainable linkages, which are also more demanding and often not immediately profitable. This tendency is maintained by a mutual mistrust resulting from past and recent practices.

Ignoring these unfavourable factors is dangerous, because they will hamper the development of contracts between stakeholders, in any case, and still be the source of many uncertainties and hesitations. They must therefore be taken into account. Helping to explain them may enable certain traps to be avoided.

3. Methodological propositions

In the following pages we first strive to take stock of the principal factors to be considered when developing and drawing up contracts, and secondly, we draw attention to certain mistakes that can be avoided.

3.1 Different types of contracts: a typology

The possible contracts between role-players may vary in their duration, contents and forms of partnerships. The following typology considers only the contracts between farmers and one or several partners.

3.1.1 Contracts between farmers' organisations

This type of contract involves farmers' organisations that establish multi-local or regional economic relationships.

- The content of such contracts may vary greatly, involving the following for instance:
- *Food security*. Organisations located in areas with excess cereal production sell their surplus to organisations located in deficit areas. These sales are usually done according to conditions that are advantageous to the buyers. On the other hand, sellers have a guaranteed market. Such exchanges often take place in the context of a food security programme (cereal banks). This is the case in Senegal between the farmers' organisations belonging to FONGS (see Chapter VI: The farmers' organisations).
- *Seed supply*. Organisations situated within irrigation schemes sell high-yielding rice variety seeds to others whose members grow rice in uncontrolled water supply conditions (outside any scheme). This possibility was initiated by CADEF in the Casamance area and AFEGIED in the delta area, both in Senegal.
- *Exchanges between organisations with complementary production*. This happens for instance in Mali, between irrigated crop farmers and organised stock keepers in the Mopti area. The latter supply bulls for animal traction and in return purchase rice. Exchanges of this kind exist in different countries in western Africa, through the NGO Six-S.

These exchanges are promising, but they can

arise and survive only under the following conditions:
- The organisations involved must be strong enough, as economic units and institutions, to engage in this type of contract. Moreover they must be fully informed of the possible complementarities that may feed such exchanges. Among the known examples there has always been facilitation by a federative organisation (FONGS of Senegal), by a NGO (Six-S for instance) or by an external operator.
- Such contractual exchanges require sound material organisation. Any unknown factors regarding planning, logistics (transport), cash management or a certain laxness from leaders can compromise the contracts.
- A quick, good bargain or an external opportunity (subsidy, timeous aid) is often more attractive to farmers than the prospects of benefits resulting from a long-term strategy. Thus, it is important to identify clearly the farmers' reasons for accepting or not direct, long-term and often less profitable exchanges (in the short run) with their peers, rather than seeking short-term but uncertain opportunities.

3.1.2 Contracts between farmers' organisations and private economic role-players
Such contracts may deal with different topics:
- Input supply

> In the Senegal River delta, the development company SAED organised forums, in which contracts were established between farmers and input suppliers.

- Product commercialisation
This concerns marketing products out of any official stream. This practice sometimes exists at national level (for vegetables, fruits, meat, etc.). Informal contracts have existed for decades in Africa between farmers and salespersons, hawkers, and the like. Improving the farmers' socio-professional network increases their negotiating power (see Chapter X: Product management). They can also adapt the offer (products, period) according to the market demand more efficiently.

> **Towards the organisation of a fruit and vegetable supply chain in Senegal**
> Fruit and vegetable growers in the Bignona area were facing a tricky commercialisation problem. A survey was conducted by the farmers' organisations, supported by a NGO (CIEPAC). This made it possible to identify the different stakeholders involved in the supply chain, from the rural communities to Dakar. The survey also made it possible to characterise their expectations and current strategies.
>
> A meeting between all stakeholders (small-scale growers, transporters, wholesalers, retailers, hawkers, tradespersons, consumers) was organised. This consultation resulted in agreements upon
> - changes to the cropping calendar in order to match offer and demand better;
> - the withdrawal of the farmers from market places, where they were attempting to settle as retailers;
> - the creation of a weekly, wholesale market.

The development of contracts between farmers' organisations and private operators faces certain constraints:
- Certain private operators are opportunistic and capable of adapting to the new context of liberalism and of the state's disengagement. They easily find ways of taking advantage of the situation. They can for instance benefit from the farmers' access to agricultural credit and undertake some fraudulent practices (e.g. diverting the use of credit, proposing cash to farmers in ex-

> In 1990, a French agro-food firm proposed a contract to the rural organisations from the Bignona area (Senegal) for a yearly purchase of 1500 tons of vegetables. The vegetables were to be produced according to certain organic-farming norms and dried in specific plants, the establishment of which would be supported by the firm. Despite the attractiveness of the offer, the rural organisations did not accept it, as they were aware of the difficulties of meeting certain terms of the contract, which included, among other things,
>
> - adopting new cropping techniques, which were not yet tested locally (organic-farming practices);
> - organising supply and transport;
> - paying penalties in case of delays in delivery.

change for an inflated demand for credit), which can result in havoc and trouble for farmers and banks.

- Other downstream private operators have sometimes established long-term strongholds (middlemen, stock dealers, wholesalers, etc.). They are not very keen to give up their advantages and it is quite difficult to interest them in negotiation.
- New economic role-players are interested in dealing directly with the producers (chemical industries, agro-food industries and exporters) and they propose contracts to the farmers' organisations. This means drawing up written, formal contracts, with financial guarantees, possible penalties, legal aspects and so forth. Although (or because) such contracts are very clear, they often scare the farmers by pointing out their responsibilities and the commitment that is required from them. Such contracts seem to be the way forward in the future, but the emerging farmers' organisations are currently seldom prepared to join in proactively.

3.1.3 Contracts between farmers' organisations and upstream/downstream economic groups (co-operatives)

The oldest and most seasoned farmers' organisations have long discovered the shortcomings in the collective management of productive equipment or of services: management committees are difficult to handle and voluntary help quickly displays limitations. The time has come to promote real rural enterprises, which will enjoy full financial and operational autonomy, and which will take over the required economic functions of production. In many cases, these new enterprises benefit from an initial donation (capital) from farmers' organisations (subsidies or credit), to which they have preferential but not exclusive links.

> Such economic entities can be, for example,
>
> - an enterprise in wire netting or fence building;
> - a co-operative managing a workshop, or a mill;
> - an economic interest group (GEI) in charge of commercialising market garden products;
> - an economic interest group involved in transport;
> - a local shop specialising in input supply or equipment spare parts;
> - an enterprise dealing in tillage operations, and so on.

The farmers' organisation signs a contract with the enterprise that it has helped to create. Farmers can join the management board of such an enterprise (co-operative-like mode of

operation). This contract generally specifies
- the nature of the services to be delivered and their remuneration (calculated in such a way that the profitability of the enterprise is guaranteed),
- the modalities of reimbursement to the farmers' organisation for initial investments (reimbursement of a loan or the payment of a percentage of benefits).

In practice many shortcomings are encountered:
- The farmers' organisation (and its leaders) may feel they "own" the enterprise created and then misuse it (the company truck being used to take the members to a meeting; the tractor being used to plough the leaders' private plots, etc.).
- If the new enterprise performs well, it may cast its services wider afield and find new clients, sometimes at the expense of the farmers' association and its members, who supported the creation of the enterprise at the outset.
- Conversely, the feeling of being closely linked to the farmers' organisation can make the managers lax and passive, which may jeopardise the enterprise's profitability.

3.1.4 Contracts between farmer's organisations and development projects

For a long time NGOs have attempted to promote forms of contractual intervention through flexible development projects that support emerging local initiatives, which are difficult to identify at an earlier stage. From the eighties onwards, large development projects, financed by international agencies, have also appeared to commit themselves in this way and have proposed framework contracts or specific contracts to farmers. Many projects operate this way all over Africa (e.g. the "communal development" project in Rwanda, the "food security" project in Burkina Faso, etc.).

This orientation is, theoretically, perfectly adapted. However, in practice, such contracts encounter several difficulties:
- External operators sometimes make initial choices that are ill-adapted to the farmers' demands. These choices may concern technical issues (e.g. in the context of a "self-managed" project, a NGO imposed manual waterpumps on farmers who did not want them). They may also concern activities which are excluded from the contracts.
- Flexible projects are sometimes actually not that flexible in their financing procedures. Delays may occur and make it impossible for the project to respect its commitments.
- Most of the time, the projects' staff come from the administration. They do not always have a "participatory" or "contractual" approach. Also, they can lack a certain "economic profitability" perspective.
- The so-called "farmer-partner" does not always exist as an institutional entity (lack of organisations or representatives), which does not favour the formalisation of contracts.

3.1.5 Contracts between local communities and public technical services

In some domains, public services strive to promote contractual relationships with villages and local communities: health, water services, school management, borehole drilling, and the like.

There are also many problems related to such arrangements (see Chapter XI: Management of collective assets):
- During the drawing up of the contracts, chiefs and local influential people represent the community, whereas their legitimacy and representativity are sometimes contested. As a result, a number of malfunctions appear in the implementation of the contracts, as those who involve and represent the community are not necessarily those who implement the work.
- Public services do not always fulfil their promises: there may be a lack of means, a unilateral reviewing of the contracts by administrative upper levels, excessive casual-

ness from local officers, a lack of skills, and so on.
- Laxness is aggravated on both sides owing to the absence of authorities to turn to in the event of a disagreement or malfunction.

3.2 What has been learnt from experience?

3.2.1 Five inescapable conditions
To set up a real contractual situation, five (permanent) conditions must be met:
- The *contracting partners* must be accurately identified: Who is involved? On behalf of whom, or from which institution?
- The *object of the contract* must be clearly specified: What is the contract about? If it is a framework contract, it must clearly specify that the different activities to be included will require specific/sectorial contracts.
- A contract is sustainable only if there is a *reciprocal interest* in it, each contracting partner gaining some benefit from the contract (win-win situation). This supposes that the contract must correspond to real needs, considered as priorities by the farmers. It must improve the current situation, otherwise contracts just become additional constraints, justified only by ideological or specious arguments.
- Concrete modalities for the *implementation* of the contract must be defined at the outset, even if they may develop after mutual agreement between all parties: Who will do what? When? How?
- The modalities of *monitoring and evaluation* must be defined from the beginning. They must be associated with penalties in case certain clauses of the contract are violated. This requires an arbitration authority, accepted by all parties.

3.2.2 Operational modalities
Contracts must be negotiated. A contract exists and is durable only when it is possible for the parties involved to refuse it at the outset. It must not be a hostage situation.

A contract cannot be drawn up in a few minutes or hours, during a somewhat extended meeting. It requires the effective participation of farmers in detailing and discussing the contract. The more organised and professional the farmers' organisations are, the easier it is for the farmers to participate (assuming a real representativity of the leaders). If the farmers are poorly organised, the discussion of a contract proposal may be an opportunity for them to improve their organisation. How can they organise themselves, in order to benefit from the opportunity offered by the contract?

Any dictatorial attitude from external partners is incompatible with a contractual spirit. Farmers may get involved because of external pressure, but they will eventually challenge or bypass the contract in the long run. It is therefore advisable to create conditions for the farmers to express their objectives, priorities and reservations.

If the contract proposal emanates from a development organisation (public, parastatal or non-governmental), it is indispensable for its promoters to avoid pre-established schemes, or some sort of turnkey contract. There must be real room for negotiation, which will allow the farmers' propositions to be taken into account.

Each aspect of the contract must be clarified and negotiated: financial implications, monitoring and evaluation modalities and incurred penalties. On this last point, one must highlight the fact that the credibility of contracts between farmers and service providers is very much linked to the reciprocity of possible penalties, which cannot only pertain to the farmers.

In many cases, facilitation and mediation are necessary. These roles are efficient only if all parties accept them. They consist of
- facilitating the access to information of all parties (pre and post-contractual asymmetric information and imperfect foresight are two key issues of contractual approaches, most

of the time at the expense of rural people);
- putting in contact potential partners and helping to establish a real dialogue;
- supporting the most fragile partners (supporting their reviews, analysis and training);
- being aware that playing a regulatory role, even unofficially, can jeopardise the contract if malfunctions or deviations occur;
- helping different partners to overcome short-term difficulties and defining mid and long-term strategies.

It should be the role of the government and its public services to establish and regulate healthy and durable relations between role-players of the civil society. However, NGOs or consultants may also play such a role.

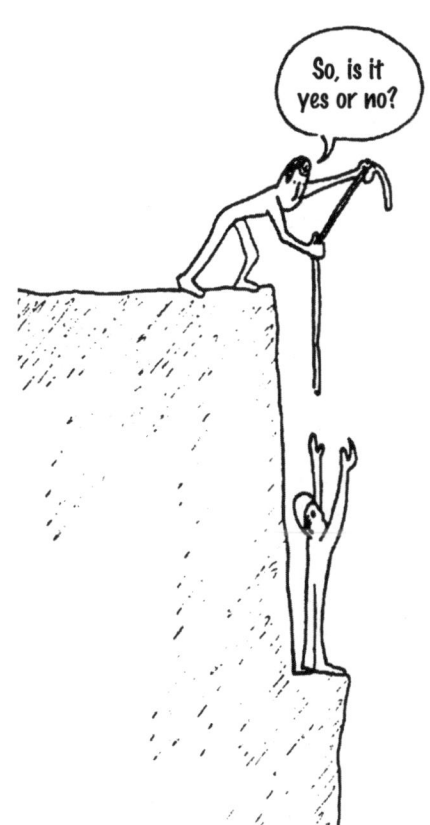

A contract should really be mutually discussed.

3.3 Many questions, several answers

3.3.1 Short or long-term contracts?

In the first stages of a partnership, it is often advisable to set up short-term renewable contracts (spanning a cropping season, for instance). This is especially true when poorly organised farmers are involved. Short-term contracts allow the different stakeholders involved to go through a learning experience, to discover one another, and to refine the terms and scope of further contracts gradually. An evaluation process, involving all parties, must precede any contract renewal. This allows one to implement adjustments and possible extensions to the contract contents.

Short-term contracts, and related discussions, may be compatible with mid and long-term development plans. Three or five-year framework contracts may be established, in broad terms, that define orientations and modalities of general collaboration (see Chapter I: Local planning). Such frameworks can then accommodate sectorial or short-term contracts, and allow accurate planning.

These two levels of planning (mid-term and annual planning) are necessary to deal efficiently with certain cross-cutting issues (e.g. agricultural credit, land management). They also allow development activities to be put into a time perspective (see examples in Chapter I).

3.3.2 Specific or multi-sectorial contracts?

The content of a contract may be limited to one activity or one function (e.g. a contract between a farmers' group and commercial agents for input supply, marketing of a specific product, etc.). It can also be broad (e.g. framework contract with public services or a NGO, land and resources management support to a farmers' organisation).

The scope of a contract depends on the size and the structure of the contracting organisation, but also on the farmers' needs. It is advisable to start with a sectorial, limited collaboration, which may be strengthened and diversified over time, if necessary.

3.3.3 Contracts between two or more partners

The diversity of problems encountered by farmers sometimes leads to ambitious multi-sectorial contracts being drawn up between farmers' organisations and numerous partners. This may be done if the organisations are strong enough to defend their interests and negotiate as fully-fledged partners. This is still the exception. In most cases, it is easier for the farmers' organisation to negotiate and to draw up specific sectorial contracts with each of its partners, rather than face all of them together.

In other instances, it may be interesting for the farmers' association to link up with another local operator (e.g. technical services, a local association or a NGO) to reinforce its negotiating power with a third party (e.g. private operator, large development project, etc.).

3.3.4 Contracts between stakeholders located at different geographic levels

A. At local level

Different types of contracts may exist:
- Contracts between crop farmers, livestock farmers, craftspeople and local businesses in order to solve certain local issues or to address conflicting interests (e.g. demarcation of grazing areas, collective use of scarce natural resources, etc.). These can be set up under the supervision of traditional authorities or of a local association (formal or not).
- Contracts binding a federated farmers' organisation (at district level for instance) to its grass-roots members. Each basic group designs its work programme, with the support of the organisation (which provides information and methods). Sectorial contracts can then be drawn up, reflecting the diversity in needs and priority. Such contracts usually define the distribution of tasks and responsibilities between the local and the federated levels. The organisation supports its local groups either through its own human or financial resources, or through external support from other partners.
- Contracts between different local stakeholders (persons, groups, institutions physically present at the local level). Such contracts usually aim to reinforce the negotiating power with external operators. The objective is often collectively to approve and design a local development plan, which will be submitted and negotiated with external operators (see Chapter I: Local planning).

It should be noted that supervising administrations or services may challenge or disregard contracts drawn up at local level. Local authorities and local governance entities can play a key role in approving, certifying and/or backing such contracts.

B. At higher geographic levels

The participation of the local people in negotiations and the drawing up of contracts relies on
- their capacity to delegate representatives (this supposes that autonomous and efficient federative organisations exist);
- the government's real willingness to promote consultation and negotiation regarding political orientations at all levels.

The examples show the necessary emergence of an inter-professional platform for discussion between stakeholders involved in the same production network or in a given commodity chain. Experience in northern countries confirms the need for such an approach, which proved realistic and efficient. It allows for both the formulation of the different stakeholders' interests and the search for acceptable compromises for all. This is possible if several conditions are met:
- Farmers must enjoy significant weight and power in the local, regional and/or national economy.
- Their representatives must carry the interests of the different categories of farmers,

Examples

It has only been since the 1990s, that FONGS (federation of local NGOs in Senegal) has participated in official technical meetings, along with major sponsors, development operators and policy makers.

Farmers' participation in discussions on agricultural prices and commodity chain management is even more recent.

In 1991, SYCOV (Association of Cotton and Food Crops Growers, in Mali) asked for a renegotiation of the contract between CMDT (cotton company) and the government, as the contract initially excluded farmers. SYCOV distributed widely the existing contract to its members so that the debate could take place at local level. The results of these debates helped SYCOV to define its position during the new negotiations between CMDT and the government.

In Senegal, a cotton dispute (la fronde du coton) occurred in 1989. Since then, growers have been represented during discussions regarding the management of the chain and its organisation. Moreover, the Ministry of Rural Development initiated a forum on the rice chain, which involves the different operators, including smallholding producers.

Towards a contractual policy in Guinea

The coffee producers themselves initiated a contractual policy. For cotton, a development project took the initiative. In both cases, the objective has been to achieve better information for all, regarding the costs, the world prices and the sharing of added value throughout the chains.

In practice, the dialogue first concerns fixing an indicative price, at the beginning of the cropping season, in order to encourage production. Modalities of distribution and of added value sharing are also discussed.

This first supposes that the professional organisations are strong and representative, at local level then in federations (groups, associations, unions, etc.). Secondly, inter-professional organisations must be set up. This also implies that a real "negotiation culture" exists, that the different operators are used to debate economic and institutional issues, and to negotiate with one another.

In the context of a coffee development project, the organisation of local groups of growers was first promoted. This started with the management of nurseries. Thereafter, these groups received support for organising the coffee harvest. Later on, they elected delegates and grouped themselves into a federation, which became the key contact for commercial agents and exporters.

which presupposes internal agreement within local associations and groups.
- Farmers must be informed on the national and global economic context for them to define their objectives and have room for negotiation and then to prepare their arguments.
- Regulation of the negotiating process between economic operators is often necessary. The government should perform this regulatory function, as it has some means of intervention (legislation and policy, incentives, subsidies, price supports, loans, minimal guaranteed price, etc.). This allows sustainable relations between role-players to be established (according to government priori-

ties, be they social equity, economic efficiency, rural development, etc.) and uncertainties and international market variations to be buffered.

3.3.5 Oral or written contract?
African rural societies are sensitive to the spoken word and there are hardly any cases where a farmer does not respect the commitments he/she has taken before his/her peers (be they social or economic commitments, e.g. marriage, land renting, loan, etc.). This sense of responsibility is due to the existence of local social values and benchmarks, including social penalties. However, when it comes to contracts established with external operators, the same sense of responsibility does not always exist. Thus, written and formal contracts are often considered safer.

However, written contracts have some limitations. Farmers may be illiterate, especially in English. They may not understand the technical and juridical jargon, which often characterises such contracts. "Paper" as such, may not be seen as a reliable medium (it may cause concern, or be considered as a meaningless object in certain cultures).

The choice depends on the circumstances in which the contract is drawn up. One should remember that a contract exists only if both parties know exactly what is at stake, what their commitment is. Also, some solemnity and etiquette can help emphasise that signing a contract is a serious matter ...

3.3.6 Strictness, not inflexibility
Implementing a (negotiated) contract requires some strictness. Each party must carry out the tasks it has committed itself to performing. However, any contract should remain flexible in its implementation, meaning that some adaptations may occur, as long as they are negotiated and that all parties agree to them.

This justifies setting up mechanisms and structures for concerted monitoring.

3.3.7 What should be evaluated?
The evaluation of contract implementation must take place in accordance with what was prescribed at the outset (signature of the contract). It must address the ways and the extent to which the clauses have been respected, and also the efficiency of collaboration. In other words, what progress has been made towards solving the identified problems.

A contract can be fully realised without generating any benefit for either of the parties involved, which proves that the objectives were probably not well identified at the outset.

All contracting parties must be involved in the evaluation, even if each party prepares its evaluation separately. Yet again, mediation or facilitation may prove useful in this regard (see Chapter IV: Monitoring and evaluation).

4. Conclusion

The farmer's integration into society and into markets is as old as society itself. However, recently this integration has taken place under conditions that were not very favourable to rural people. Farmers have had to deal with agricultural pricing policies in which their interests were often sacrificed for the greater interest of urban consumers. They have had to adapt to external decisions made without their participation.

Farmers can be motivated to take the initiative and to innovate only if some corrective measures, which will take into account the interests of the different stakeholders, including those of the farmers are implemented. Contracts between role-players can help define this new balance. However, two conditions are necessary:
- There must be an emergence of social and economic operators capable of defending their interests and fully aware of their interdependence and complementarities. This supposes that smallholding farmers must no

longer be considered as mere beneficiaries or "targets" of development projects, but rather as fully-fledged partners, capable of taking decisions and actions as responsible people.
- Durable relationships between the different role-players must be established, such relationships being negotiated on the basis of common, negotiated and long-term interests.

To realise these ideas (or to prevent them from being bypassed by certain stakeholders), some measures are necessary:
- Farmers' organisations must be supported (see Chapter VI) for them to identify potential partners, possible alliances, as well as potential competition.
- Mediation and facilitation are often necessary, at least during a transitional phase, to help the emerging farmers' organisations negotiate with other operators.
- There must be a set of rules (institutions) to limit laxness, irresponsibility or free-riding behaviour.
- There must also be regulating mechanisms that protect the operators and their relationships against global market fluctuations, for instance.

This is possible only if the agricultural sector is recognised as a full social and economic sector, with its own demands and requirements, its constraints and main interests. It should no longer be a place where wealth and added value is created only for the benefit of the urban sector.

5. Recommended literature

Chambers, R. & Pretty, J. (1994). Towards a learning paradigm: new professionalism and institutions for a sustainable agriculture. In: Scoones & Thompson (editors), *Beyond Farmer First*. Intermediate Technology Publications, London, UK.

Furubotn, E.G. & Richter, R. (2000). *Institutions and economic theory. The contribution of the new institutional economics.* The University of Michigan Press, 556p. (pp. 179–264, Contract theory).

Fafchamps, M. (1996). The enforcement of commercial contracts in Ghana. *World Development*, 24 (3):427–448.

PART THREE:

THE FIELDS OF INTERVENTION

- VIII. Extension services and farm management advice
- IX. Natural resources management
- X. Product management
- XI. Management of collective assets and facilities
- XII. Financing local development
- XIII. Women and development
- XIV. The non-agricultural sector

Chapter VIII: Extension services and farm management advice
Or how to support technical changes
M.-R. Mercoiret, D. Gentil, J-F. Bélières, J. Marzin & S. Perret

Chapter VIII may usefully be connected to the following chapters:

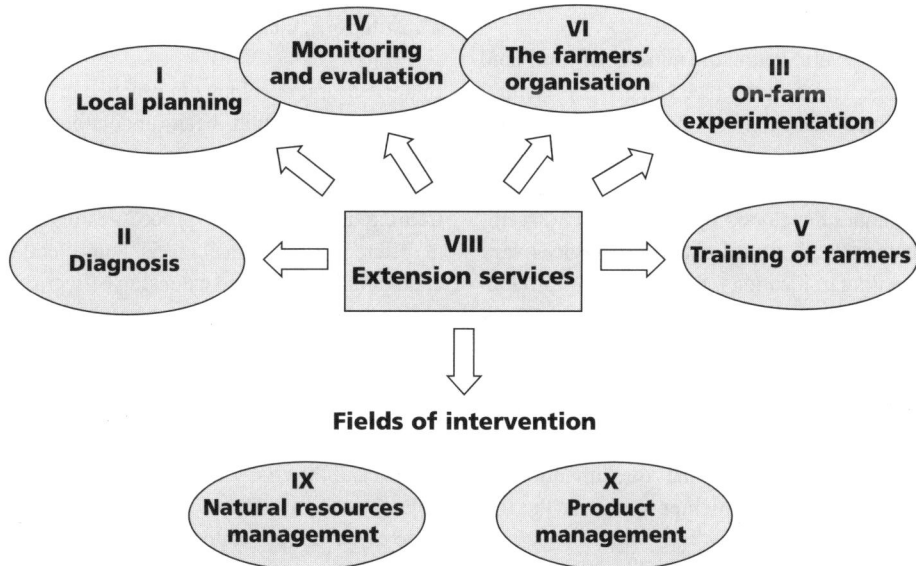

1. Introduction

1.1 Necessary technical changes

In sub-Saharan Africa, *the need to improve and secure land productivity* (i.e. yield per hectare), as well as *labour productivity* (i.e. yield per working day) is quite obvious. It results from demographic growth, the reduction of available land, and from the degradation of the environmental and economic context.

Not only do external operators (decision makers, technical services, financial sources, NGOs) agree on this point, but also independent smallholders' organisations and producers themselves who face serious problems (fear of food shortages, drop in income). Another important factor is *the need to preserve natural resources* while promoting sustainable, reproducible agricultural practices so that future generations may inherit sound ecological capital that they will in turn protect and put to good use.

In all three cases, producers must change their agricultural practices in order to achieve these objectives.

1.2 Enormous efforts for disappointing results

For decades, considerable human, material and financial means have been invested to reach the afore mentioned objectives in sub-Saharan Africa.

Agronomic national research institutions have been created (and supported by scientific research institutions from northern countries and sponsors) to perfect new techniques.

Countless operations, projects and programmes have been set up with the aim of transferring the results of research to producers through extension services.

For several years, the number of rural land management projects (see Chapter IX: Natural resources management) has been on the increase.

The smallholders' reactions to the new techniques developed by research and disseminated by agricultural extension services vary according to locations, the category of producers concerned, the type of techniques disseminated and the products to which the techniques are relevant. On the whole, the adoption rates are disappointing and not up to expectation.

This chapter will rapidly review diverse approaches related to the dissemination of techniques and will attempt to answer the following questions:
- What kind of support can be provided to producers to facilitate or accelerate technical changes within the production units?
- How will these changes be implemented?

2. A framework for reflection

2.1 Short historical review: from extension services to agricultural advice

The approaches and methods used to support changes in agricultural practice have varied according to periods, locations and the operators. Known as extension services, they have sometimes developed to become agricultural advice systems.

2.1.1 Agricultural extension

The dictionary defines "extension" as
- the act of spreading into the public (it refers back to the concept of prolongation of agricultural research, dissemination, propagating);
- the act of adapting technical and scientific knowledge to bring it within the reach of the non-expert reader (it also refers to scientific extension, extramural training by academic institutions, etc.).

In agricultural development practice in Africa, extension has often been understood as a means of having producers adopt techniques developed by agronomic research through to training and supervision procedures organised at different geographical levels. For several decades smallholders and extension officers have been exposed to various extension methods.

A. Coercive extension and extension disseminated by agricultural services

During the colonial period (agricultural supervisors) and the first years of independence, the objectives of extension activities were to introduce export crops (cotton, groundnuts, coffee) and sometimes complex innovations (animal traction for instance was already well developed in Guinea in 1930). At the beginning the methods used were sometimes coercive (mandatory crop), but later became more flexible.

B. Close supervision

This method expanded especially with "operations productivity" (e.g. in Senegal, 1964), cotton cropping and other projects. Its distinguishing features were the sheer size of the extension activity, the importance accorded to hierarchy, good organisation and strict control of personnel. The pyramid relied on a large number of grass-roots supervisors, who often had poor qualifications, transmitting generally simple messages (high-yielding seeds, densities, fertilisers, plant protection), though these were sometimes more complex (like animal traction).

Although expensive, this method has achieved uncontested results especially for cotton crops, even though it may be challenged with regard to producers' responsibility and real autonomy.

> In many cases, this integrated approach is based on cash crops. The cotton industries, especially in Mali at CMDT, provide a good illustration. The approach is based on three principles:
> - The extension system is based on approaching smallholders as a group at village level. There are no pilot farmers or contact villagers. Extension officers always talk to a group (even though some demonstrations are carried out on an individual level).
> - Besides his/her extension duties, the extension officer is also responsible for supplying services, agricultural credit and marketing.
> - The extension officer only passes on information tried and tested by scientific research. All interested farmers can apply this information since they possess both the equipment and inputs required for the implementation of the recommended tasks.

C. Training and visits

From the early 1980s and for about ten years, the World Bank has been proposing this system in all the countries in which it is involved. It is based on a number of simple principles:
- Extension officers must focus on extension only (and not be involved in supplying services in credits, in marketing or in statistics).
- They must regularly visit, at fixed dates, a limited number of contact smallholders.
- They must be regularly sent on refresher courses and must be supported by technical services that remain in close contact with research.

The model proposed in Africa has often resulted in "dynamising" programmes towards agricultural technical services, so that they become national agricultural extension services. This system has some strong points:
- Owing to its strict organisation, it has enabled agricultural officers, who were for a long time marginalised by development projects and their specific framework, to go back to work.
- It makes it possible to use the results from research.
- It improves the extension officers skills through regular training.
- It allows regular contact between extension officers and smallholders.

In practice, however, the efficiency of this system and its costs are subject to controversy:
- Even if texts recommend taking into account farmers' demands and reactions (feedback effect), in practice, costly hierarchical national services are created or reinforced that are similar to the system of close supervision.
- Does this system really allow for information to get back to the research centres and to orientate research according to the needs of the producers? To date, there have been very few convincing examples of success in this field.
- This system does not seem able to take into consideration the diverse agricultural situations existing in the same area.
- It seems unable to disseminate the most complex messages.
- Finally, setting up such a system supposes that, from the outset, existing innovations are adapted to the production systems and can be immediately transferred, and also, that the environment allows the producers to adopt such innovations (organised supply and marketing, access to credit). This was the case in certain countries in Asia where the system obtained good results.

2.1.2 Advice to farmers

The dictionary defines "advice" as "an opinion given to somebody about what he/she is supposed to do". *Agricultural advice* excludes any idea of coercion, or of imposing. The farmer is free to make decisions once he/she has been given an "opinion" (information, analysis of the various options open to him/her with their advantages and disadvantages) that may throw light on the choices. It is an aid to the decision-making process.

In Africa, agricultural advice is still rarely practised, even if technicians in some developmental services are now called agricultural advisers.

> However, there have been some interesting experiences. In Senegal, managerial advice was used from 1974 in the experimental units in the Sine-Saloum region, followed by a more simplified formula (equipment advice) used by SODEVA in the groundnut production area. It is also used by CMDT in certain farmers' associations in south Mali.
>
> Many research-development projects[1] have resorted to this system. A project is then carried out with farms selected for representing different situations existing in a given area. They are monitored and benefit from the advice. These farms, often known as *reference farms*, are present in Madagascar (Lake Alaotra, the highlands), in Rwanda, and so on. In certain cases, research-development projects have achieved significant results in disseminating new techniques through exchanges between farmers and through training (for example in Haiti, on market gardening at Madian-Salagnac or on irrigated rice in the northern flat lands).

2.2 Limits of extension

Thoughts on extension methods have often been limited to debates concerning the educational aspects of dissemination and the organisation of the advisers' and the extension officers' work. Without undermining these aspects that will be discussed later, it should be underlined that the changes introduced by producers in their practices are not only due to extension and agricultural advice.

2.2.1 Technical progress is as old as the world

For centuries, the farmers themselves brought about the essential part of technical progress in agriculture:
- They created techniques that were later distributed through imitation and informal exchanges.
- They borrowed techniques from other societies they were in touch with and later adapted them.

The following should therefore be remembered:
- Agronomic research does not have the monopoly of creation, even if, nowadays, it has an increasing impact upon technology development.
- Systematic support to technical changes (research institutions, extension and advice services) are historically recent (a few decades). In Africa, it has often been linked to two factors:
  - The more or less compulsory introduction of new crops
  - The intention of states and external operators to *accelerate* technical changes in agriculture either for macroeconomic rea-

---

1. Research-development is an approach which, at an experimental level and based on sites representing the diversity of an environment, attempts to improve techniques and methods of work that will later be extended to larger areas. Research and development projects regroup researchers, producers, farmers' organisations and extension officers, etc.

sons (reducing imports, need for foreign currencies) or to improve or preserve the farmers' incomes threatened by changes in the overall context (drought, decline in soil fertility, degradation of exchange terms)

2.2.2 Changes must be of interest to the producers and within their reach

The desire to accelerate changes in agricultural practices is often justified. However, the propositions are often not adapted to the producers' expectations and strategies.

For farmers to adopt a new technique, it is not enough just to make it known. Many other factors are involved.

A. Three main conditions for farmers to innovate and change their practices

- Producers must know the new techniques. To achieve this, a system of dissemination is useful as it facilitates the farmers' access to information and to technical training (see Chapter V: Training the farmers).
- They must be interested in changing practices, either because the new techniques are *more efficient* (e.g. it takes some of the tediousness out of the work, saves time during ploughing and sowing), and/or because they consider the new technique to be *more profitable*, resulting in benefits (increased yield, increased income) and/or better security as it contributes to reducing the risks related to rainfall patterns (e.g. in the case of early ripening varieties).
- They must have the means to adopt the new techniques:
  - Material means: an organised supply system, an adapted system of credit if the new technique requires external inputs, and so on.
  - Social means: new techniques must be compatible with the social rules in force in the farmers' society (land management, for example) or must go along with the amendment of these rules. This supposes an agreement within the social group or at least among those who make sure the rules are respected.

When the economic profitability of a technique is demonstrated, it is not unusual for it to generate changes in the social rules (e.g. land management, division of labour based on gender) sometimes without any outside intervention being necessary.

> In the Sahelian area, the quick planting and development of crops is crucial, since it protects crops in a region very exposed to climatic hardships. Hence, the rapid adoption of a sowing drill pulled by a donkey or a horse is not surprising. It has been combined with the use of the traditional Sine hoe (in Senegal for example) and great benefits are gained and the tediousness of weeding with the hoe is greatly reduced.

> As state-managed agricultural programmes (which guaranteed material supplies and credit input) have disappeared and been replaced by private structures (sometimes unreliable and offering less advantageous conditions), emerging intensification techniques have been disappearing in certain regions: that is the use of fertilisers has greatly declined, agriculture equipment is rarely renewed and so on.

> In Guinea, significant technical improvement in lowland rice cultivation, an activity traditionally reserved for women, has attracted men who have tended to replace their wives.

B. Variation in smallholders' reaction

The smallholders' reactions to technical changes proposed to them also vary according to certain conditions.

- Their reactions are affected by the degree of complexity of the proposed technique.
- The cost (in labour and/or in money) of a new technique and the risks involved also affect the decision to adopt it. A tough farmer in a savannah area will be more tempted by changes than a poor Sahelian smallholder.
- It seems that a technique is easier to disseminate when it responds to a problem affecting the majority of the producers, even if this implies a change in individual practices. Perhaps it is the fact of embarking on something together that reassures the farmers.
- An already known technique ("we have heard about it", "the neighbouring village is using it ...") is easier to distribute (if it is adapted) than a totally new technique. In the latter case a few daring people need to get started for others to follow, hence the relevance of experimentation in rural areas (see Chapter III).

As a general rule, an innovation is better adopted if the smallholders themselves participate in the setting up process.

A system of dissemination, no matter how strict and how adapted it is, cannot encompass all the necessary conditions for making producers adopt new techniques. It can only facilitate the adoption of a technique or accelerate changes in practices, but the latter is usually a result of a multitude of factors and is not predictable.

3. Propositions for action

3.1 Three main orientations

The previous remarks lead one to propose three main orientations to support technical changes inside production units:

3.1.1 A need for diversified propositions

Diversified technical propositions should be adapted to the diversity of production units and to the diversity of producers' strategies.

In many situations in Africa it has been observed that the differences between farms and the technical and economic have reached a level for which universal formulas for extension have now become obsolete. In fact, in many cases, repeating at simple and uniform messages to producers, sometimes for decades, has not corresponded to the producers' expectations and consequently they have been rejected. "Their book is finished, they keep on repeating the same old story," said an old Senegalese smallholder about extension officers in the early 1980s.

3.1.2 Need for a renewed attitude, open to dialogue

Experience has demonstrated, sometimes in the short term but always in the mid and long term, the inefficiency of directive methods based on a one-way transfer of knowledge from the "ones who know" (technicians, researchers) to "those who do not know" (the smallholders). Smallholders always have good reasons for doing what they do. Practices seen as irrational are seldom so when one tries to

> One can differentiate between *simple innovations*, which bring about very few changes in farms; *radiating innovations* (animal traction), which, when adopted for a sectorial problem (e.g. improvements in tillage, weeding), have an impact on the whole farm's operation; and *systems of innovations*, whereby various techniques are adopted simultaneously (for example the fight against the decline in soil fertility).
>
> One can also classify innovations in a similar way: *additive innovations* (one practice is added without changes being made to the whole); *adaptive innovations* (in which progressive and more or less important adaptations are made to the running of the whole farm) and *transformations* (which means changing the running of the farm).

understand the factors which motivated them. This does not mean that opposition to change should be promoted. It is only a way of saying that researchers, extension officers, advisers and smallholders all share a portion of the truth, and that development is not a match where one side has to win. It is in the interest of everybody to work together from the very beginning to improve the performance of local agriculture.

3.1.3 Need for the farmers' involvement in the different stages of innovation processes

The innovation process, with its three main components (creating a technique, distributing it and adopting it), forms a whole. It cannot be chopped up and given to different role-players who do not communicate.

In the classical scheme of agricultural extension, the three components are more or less strictly divided and attributed to three major role-players: research design to creates new techniques, extension services to distribute them, and farmers to adopt them. Feedback is seldom implemented. (see figure above).

By including all smallholders in the process of innovation, conditions are created for them to participate more actively in the creating/setting up of techniques, in the definition of the means of distribution and in the evaluation of its adoption and the effects incurred.

The following figure and table show the main principles of the so-called "bottom-up" approach.

A typical top-down approach to extension

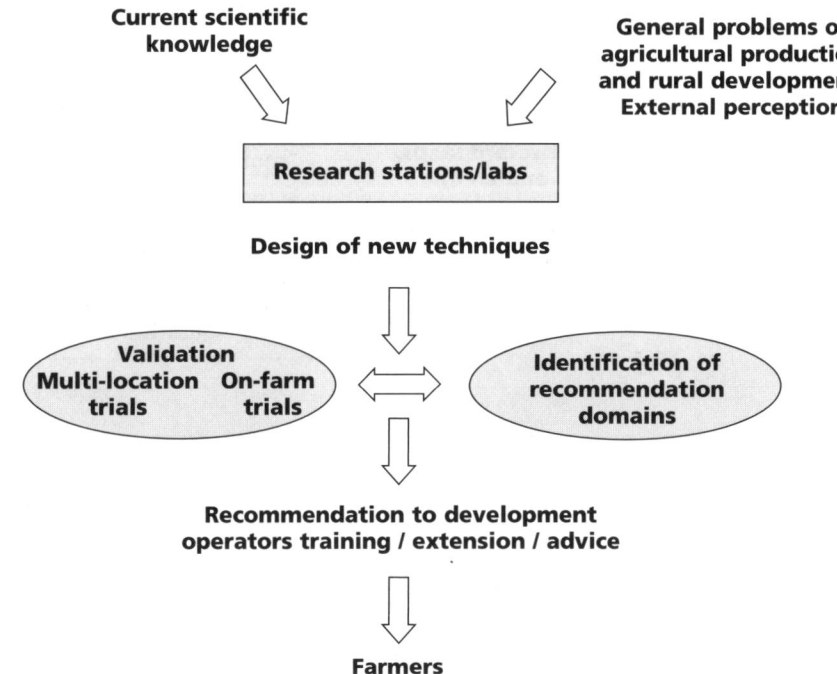

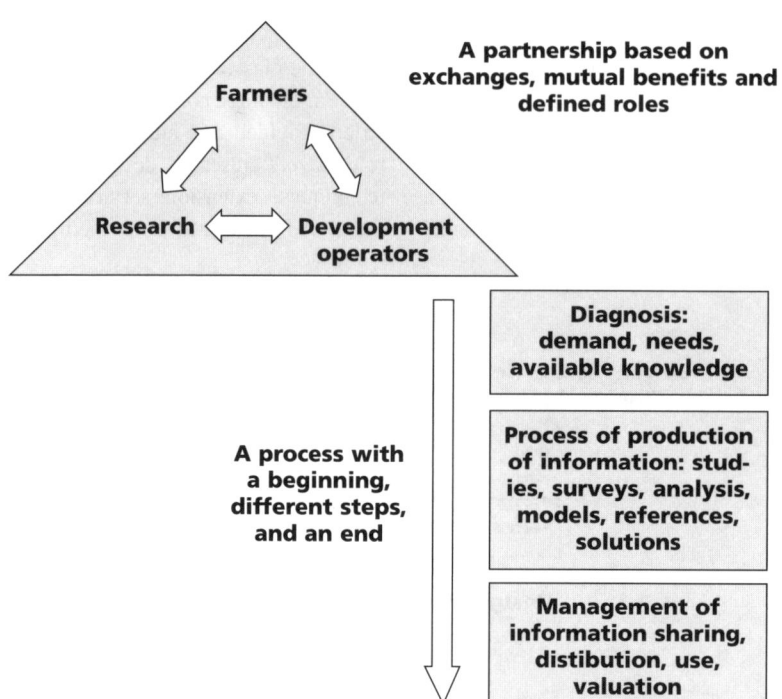

Role-players Phases	Farmers	Research	Development operators
Diagnosis	Expression of needs, problems, knowledge models (know-how, representations). Exchanges between farmers	Collection, reformulation of problems into questions, demand analysis through specific tools, collection of available data (scientific knowledge, generic references)	Expression of knowledge of the situation, collection of available data (local references)
Production of information	Participation in on-farm experiments, surveys, feedback of results, role of validation	Proposals of methods, setting up of experiments, surveys, sampling, additional labs/station experiments	Contribution to sampling processes, experiment setting up and monitoring, role of validation
Management of information, products	Enrichment of the information systems, of the knowledge models, sharing with the socio-cultural and professional environments, distribution of knowledge	Storage of data, production of generic knowledge, models, methods, sharing with the international scientific community	Local knowledge and references, sharing with professional environments, capacity building, dissemination of knowledge and methods

There never is a unique solution to a problem.

3.2 A procedure

A procedure includes four components: diagnosis, experimentation, dissemination and monitoring/evaluation.

3.2.1 Diagnosis (see Chapter II)
The importance of the following should be noted:
- Consensual diagnosis, which allows the views of both smallholders (internal) and researchers, technicians, and similar, (external) to be expressed
- Historical analysis (based on the elders' memories, archives, comparisons made between aerial photographs) in order to understand the development and main tendencies towards change
- Taking account of the different levels of decision making (the plot or the herd, the various types of farms, the area, the village, the sub-region and the region)

The progressive nature of diagnosis: there is no need to know everything before starting.

In other words, diagnosis is endless. In fact the best way to learn is by observing dynamics in action through monitoring and evaluation. A quick diagnosis (one to three months, if possible during a cropping season) is enough to initiate the first actions provided they are permanently monitored and evaluated.

As soon as a local society is broadly understood, it is necessary to identify the main hindrances and major problems expressed spontaneously by smallholders or raised during discussions or surveys. As time goes on, more detailed knowledge of the society will be acquired.

The above problems must then be classified according to different criteria.

A. Clear distinction between farming issues and general issues
A clear distinction must be drawn between problems stemming from agricultural techniques and those of a more general nature: remoteness and accessibility, price ratios, supply, credit and consumption systems, land issues,

and such like. Of course the general problems must be addressed, but they must be given specific treatments (see Chapter I: Local Planning, and Chapter VII: Contracts between role-players).

Technical problems are only one of the numerous reasons that explain the adoption of new techniques. It is often useless to insist on the promotion of certain methods when the environment is inadequate (for example, recommending fertiliser if the supply system does not work).

From a technical point of view, it is necessary both to work in a given environment and to improve that environment.

B. Classification of problems
Problems can be classified
- according to their complexity. Does the solution depend on a single producer? Is an agreement from all those who work on the farm needed? Are higher levels of decision making involved (the area, the village)?
- according to the general nature of the problem. Are the majority of producers affected or is this a problem encountered only by a minority?
- according to the urgency of the situation. This is evaluated by noting what sort of priority the smallholders give to the problem.

The problems, once identified, can then be put into an order of importance (a sort of hierarchy) and a time scale can be established to resolve them.

3.2.2 Experimentation (see Chapter III)
Two kinds of situations are possible:
- The problems have been identified and some smallholders in the area or in the surrounding area have already tested one or more solutions (for instance, alternative methods of preparing nurseries or new varieties). However, opinions differ concerning the conditions of use and the results. Moreover, the solutions tested in these on-farm trials are available but have not been distributed to smallholders. The experimentation must therefore be organised with several volunteering smallholders while other smallholders look on, following the trial because they are interested in the results.
- The solutions are hypothetical, coming from research stations or studies conducted in other countries (in certain cases, the solutions are not yet available and require more in-depth research in the stations). A twofold operation is therefore often necessary: local experimentation in a controlled environment and on-farm experimentation closely monitored by national research.

After several years of trial and error, experiences and written works concerning this matter abound (see references on Chapter III: On-farm experimentation).

Particular attention should be paid to
- finalising a set of reliable but also diversified techniques that should provide a range of answers to the problems identified. It is rare that only one answer to a problem exists. Usually there are many answers, each one having its pros and cons. Traditionally, extension often tended to select one answer and to distribute it as the one and only answer to a problem. In reality, however, smallholders' reactions are not predictable, and it is risky to select for them the technique that suits them best.
- finalising techniques that take account of the general perspective of the context. A particular technique may seem suitable at plot level (e.g. increased yield owing to the use of high-yielding variety seeds and fertilisers) while it is less suitable at whole farm level (the farmer may be keener to invest his money in pastures to fatten his sheep than to buy inputs). Realising this has paved the way for improvement schemes at farm level (cf. below management advice or farm advisory system).

3.2.3 Dissemination

If the techniques for distribution are based on a sound and consensual diagnosis and have been tested with the smallholders, the chances are that they will be adapted to the smallholders' expectations. In fact, when innovations are really adapted and profitable, they spread rapidly through informal visits, discussions at the market or in the shebeen, and also through the social networks smallholders have at regional and sometimes national level.

Nonetheless, the dissemination of techniques on a large scale can also be accelerated or facilitated by particular means. The possible tools are discussed later. At this stage, the idea is to highlight some inescapable features of distribution.

A. Differentiated messages for different people and situations

In the same village or the same region, very different types of farms can exist (because of land surface, the type of labour, equipment, livestock count). Diagnosis brings this diversity to light, experimentation takes it into account and therefore it must surface again during the dissemination / distribution process.

B. Using adapted codes for communicating with farmers

Permanent attention must be paid to communicating with the producers. Technicians, extension officers and smallholders must speak the same language. This is not always the case. Above and beyond the problems encountered owing to approximate translations from English or French into national or regional languages, efforts must be made to adapt the language and its meaning for technicians and smallholders to understand one another.

The same applies for other media (drawings, sketches) which, although clear to those who are used to them, are not necessarily so for those for whom they are destined (an arrow does not automatically mean a relationship between cause and effect for a Bambara, a Dioula or a Zulu smallholder!).

C. Using adapted media for communication

Information travels through a wide range of channels (demonstration meetings, visits, radio, local press, booklets, TV). In principle there is no such thing as a good or a bad channel. The appropriate channel all depends on

- the nature of the information (e.g. it is difficult to teach people to adjust a plough through a radio programme);
- the ability of the smallholders to access different facilities (Do they have a radio? Can they get to a meeting? How far away are they?);
- the credibility of the medium in the smallholders' opinion.

D. The role of smallholders' organisations

It would be better if the distribution of techniques was done by the smallholders' organisations where they exist and where the smallholders trust them. The advantage of this method is that, as men and women from the rural society distribute the information, it is much easier to find the appropriate language and means of communication. This supposes that specialised farmers or villagers selected by the community have been trained (see Chapter V: Training the farmers).

3.2.4 Monitoring and evaluation (see Chapter IV)

It should be remembered that the evaluation-monitoring procedure enables operators and stakeholders to learn from actions. It enriches their initial knowledge of the local environment and also becomes a tool for permanent diagnosis. It is interesting only if it results in discussions and feedback from the smallholders and if they do not get tired of it.

Example: list of possible indicators of performance for extension (non-exhaustive)

1. Realisation indicators

Actions realised / planned actions

Causes of discrepancies

2. Impact indicators

a) Percentage of smallholders who use the whole package of proposed techniques

b) Percentage of the surface affected per farm

c) Percentage of smallholders using recommended inputs (fungicides, high-yielding seeds, phosphate, urea, animal traction)

All these indicators should be measured, as well as their development in time (if possible by going back a decade). If possible the figures should be compared with other projects or with areas outside the project.

3. Efficiency indicators

a) Average yields and variations in yields, from those who use and those who do not use the technical package

b) The explanations of yield differences

- in connection with recommended features (e.g. respecting the fertiliser doses)
- in connection with other causes (e.g. date of sowing, soil preparation, correct weeding)

4. Indicators of smallholders' reactions

a) Reasons given by the more advanced smallholders for not using the techniques over the entire area of their farm

b) Reasons given by smallholders who do not apply any of the techniques

c) Reasons given by smallholders who have partially applied the techniques

d) Other problems raised by smallholders, and not included in the extension programme

5. Cost indicators

a) Direct cost of one extension officer per smallholder concerned

b) Total cost of extension personnel (all levels) per smallholder concerned

c) Total cost of the project per smallholder concerned

3.3 Tools for distribution

3.3.1 Visits and exchanges between smallholders

These remain the best means of distribution. Smallholders are more convinced by what they see on other smallholders' farms than by the words of an extension officer. They can, however, still be interested in trials or in certain demonstrations at stations if the latter are not too different from their own situations. Visits and exchanges are efficient only if they have been very well prepared materially (careful preparation of transport and catering) as well as pedagogically (precise and limited definition of objectives, preparation with the host farmers, someone responsible for facilitating the exchanges).

3.3.2 Local information centre

The following formula has proved relevant and efficient: a competent officer is on duty for farmers' information and advice, he/she is available in a given place, on a specific day and at a time in order for the smallholders to talk to him/her about their particular problems. This method is widely used in many countries, where the agricultural adviser to a particular area regularly opens his/her doors to farmers (for instance on market days, which take place on a weekly or monthly basis).

3.3.3 Training of specialised farmers

These farmers, acknowledged and chosen by the community, become experts in certain themes. It is an efficient way of distribution. The duties they will be assuming, as well as the material and financial advantages associated with them, must be made very clear. These must be negotiated with the community and all decisions must be made in the open. The objective is not simply to train people to relay messages from the technical services, but really to transfer skills to the community or smallholders' organisations (see Chapter V: Training of farmers).

3.3.4 Group meetings

Group meetings remain efficient as long as
- they are based on specific themes of real interest to smallholders (which means they correspond to their needs);
- they combine demonstrations with discussions during which the smallholders can openly express their viewpoints.

Who should attend these meetings? And where?

Meetings can take place at village or district level, or even at a specific local area level. The public is often large, and heterogeneous as not all the producers have the same kind of farm nor the same possibilities and constraints. This kind of meeting is usually necessary at the beginning of a project or for matters that need the involvement of all producers (grazing areas for example).

Meetings may also involve smaller and more homogeneous groups. This type of meeting is only efficient if the participants acknowledge the criteria that are used to justify the grouping (typology). If the smallholders do not understand or do not accept this, the homogenous group approach can be seen as a manoeuvre by the extension agent to divide the village or favour certain people. Moreover, certain projects classify the villagers as "progressive", "serious", "with potential", "semi-traditionalists", "traditionalists", and so forth. It is then understandable that smallholders do not wish to be in some of these groups. If the typology is based on objective criteria (e.g. the existence or not of animal traction, herds of cattle, size of the family, on-farm food processing, etc.), it will be more easily accepted.

Dissemination can also take place with smaller heterogeneous groups who are motivated by a common interest. Participants do not necessarily have the same kind of farm, but everybody wants to progress by exchanging and sharing the results of experimentation, for instance. The various problems encountered by different types of farms can then be discussed.

The following example demonstrates an approach at village level in south Mali. It is a good illustration of a differentiated approach to different types of farms in the same village.

A village approach in south Mali

The aim of this extension method was to distribute differentiated messages, adapted to the particular constraints of each category of producers. The programmes covered a calendar year according to the following steps:

1. Awareness and facilitation meeting. The aim of this meeting was to explain the process, identify the general constraints and launch a simple typological survey.
2. Second meeting to give feedback on the results of the typology survey. Two target groups were chosen for the first year.
3. Analysis of one or two farms from each group was carried out in the presence of all concerned. The farms were chosen on a voluntary basis. The analysis, which was based on the results of a survey previously carried out by the extension officer, was done to show the specific constraints for this type of farm, and propose solutions.

A simplified example: Poor yield

Causes: Delays in ploughing and sowing as a result of: lack of adequate and available equipment, poverty. Solution: Credit system for basic equipment.

4. Demonstration sessions and technical training. At specific times the group would meet at a member's property (on a rotational basis) to participate in a talk or demonstration about specific techniques.

For example, in the case of a first-time credit/loan for equipment, the programme was composed of

- a session in calculating the profitability of the loan,
- a training session on breaking-in bulls (3 weeks),
- dry soil ploughing (refresher course),
- wet soil ploughing, and so on.

5. Strict monitoring of work was carried out. To avoid being weighed down by paper work, monitoring was carried out on one or two farms only and relevant data recorded: for instance, the surface ploughed by a pair of bulls and per cropping technique, the inputs used, the surface cultivated and the yield harvested.
6. Evaluation and reporting back on results. Through to the monitoring, albeit limited, the results obtained at the end of the crop year could be reported back to the village general assembly. The discussions that followed enabled a programme to be set up for the following year.

The following table shows the four categories of farms chosen and the technical proposals for each of them to integrate crop and livestock farming.

TYPES OF FARM	TECHNICAL PROPOSITIONS
A A farmer equipped with at least a pair of bulls, a plough, a multi-purpose cultivator, a sowing machine, a donkey or a bull cart, plus at least ten cattle including those for animal traction and a tractor.	Construction of improved paddocks (kraals) with abundant litter (cereal straws) Intensification of cropping systems + organic fertilisation + ploughing High-performance equipment Improvement in animal productivity + generally same proposals as in B.

B. A farm with at least one pair of bulls, a plough or a multi-purpose cultivator (one unit of animal traction), but which is still under-equipped	Forage production of cow pea or Mucuna Training of bulls for animal traction, building stall + litter Matching the potential cultivated area and the bulls' working capacity Amelioration of tillage techniques, dry soil weeding, ridging, tine cultivator use Groundnut mechanised cropping (density, hoeing/weeding techniques) Composting with cereal straws Erosion control programme, fencing with hedge trees
C. Non-equipped farm or having an incomplete animal traction unit, but where animal traction is known (since it has been practised)	Basic equipment credit: priority to a multi-purpose cultivator (hoeing/weeding) Mastering tillage techniques (hoeing/weeding) Limitation of cultivated surfaces (especially with cotton) or borrowing equipment
D. Non-equipped farm practising traditional manual agriculture Animal traction is more or less unknown	Organisation of a provisional stock of staple food cereal by the village community in order to help the most needy persons Partial intensification: introduction of intensifying techniques by manual cropping, first on a small surface (0,25 to 0,5 ha)

3.3.5 Using audio-visual equipment and media

This includes slides, posters, post-literacy booklets, radio programmes, and the like.

These supports are useful and efficient as long as they are adapted. They cannot replace the need for direct contact or meetings in small groups, but they can facilitate or help in the preparation of the latter.

They call for some operational observations:
- The least expensive and least sophisticated supports are often found to be the most suitable. Wind, sand, and poor maintenance in the harsh environment of rural areas usually get the better of videos, projectors and similar equipment. Moreover, not all the supports and media have the same purpose.
- Posters can be used to focus attention on a particular problem and/or to remind farmers about a theme of general interest, provided they are visible and attractive. This requires financial means that are not always available.
- Slide projection can be used to introduce a debate on a general problem (e.g. soil degradation) or to review all the different techniques that are likely to solve a problem (e.g. water management at village level).
- Video programmes are an excellent way of replacing visits to locations that are difficult to reach.
- Technical booklets are efficient for disseminating techniques and for post-literacy

learning. Their efficiency is increased when they are distributed straight after a meeting or a training session. It enables smallholders to go back home with something that shows what they have learnt and seen, and that they can discuss further. A variant of this kind is the technical pamphlet distributed after a demonstration. In both cases, they must be made with care and illustrated in a very understandable manner (for example, by reproducing in a reduced manner the drawings used during the meeting).
- Strip cartoons can also be used. They are interesting for complex themes, but must only be distributed after a meeting or seminar. The caricature aspect attached to certain cartoons can sometimes limit the rural people's understanding.
- Drawn pedagogic media (illustrations on boards) are particularly appreciated during meetings and seminars. The drawings, however, must stick to certain rules, as they must be simple for example. Their use also requires some explanations (especially in the beginning when the small-holders are not familiar with drawings).
- In a dissemination programme, it is often a good idea to combine different pedagogic methods according to the objective targeted and also to avoid monotony.

3.4 Farm integrated (management) advice

The example of south Mali mentioned previously shows the importance of adapting disseminated technical contents to the needs of the different categories of farmers. This development, linked to the awareness of the diversity of farms, has resulted in some cases, still rare in Africa, in an individualised approach to agricultural advice (farm integrated management advice).

3.4.1 Some definitions

A farm is relatively easy to define in developed countries, for instance where a family (usually limited to one household) produces on a given surface (rented or owned land), consumes and accumulates what it produces. It is more difficult to define in Africa where several households may live on the same homestead or concession, exploit the same land and/or the same livestock (collectively or not), possibly share the same kitchen and may have separate or joint accounts. That is why the term "farm" will be used to mean a *basic socio-economic unit*. In other words, it represents the smallest level where production, consumption and accumulation take place. There might therefore be many "farms" on the same plot or concession (see Chapter II: Diagnosis).

Advice can be purely *technical*: it often aims to modify the producers' farming or cattle-rearing practices. Advice can also be of a *technico-economic* nature, the aim of which is to help the farmer to select technical practices that take into account the economic effects they will have. *Managerial advice* (or farming integrated advice) is advice given to the farmer to help him/her in the decision-making process. It can be defined as a method taking into account the entire situation of a farm, seeking a dialogue with smallholders and aiming to make improvements over a period of many years.

3.4.2 Procedures for farm management advice

These are based on the idea that farms can improve their technical or agronomic results, but that the road to be followed is not the same for all.

Each farm, each production and consumption unit has its own characteristics (land, equipment, livestock, labour, etc.). It also has its own objectives (independant food security, marketing and capitalisation, accumulation of land, of livestock, of money, etc.).

The farmer uses his/her material resources and available manpower in the most adapted manner to achieve his/her objectives, given the possibilities and constraints of the ecological and economic environment. This includes

many components, for example soil condition, weather risks, organisation of sales and supplies, products and inputs prices, information availability, and the like. In such a complex environment, the farmer's decisions are motivated by three factors:
- The objectives targeted
- The means he/she disposes of or to which he/she can have access
- How much he/she knows about "what could be done", taking into account his/her objectives and means.

Farm integrated management advice strives to
- clarify with the farmer the objectives he/she is trying to achieve (which are often implicit);
- analyse with him/her the means at disposal;
- analyse with the farmer the way he/she is using these means to meet his/her objectives and how successful he/she is;
- seek with the farmer possible improvements.
When it is necessary, technical and economic information on management should be produced.

Managerial advice presupposes there is a relationship of trust between the farmer and his/her adviser. The only way to work with is the cards on the table, to monitor effectively (recording data), to analyse the technical, financial and economic results and to create adapted scenarios for improvement and evaluate the results.

In terms of practice, management advice often occurs as follows:

- The collection of data, concerning the farm and its environment, enables an initial diagnosis (or pre-diagnosis) to be made and gives an indication as to how a more in-depth procedure for monitoring should be conducted.
- Farm monitoring by the farmer him/herself or by an external officer aims to record all the technical operations performed (dates, work times, manpower and tools used), as well as the circulation of goods and money (consumption, purchase, sales, loans).
- The farm diagnosis shows the farmer's constraints and assets and the strategies used to achieve his/her objectives.
- The definition of a provisional plan for the cropping season aims to modify the farmer's practices and improve his/her results while taking into account his/her constraints and objectives. The plan is consensual and is based on technical propositions formulated by the adviser (or farmer himself).
- Technical monitoring is done of the plan's implementation, and of evaluation at the end of the cropping season, of the modifications made during the year, of the technical and economic results recorded and of the problems encountered that explain any differences.
- There is a definition of a new plan for the next cropping season, which, in certain cases, means defining a plan of improvement extending over a period of three to four years.
- Technical monitoring and evaluation is also done.

An example: farm integrated advice in south Mali

Managerial advice complements the village approach described previously and uses the same typology of farms. The smallholders in question are literate in the Bambara language. A technico-economic analysis of a number of farms enables group discussions on agricultural development on farms to be conducted between the smallholders themselves and between the smallholders and the advisers.

From the analysis of production factors from each farm, it is possible to propose a production system covering monetary and cereal needs by using labour, agricultural equipment, harnesses, livestock and the like in a rational way, while ensuring at the same time that the soil is protected and fertility maintained.

All group sessions are plenary and other villagers are encouraged to attend. It is supposed that the participants belonging to one type of farm recognise themselves in the case studies. The technical actions proposed are open to all participants, on a voluntary basis. Technical monitoring and evaluation of these actions apply to all the volunteers.

The method revolves around the following:

1. Data is collected on the environment and the farm. A fact file on each farm is thus established from surveys conducted by the participants and includes information made available by the village association.

2. The situation is analysed. A diagnosis of the constraints and opportunities is made. Discussions with the smallholders enable their objectives and strategies to be clarified and checked. This procedure should be supported by technical demonstrations of the proposed innovations that have been carefully chosen by the producers themselves.

The analysis is mainly based on the following:

- Balance of crops
- Cereals produced according to the number of people to feed within the farm
- Gross margin on cotton production
- Food and monetary needs

3. Provisional plan for the cropping season. The conclusions of the analysis are compared with the propositions made by the smallholders concerning the surfaces, the levels of productivity to be achieved, the necessary inputs, investment in material, and so on. Precise technical propositions are formulated. Based on the outcome of the discussions, the smallholders can then determine their definitive plan.

4. Technical monitoring and evaluation during and at the end of the cropping year. The participants write down in their notebooks the main events of the cropping season: intervention dates, the quantity of inputs brought ... according to standard tables. Group excursions are organised to visit the demonstration fields/actions. Estimates of land coverage and productions (yield) are essential.

This method has been successfully used with literate producers in south Mali. Experience shows that the application of these methods in Africa is difficult because managerial advice is inseparable from a tendency to individualise the given advice, which makes it very expensive. It is therefore not an alternative to group extension but, by concentrating on some representative farms, it may be possible to establish ways of improving the running of farms generally. References and typical case studies are obtained that help to define the contents to be distributed to larger groups.

3.4.3 Farm integrated (management) advice: the tools

Much research has been done in Europe on farm management. The studies are linked to a variety of approaches that are briefly discussed below:

The tools used for managerial advice can be classified in various ways. They can, in fact, be considered according to the level of complexity (or refinement) of the method, the individual or collective use of the tools, the end results (financial, technical or fiscal ...) or the support, be it manual or computerised. We will distinguish between the tools used for technico-economic management, the tools used for prospective management, those for financial management and finally the tools used to help in the decision-making process.

A. *Technico-economic management*

The aim of the various methods of technico-economic management is to define the observations made regarding farmers' practices, identify the bottlenecks and suggest solutions by using references that generally come from outside. Technical analysis is backed up by an analysis on sources of income that is based on the use of three main methods:

- *Group analysis* consists, in the context of extension, of analysing, classifying, comparing and advising a group of homogenous farms based on a set of normative references. The latter can be endogenous (practices used by other farmers in the group) or exogenous (techniques derived from research or from extension). Data on structure, results and more incidentally on operations are necessary to try to define the relationship of cause and effect between production factors and the results obtained. The main limitation of this method is its normative aspect. This makes it difficult to take fully into account the coherence between the farmers' objectives and the strategies he/she uses to create income. This kind of analysis can be manual or computerised.
- The *gross margin method* is individual and concerns the analysis of a farm. It consists of allocating their operating costs (production costs or inputs, manpower, that are renewed each year) to the different products on the farm, and therefore the gross margins (raw products, operational charges) can be calculated. By comparing these margins and identifying production factors, managerial advice can be established based on optimising gross margins. One can estimate how the various products contribute to cover the fixed charges (fixed charges may be the reimbursement of annual payments on agriculture equipment, equipment depreciation, the general running cost of the farm, etc.). This method can be very interesting in diversified systems of production as it enables farmers to choose between products (real or potential) and to look for a production plan. Its main limitation is the fact that it does not directly take into account the fixed charges and therefore it is difficult to make estimates including investments. Moreover, it is impossible to compare the gross margins obtained from one farm with another owing to the differences in existing fixed charges that affect income.
- The *cost price method* is based on accountancy techniques. It consists of taking into consideration all the charges (operational meaning variable and fixed) related to one activity and reducing it to a technical production unit (i.e. cost per ha, per head of animal) or to a unit of goods produced (per kg of wheat, per litre of milk). This represents the unit-total cost. This method, which is more demanding in terms of the information required, is essentially used in specialised farms where the concept of cost price can be a criterion for management. In more traditional farms where the remuneration of all the productive factors (including land and labour) at the level of the market is incidental, this kind of method can be avoided.

When European agriculture was characterised by production growth and good product sales, the tools for technico-economic management were essentially based on technical aspects: increased production would result in better income than would the refining of resource allocation. The tools were therefore basically normative, partial and incidentally economic. Accounting did not play an important role. Management was more a diagnosis of the past than a plan for the future.

B. Prospective management

However, the need for better forecasting has become obvious, on the grounds of the hypothetical modification of farm structures. As the previous tools are insufficient or difficult to use, it has become necessary to search for new methods of management that attempt to evaluate the impact of structural changes while optimising activities. Inspired by the methods of *operational research*, one can distinguish three methods: the partial budgets method, linear programming and simulation tools.

- The *partial budgets method* attempts to simplify the prospective approach by using the concept of *opportunity costs* (decision results in a different allocation of production factors and, therefore, other possible allocations and their advantages, be they monetary or not, are abandoned[2]). The method consists of asking oneself what is to be gained or lost in carrying out certain change, firstly at a qualitative level and secondly in quantifying as much as possible. It is similar to the gross margin method in that it does not take fixed charges into account either. It is therefore a *simple tool* to solve everyday problems that can also be of great help in explaining the farmers' objectives. In fact, it is a frame for review aimed at identifying and organising the changes envisaged. However, it is difficult to make predictions using this simple method in complex systems.

- *Linear programming* is a "refined" technique that combines all the elements that make up the running of an enterprise. However, the process of optimising the allocation of resources is carried out in a context of technico-economic normality, given price systems and hypothetical behaviour. In order for it to be used in management, the model needs to be parameterised or calibrated, which aligns this approach with *simulation*. The obvious limitation of this method is its cost in terms of time and instruments: preparation is time consuming and must be computerised. This explains why its use has been very limited in farm management. Today the main users of the system are research centres and planning organisations. It should be stressed that an exaggerated interpretation of linear programming has resulted in a tendency to accentuate its normative character, whereas it should be considered as *an aid to decision making*.

- *Simulation* increases the possibilities of linear programming as it allows for the intensification of different projects. The advantage simulation offers is that by firstly analysing the coherence of the technical choices, then evaluating the economic or financial consequences, forecasts are set between limits making it easier to appreciate the risks (climatic changes or price fluctuations). For these tools to be used a great deal of reliable information must exist. A high level of expertise is also required to set up the simulation computer software and to parameterise hypotheses. The results obtained from the hypotheses, developed for the purpose of simulation, greatly demystify the previsions. Computer software programmes are available on the market and are sometimes used for investment plans.

C. Fiscal and accounting management

In Europe, the conditionality of certain aids and fiscal requirements (VAT, real benefits) has resulted in the distribution of accounting tools, which has largely contributed to the confusion between management and accountancy.

Based on the farm accounts and the balance sheet, these methods have reduced man-

---

2. For example, if one has an abundant labour force, the choice might be either to improve the maintenance of the main crop (several people hoeing for weed control) or to use the labour to diversify activities, be they related to agriculture or not.

agement to the rapid interpretation of some criteria, essentially financial (cash, debt rate, agricultural income, depreciation). And yet, these tools have been used mainly to obtain financial aid, or to pay less tax. Moreover, the persistent decrease of agriculture tends to devalue the activity in the balance sheets tremendously. Variations of prices sway the evaluation of stock. Under these circumstances the relevance of these accounting tools must be questioned and their utilisation limited to the limits of their validity: fiscal and tax aspects.

D. Fields of investigations

Such an overview would not be complete without mentioning some fields of research that will undoubtedly emerge as managerial tools. In France a method known as the *integrated approach* (*approche globale*) attempts to reconcile farm management advice with the farmers' way of making decisions, by using certain previously mentioned tools. It combines a systemic approach with attention to the coherence of objectives, strategies and means used.

The tools used in Europe over the past few decades seem difficult to extrapolate for use in Africa in the current economic and institutional conditions. Under such circumstances, it is better if *management advice* aims at farmers' groups (or types), and is a means of helping them in the decision-making process (organising information on techniques, the running of farms and the environment) rather than being normative advice which is inevitably individual.

However, the tools are still useful and are increasingly used in research-development programmes that finalise technical contents and work methods in representative but geographically limited situations before transferring them to larger areas. In this framework, recourse to reference-farm networks constitute a compromise between the usual mass extension and sheer individual management advice. It has been implemented successfully in France, Madagascar, Rwanda, Reunion Island, for example.

3.4.4 Future of managerial advice

Farm management advice is currently at an experimental stage in many regions. And yet it is very important to encourage it if the mainly short-term strategies of farmers are to be overcome. Good management is necessary, not only for the farmer, his/her farm and the rural organisations to which the farmer belongs (see Chapter IV: The farmers' organisations), but also for the preservation of natural resources (see Chapter IX: Natural resources management).

He/she who manages his/her own business well will be in a better position to manage those of the community.

4. Conclusion

The methods presented here are inseparable from the institutions and the officers who implement them. Hence, some questions need to be raised.

4.1 Who should take responsibility for agricultural advice?

In sub-Saharan Africa, for many years, agricultural extension was a field reserved for development organisations (commodity companies) and big projects. From the 1980s, national extension programmes supported by the World Bank have appeared, along with the transfer of the responsibility for extension to technical agriculture services, thereby becoming increasingly dynamic.

There are also alternatives, raised by the recent socio-economic changes (state withdrawal, decentralisation, privatisation, local empowerment).

The empowerment of federative farmers' organisations (see Chapter VI: The farmers' organisations) opens new perspectives. It would seem logical that progressively farmers' organisations should take over the responsibility of providing agricultural advice to their members.

This should undoubtedly guarantee that the advice given is best suited (in terms of contents and distribution methods) to the diverse local and regional circumstances.

Decentralising agricultural advice and extension services does not exclude a *national policy on extension*. This should be developed in a consensual manner with the different role-players involved. Negotiated at regional and national level, based on the expectations of the farmers and on macro-economic objectives, the national policy on extension could be implemented through *regional intermediaries* (managed or co-managed by organised farmers). Through this means it will be possible to

- define orientations and procedures,
- look for overall coherence,
- organise support for a variety of experiences and exchanges between them,
- train advisers,
- liaise with research and other developmental institutions (credit).

4.2 What status for agricultural advisers?

Extension officers must not necessarily be civil servants or public contract employees. Many other statuses are possible. Such a post could be:

- a part-time job for someone who has recently learnt to read and write, and who receives an allowance from the farmers' organisation (case of village associations in south Mali);
- a full-time adviser, employed by a professional organisation (even if the government has to subsidise a large part of the salary–extension, human capital and capacity building should remain a public service duty and mandate), and the like.

Whatever the institutional context, there should be coherence between the technical support network (advice, training) and the economic support systems (organisation upstream and downstream of production). Farmers' organisations must be partners and play a key role in this coherence.

To carry out the responsibilities inherent in agricultural advice, the farmers' organisations need

- financial and material support,
- pedagogic support and personnel (allocation of technicians for example to organisations).

Setting up semi-public structures (like the Chamber of Agriculture) is a possibility that perhaps needs to be investigated. The idea is not to transpose European models, but to develop an original organisation inspired by experiences that have proved efficient.

> The Chamber of Agriculture in France is a semi-public organisation as opposed to the economic organisations set up by producers that are private structures. The chamber operates on a regional and local basis (even though it forms a national council as well); it is managed by the farmers (through elected boards, with farmers' unions representatives), but receives public funds that cover a variable part of its running costs (salaries of the agricultural advisers, who remain salaried employees of the chamber). Local officers are in charge of most training, development projects, extension and advice.

4.3 Can the adviser hold several positions?

It is highly recommended that advice (technical and managerial) be his/her first priority. But when the crop season is short and there is no other efficient system (private or public) to distribute inputs or market them (duties that are normally conducted outside the productive season), holding several positions can be justified. It is nevertheless recommended that credit be separated from extension and that statistics-related duties be reduced. Instead, some specialised officers may conduct surveys based on samples.

5. Recommended literature

Benor, D. & Harrington, J.A. (1977) *The training and visit system of extension*. The World Bank. Washington D.C., USA.

Bosc, P.M. & Hanak-Freud, E. (1995) *Agricultural research and innovation in Africa*. CIRAD-SPAAR, Paris, France, 134p.

Cernea, M. et al. (1985) *Research, extension, farmer: a two-way continuum for agricultural development*. The World Bank, Washington D.C., USA.

Gentil, D. & Devèze, J.-C. (1988) *Organisation paysanne et vulgarisation*. CFD, Notes et Etudes No.10, Paris, France.

Mtshali, S.M. (2000) Monitoring and evaluation of women's rural development extension services in South Africa. *Development Southern Africa*, 17 (1): 65–73.

Merrill-Sands, D., Kaimowitz, D. et al. (1990) *Production and technology transfer*. ISNAR, The Hague, The Netherlands.

Papy, F. (1994) Working knowledge concerning technical systems and decision support. In: *Rural & farming systems analysis. European perspectives*. Dent & McGregor Edit., CAB International Publ., pp. 222–235.

Roling, N. (1994) The interface between farmers' and research workers' knowledge. In: *Systems-oriented research in agriculture and rural development*. International Symposium, Montpellier, France, 21–25, Nov. 1994, pp. 315–319.

Roling, N. (1996) Towards an interactive agricultural science. *European Journal of Agricultural Education & Extension*, 2 (4): 35–48.

Scarborough, V. et al. (editors) (1997) *Farmer-led extension: concepts and practices*. ODI Intermediate technology, London, UK.

Chapter IX: Natural resources management
T. Gillet, J. Mercoiret, J. Faye & M.-R. Mercoiret

This chapter may usefully be connected to the following chapters:

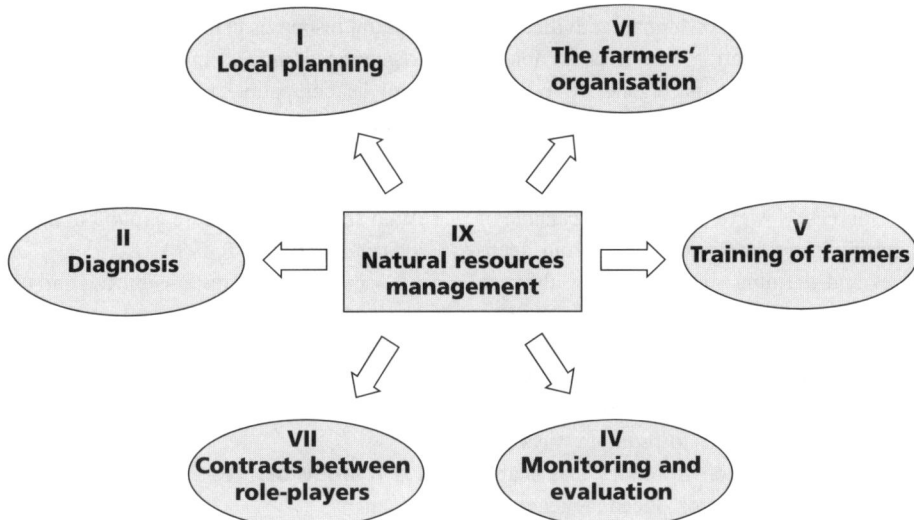

1. Defining some key terms

By *natural resources* we mean the physical elements of the environment: plants, animals, water (in all its forms), soil and air that are used by human societies to satisfy directly or indirectly their dietary, domestic or monetary needs. They constitute the *ecological capital*.

Exploitation of natural resources means using the available resources to derive the maximum benefit from them. This is as old as humanity. As the population grows, so the exploitation of natural resources intensifies in order to keep up with increasing dietary and monetary needs and because the technical knowledge that facilitates and/or accelerates this has also continually increased.

The *management of natural resources* of a given place is the whole process as of deciding about the exploitation of natural resources, the access to them, and the means used to collect and develop them. These decisions are made individually and collectively (depending on the objectives and the constraints) by those who live in the place, who have access to it or who have a right of say.

When the ratio of humans to available resources in a given place is low, the place and the resources seem inexhaustible; it is only necessary to follow a few simple rules for the resources to be naturally regenerated (shifting cultivation on slash and burn systems, spreading out scarce human settlements). When the ratio of humans to available resource ratio increases, more sophisticated and stricter measures become necessary. Gradually there is no more space to conquer and the development of towns cannot absorb an ever-growing population. It is often said that space is finite. This is when the process of deteriorating natural resources begins, when the resources lose their quality and their potential (a decrease in soil fertility, reduction in animal and plant resources, air and water pollution). This type of degradation must be differentiated from that resulting from natural phenomena (short-lived like a cyclone; or long-lasting like drought which has aggravating effects in sub-Saharan Africa). It is because of human activities and methods of exploitation that it is no longer possible to reproduce and regenerate the ecological capital.

It is, therefore, the population increase in a given space that is responsible for deteriorating resources. However, humans are capable of increasing their knowledge, inventing techniques and defining rules that allow them to get the best out of the natural resources. Human abilities vary according to time and space.

The diagnosis of accelerated degradation of natural resources gave rise to the notion of *sustainability*. Its aim is to promote ways and means of using natural resources that enable the resources to be reproduced, regenerated and, if possible, improved in order to leave future generations with a sound ecological capital.

2. Identifying key issues

In many African countries, smallholders, their organisations and external operators all have noted the accelerated deterioration of natural resources and its consequences (reduced productivity, sterilisation of certain areas, increased scarcity of grazing areas). This can only get worse in the future, if the current development tendencies persist.

How to react? A number of local initiatives are emerging with or without the support of NGOs (mostly in Sahelian countries). In some of these countries national policies have been defined (i.e. in Burkina Faso and Benin, national projects on soil management have been created), in others the state has taken strong measures (works of general interest in Rwanda). Projects on soil management have been set up both at local and regional level.

The problems encountered raise several questions:
- How can one promote methods to exploit natural resources that will enable them to be regenerated, indeed improved, and that will still be compatible with the mostly short-term strategies adopted by smallholders who are themselves confronted by so many constraints?
- Who should define these exploitation methods and how?
- How can the roles, responsibilities and charges, between the state, rural communities and external operators be shared?

3. A framework for review

3.1 Obvious signs of degradation but also important attempts to adapt to constraints

Human intervention is not necessarily synonymous with degradation, because humans can also maintain and improve physical conditions. However, it has been observed that, increasingly often, human activities provoke and hasten the deterioration of natural resources.

Certain human practices can cause or accelerate this degradation process: deforestation or land clearing, unsuitable agriculture practices (a badly run irrigation system in certain areas can cause soil salinisation) or overgrazing.

The soil as an exhaustible resource

The soil loses some of its solid layers (sand, silt, clay, humus) by these being swept away from their original site and transported a distance that can vary from a few hundred metres to thousands of kilometres.

Erosion

Erosion can be caused either by wind or by water; the latter is more frequent in Africa where erosion occurs as soon as the annual rainfall exceeds 300 to 400 mm in any given region. As the rain falls heavily on the land, earth particles are swept away by gravity. The erosion is either diffuse or linear. Diffuse erosion is a generalised process that gradually removes the superficial top layers of soil where the essential elements of fertility are concentrated. This always occurs in conjunction with human activities. In the case of linear erosion, the streams of rain water converge into channels that form ravines or gullies many metres deep. This is frequent in the savannah and is often found in the clay soils along water channels; from there, the erosion tends to spread out like tentacles upstream of the catchment areas.

Soil degradation

In this case, the soil does not lose its solid layers but some of its essential properties deteriorate on the spot.

This can be of a chemical or a biological nature (loss of nutrients and humus, salification, acidification, diverse pollution factors) or of a physical nature (soil compaction, soil conglomeration, crusting of the soil's surface, changes in the soil water economy).

3.2 A multitude of causes

The degradation of natural resources is a vicious circle (or spiral).

3.2.1 Demographic pressure and the deterioration of the general economic context

Demographic pressure and general economic deterioration usually result in a drop in farmers' incomes owing to a decrease in prices for agricultural produce. Farmers are pushed to over-exploit natural resources. Confronted with an ever-increasing need for food and in an attempt to maintain their financial income, producers have, for instance, increased areas under cultivation (including sensitive areas that were previously protected) and exploited and marketed more and more forest products (charcoal). In so doing they have reduced the land reserved for livestock and therefore the availability of organic manure.

From a technical point of view, certain traditional practices are unsuitable as available space is dwindling (long-term fallow land for instance); the technological packages disseminated through projects have in certain cases increased erosion (e.g. the removal of stumps to facilitate mechanisation).

From a social point of view, there has been a weakening in the traditional rules for managing natural resources, while no new respected rules have taken their place. As a consequence, the land is used without consensus. This results in tensions and competition (between native and migrant smallholders on the one hand, and crop farmers and livestock farmers on the other) and individualised methods of resource exploitation that are not subject to any social controls, be they external or internal.

3.2.2 Traditional techniques becoming inadequate

The traditional methods used to derive the most from natural resources and those disseminated by development interventions have often been inadequate for the new socio-economic conditions.

3.2.3 Globalisation and liberalisation

Rural societies are increasingly being integrated into the national and international market for which they produce and from which they receive goods, agricultural inputs and services that have become a necessary part of their lives. They have no control over product demand (groundnuts that were in great demand in the past have now been replaced by sunflower seeds or soybeans) or prices (that are often in decline for agricultural products and on the increase for production factors).

The societies have therefore to adjust to the new socio-economic environment.

3.2.4 Climatic disturbances

Constant serious climatic disturbances (drought) have revealed existing imbalances and accelerated them. At present, certain measures linked to state disengagement (i.e. reduction/withdrawal of subsidies for inputs) have aggravated the situation causing already modest intensification practices to be abandoned. These different factors are linked together, each reinforcing the other in a cumulative process that is at the origin of the vicious cycle and descending spiral effect.

3.3 Solutions of varying efficiency

Faced with the seriousness of the situation, various external interventions have attempted, over the past few years or sometimes decades, to find solutions to the problems.

3.3.1 State regulations on the use of natural resources

State regulations have resulted in laws, codes and decrees, as well as the reinforcement or creation of departments responsible for enforcing them (in some countries, the water and forestry officials have a fearful reputation).

However, many experiences have shown that, although official regulations are necessary, firstly they are not sufficient when it comes to changing practices (that have a multitude of causes) and secondly, they are sometimes not adapted to local realities.

3.3.2 Awareness campaigns

The alternative and often the complement to external official regulations is a social awareness campaign that targets smallholders. The official line states that smallholders destroy their natural resources because they are not aware of the consequences of destroying them. And yet, it only takes a serious discussion with the smallholders to notice that the opposite is true. Smallholders have a sound understanding of the consequences and their analysis of the causes is quite pertinent. However, they consider that with things as they are at present there is not much they can do, unless they sacrifice their immediate survival for a hypothetical future.

3.3.3 Encouraging conservative environmental practices

This involves erosion-control programmes, reforestation, and the like that offer material benefits (food, money). Generally this method produces immediate results (general mobilisation for concrete results) and sometimes creates new habits. (The Mossi smallholders from Yatenga, Senegal, have for instance integrated the construction of rocky enclosures into their practices.) In other cases "popular efforts" are of short duration. Generally speaking these incentives concern punctual programmes (although important). It is difficult to duplicate them on a national scale and they cannot claim to modify all the practices responsible for the degradation of natural resources.

3.3.4 Large land development operations

Important developments that require considerable external means (dams for example) are useful in many cases even if they sometimes have consequences both on a socio-economic and environmental level that were not entirely foreseeable or controlled. Indeed, in certain cases, they entail the development of undoubtedly less sustainable agricultural practices. This was the case in the delta of the Senegal River where the construction of the Diama Dam resulted in a rapid increase in the rice cultivation area where there were only basic facilities and no drainage.

3.3.5 Specific regional operations

Historically there have been two types of operations:
- About 20 years ago, operations concerned with the *conservation or regeneration of the environment* appeared (reforestation operations, land development projects).
- As those who were supposed to benefit from these operations were not very enthusiastic about them (in spite of interesting results), external operators (governments, sponsors) defined a more *integrated* type of intervention to take into account the short-term concerns of smallholders (improvement of production and living conditions). This was done through the implementation of *a development strategy aimed at the mid to long-term management of natural resources* and was known as community land *management projects* (*projets de gestion de terroir*). These interventions have showed varying results up to now, but some seem to be promising (Burkina Faso, Niger, Benin, etc.).

3.4 Lessons

From the past and current experiments, the following lessons can be learnt:

In many regions of Africa, environmental degradation is the most visible (or the most alarming) sign of a general decline in both production conditions and the rural people's existence as well as of the decapitalisation[1] of farms. Faced with a multitude of constraints (an increase in food and monetary needs, a decrease in rainfall, stagnation or reduction of agricultural product prices, an increase in input prices) the majority of which are beond their control, the smallholders try to adapt. Left with so few options their only possibility is to draw from their ecological capital, thus jeopardising the future. The general weakening of traditional managerial rules and the difficulty of substituting them with new and efficient ones has led to the emergence of individual practices. This applies not only to smallholders but to other local and external role-players also (charcoal makers for instance). All of these practices are dictated purely by short-term concerns (survival or the search for maximum profit) that aggravate the deterioration of the natural environment and in turn jeopardise the collective measures still in force.

From this observation, it is possible to deduce that changes in exploitation methods cannot be dictated from the outside, nor can they be reduced to the decreeing of a *regulation* or to actions in a particular sector.

Significant and long-lasting changes can only occur if there is consensus between all those who play a role in using or managing natural resources (rural or urban people, individuals or social groups, developmental institutions, local communities and the state).

In order to slow down the process of dete-

---

1. Capital consumption, e.g. when smallholders sell their animals to buy food during drought period.

riorating natural resources (and if possible to reverse it), it is necessary to see the actions that have been carried out in a broader context and not separate them from a socio-economic developmental strategy. For there to be sustainable development, "development" must take place. For there to be long-term viable methods of exploitation, farms must build an asset base. In other words, specific actions that target the conservation and management of natural resources are necessary. However, they must be part of a *real development strategy*
- that simultaneously takes into account short-term priorities and mid to long-term orientations and objectives;
- that is based on convergence or compromise between the interests of the local role-players and those of the role-players at other decision-making levels. (See Chapter I: Local planning, and Chapter VII: Contracts between role-players.)

4. Orientations for actions

From now on we are concerned with orientations that directly affect the population at a local level and the role-players with whom the population work. The orientations must, however, be associated with both a population policy aimed at controlling demographic growth and a land development policy that aims to improve population distribution with regard to available resources. Both should be long-term policies.

Population policy

Population growth is not a bad thing as such. One can for example consider that Africa is underpopulated compared to Europe and Asia. In certain regions of these continents, there is a high population density, and yet, farmers enjoy a high standard of living.

The problem in Africa is that the rate of population growth, which is around 3% per annum, is greater than the national production growth rate and climatic conditions are worsening (a drop in rainfall).

Although more rigorous developmental policies are required, it is also necessary to control demographic growth or at least slow it down through a population policy. In general, several actions are being combined: better health care is being provided for mothers and children in order to reduce infant and perinatal mortality; education for the population and easier access to contraceptives are being promoted (thus enabling births to be spaced out, etc.). These actions are effective only in the long term (20 to 30 years).

Land development policy

A way of facilitating and accelerating socio-economic development is to allocate and localise infrastructures (roads, railways), equipment (schools, community clinics, hospitals), factories and retail outlets, for example, according to the whereabouts of the population and natural resources. This is the objective of land development policies. A road, for example, may be built to facilitate the transportation of agricultural products between the capital and a cereal-growing region. Boreholes can be sunk in sylvo-pastoral areas to make it easier for livestock farmers to exploit the land.

4.1 Three factors to be considered

4.1.1 A multitude of role-players

Many role-players use the natural resources and intervene in their management.

Smallholders and livestock farmers are the privileged users of natural resources. However there are other users: craftsmen/women (bricklayers, carpenters, potters), hunters and fishermen and external role-players involved in the exploitation of local forest resources (coal producers, retailers interested in picked crops, the timber industry, seasonal and pioneer migrants) or forage resources (transhumance or nomadic livestock farmers).

Smallholders have a role to play in the management of natural resources but they are not the only ones. The *state* through its legislation and administration intervenes in management policies, as do the sponsors who at present are making an important issue of North-South co-operation. Sometimes the state delegates certain responsibilities in this field to the local community as part of a decentralisation policy (this is the case in Senegal where rural communities are responsible for the land management in the territory under their administration). Depending on the perception each role-player has of the natural resources and their interest in them, the strategies employed may be convergent or contradictory. Before any action, *one should first identify the different role-players involved as well as their objectives and strategies.*

4.1.2 Local spaces and their management: interweaving, differences and interdependencies

At local level (see definition in Chapter I: Local planning) there is no such thing as a space regulated by unique rules, but rather many spaces, regulated by many specific decision-making centres (see also Chapter II: Diagnosis).

The decisions made at each level are influenced and sometimes regulated by the decisions made at other levels; the room for manoeuvring (initiatives and autonomy) varies between the decision-making centres.

At field level, the person in charge decides (within the limits of available means and information) on the cultivation practices that will be used.

At farm level, the farmer makes the most important decisions: clearing new lands, whether to leave land fallow or clear an area for an orchard, and so on.

At village or district level, in many places, there are decision-making (or arbitration) authorities that give rulings on issues such as the designation of areas for crops and those for cattle or on the routes cattle may take.

At inter-village level, meetings of a similar nature may take place to discuss the management of pastures, the exploitation of a communal forest or lowland or the construction of a dam in a valley.

The decisions made by farmers take into consideration those made at village and inter-village levels; the farmer can, however, influence the latter. Although he/she can impose choices and practices on those who depend on him/her (i.e. women and youths to whom he/she attributes the land), he/she also takes into account their aspirations.

The local level is part of a more extended geographic and political whole (region, country) where decision-making centres influence development and management practices at lower levels.

Before proposing changes concerning the use of natural resources, *it is advisable to identify the spaces concerned and the decision-making centres that regulate them* as well as the relationship between these spaces and the decision-making centres.

> The reduction of subsidies for fertilisers was the result of a structural adjustment concluded by the government and "strongly advised" by the IMF. The decision took into account the prices of fertilisers as well as, in the final analysis, the price of petrol on the international market and transport costs.
>
> A national policy on land development supported by foreign sponsors has a considerable impact on the nature of developmental interventions and can result in the creation of new village authorities responsible for the policy's implementation (case of Burkina Faso for example).

4.1.3 Transitional land tenure systems

The different levels of management, that is the distribution of the rights and functions linked to each space, constitute the land system that can be totally or partially formalised (written as a rural code for example) or not formalised but written into tradition.

Land tenure systems in western Africa

African land tenure systems are very varied and many experts still wonder if these systems are beneficial to natural resources preservation or, on the contrary, if the ambiguities and contradictions inherent in the system rather favour the degradation process. In many West African countries the land tenure system is characterised by the conjunction of two legal systems. On the one hand there is "modern law" inherited from colonial law, the references for which are found in western culture, and on the other hand there is common or traditional law linked to the political, social and family structure of pre-colonial African societies.

Traditional land tenure systems

In many cases, land is appropriated by lineage, by families who were the first dwellers or the oldest settlers at a place (they are said to be the eminent owners or the landlords). However, this appropriation has nothing to do with "private property" in the western sense.

On the one hand, land is not seen as property with a commercial value; it belongs in fact to the spirits who occupy the land. The first settlers allied themselves with the spirits by offering sacrifices so that they may have the right to benefit from the land's resources and make good use of them. From this point of view, the land is inalienable. It cannot be sold and neither can it be refused, if it is available, to those who ask for it (authorisation for clearing or renting) and who have been accepted within the community.

On the other hand, different farmers can exercise different customary rights on the same land providing they do not have conflicting interests; the use of the soil, the trees, the water, the clay huts and game are the concern of many different farmers.

The traditional land tenure system is therefore a system that accommodates both collective and individual management of land. With the transition from subsistence farming to commercial farming, the traditional land tenure systems are too slow to adapt to competition (competition linked either to more global demographic pressure, or to urban, governmental or individual pressure that is based on modern law).

Colonial land tenure system

To simplify, it can be said that this system introduced the concept of private property and state-owned property. This was supposed to enable the administration to be free in as far as land appropriation issues were concerned and by the same token to favour the emergence of a class of "agricultural entrepreneurs". In real terms these two concepts were implemented through two legal procedures dating from the beginning of the century. Firstly, the registration process which officialised the notion of property, and secondly, the concept of "vacant lands without owner". This meant that in the absence of any other recognised rights, the state could proclaim itself eminent owner. This last concept was widely contested by the natives to whom it was finally conceded that only the land unused for ten years should fall under this law (1935).

Modern land tenure system

Generally this system was adapted from the colonial land tenure system. After independence governments modified it, giving more emphasis to common law without, however, questioning the basis of the system. Most countries attempted to reconcile the concept of private property, collective management inherited from tradition and state-owned property.

Property: This concept in its broadest sense is still alive, but generally speaking states have appropriated it for themselves. Depending on the country, the land belongs to the people, to the state or the nation. However, by reintegrating the concept of state ownership of lands without owners, the state affirms its eminent rights over the territory as a whole.

Full and entire private property: This concept is recognised and approved in urban areas. Depending on the rural area, private property can either be excluded (as in Burkina Faso where only users' rights are considered) or officially recognised and guaranteed through registration. (This is the case in the Ivory Coast where 1% of land is concerned and to a lesser extent in Senegal or in Mali where it is partially used in peri-urban areas.)

Customary and collective appropriation is recognised everywhere but the degree to which local communities are truly empowered to use the land varies. Generally speaking the logic is "the land belongs to he/she who is working on it" or "who is developing it". This concept clearly applies to land that is built on or cultivated (fields and a fortiori perennial plantations) or similar (recent fallow land), and therefore to the concept of "work". However, it still remains very unclear when it comes to lands used on a temporary basis and especially bush lands that are used for livestock farming, forestry, crop picking and hunting.

Although the current legal framework generally recognises the concept that "land belongs to he/she who works on it", it has difficulty in recognising that "land belongs to he/she who occupies it". In practice, apart from land that has been directly appropriated (cleared or cultivated), all the "interstitial" spaces depend on vague laws juxtaposing the "eminent right of the state", which is difficult to exert, and traditional rights, which are difficult to defend. (The village can increase its cultivable area but can hardly refuse to let a migrant settle on it.)

Thus, the current legal framework generally recognises two types of domains in rural areas. On the one hand there is the village domain (cultivated land) that is fairly well recognised but which may be re-appropriated by the state for works of public interest (especially irrigated perimeters). On the other hand there is the classified domain that clearly belongs to the state (classified forests, parks and reserves) but which is often badly defended. Between the two extremes exist domains the appropriation of which is less

> clear and where opposing powers are exerted. This is the case of the protected domain in Mali (these levels of state appropriation can be found regarding the classification of forests in the Ivory Coast where a distinction is made between the two types of domains: the National Forest Domain (NFD) and the Rural Forest Domain (RFD).)
>
> The dividing line between the types of domains, "village" and "protected", is unclear and is sometimes challenged by the administration. Recently in Mali, at Koutiala, the administration prohibited all new land clearance for three years in order to counteract the effects of excessive agricultural pressure on the remaining land. In this context, it was decided that all land that had been lying fallow for three years should be subject to this decision. In other words, the powers of decision concerning this part of the land were taken away from the villagers. One can imagine the disastrous effects such a decision has had, as lands that have lain fallow for one or two years only are systematically put back into use.
>
> The individual appropriation of land is exercised within the village system according to traditional references that recognise users' rights. The law in this respect is very close to that regulating property rights, as it guarantees the perennial nature of the land occupied and, in general, the transmission of this right to any descendants. The difference however, resides in the "non-value" of the land that cannot really be sold. Behind this global theoretical scheme, two very different situations and status exist and it is important to distinguish between them. On the one hand there are the people who have a guaranteed right to family land that belonged to their original lineage, who are sort of "eminent owners". On the other there are the people who only have the right to occupy and cultivate land after an agreement with the first group of people who gave them a portion of the land they control. This second situation, which is more delicate in terms of rights, is reminiscent of tenant farming owing to its temporary character even if, generally, there is no monetisation involved in accessing the land. The precariousness of the land situation for an appreciable number of farmers is often an important limiting factor in the management of natural resources in the long term.

In many regions, the land tenure systems in force seem to favour practices that do little to preserve the natural environment.

In villages, farmers who cultivate the land on a casual basis cannot become involved in long-term actions such as planting trees, for example, as the legal "owners" of the land want to avoid any possible future claims. The implementation of land laws based on "land for he who works on it" has shortened even more the loan period and has made land improvements impossible. (A Senegalese smallholder said to us recently: "They lend land to those who don't have any, outsiders – even if they have been in the village for two generations but now they make them change fields every two or three years so that they don't get used to it.")

The co-existence of two forms of appropriation (by the state and by common law) frequently results in evaluations that seem contradictory. On the one hand, common law is gaining ground on all the other traditional rights and tends towards a more western style of property rights (sale and pledging of lands in peri-urban areas, for example). On the other hand, the authorities and rules regarding clearing and users' rights (crop picking, hunting and grazing) on common land are not respected. As the state does not have the means to enforce the law, abusive practices are becoming increasingly common on the land.

This lack of responsibility has resulted in deregulated management of natural resources as well as an over-exploitation of the environment. This is true not only for the grazing areas, around cemented boreholes and for state-financed drilling, but also in farming areas where it contributes to the abusive exploitation of forestry resources.

It is not the duty of field operators (government officials or those in charge of smallholders' organisations) to modify the existing land tenure systems. They must, however, know them, understand them and take them into account in the proposals they make, as well as highlight adjustments and revisions that would seem necessary.

4.2 Methodological orientations

4.2.1 Seeking a new equilibrium between rural communities and their physical environment

Owing to their legitimate concern for the environment, external operators are sometimes led to place more value on the environment than on the people who live in it. We are then faced with a rather deviant situation that focuses on "protecting the environment" against the activities of the population who are seen as "predators". This approach ignores that

- rural societies are obliged to exploit natural resources in order to fulfil their dietary and monetary needs, to take care of themselves, dress themselves and build their houses;
- an *over-exploitation* of natural resources means that *the ancient equilibrium is no longer adapted* to the new conditions created by demographic growth, the increasing need for money, and so on.

The question concerning the conservation and management of natural resources should be looked at more in terms of environmental and social dynamics. It is not about preserving or attempting to return to the ancient equilibrium, but about finding a new equilibrium based on ways of developing and exploiting resources that will enable them to be regenerated and, if possible, improved.

> It should be noted that this new equilibrium cannot always be found within the limits of a given space. In fact, in a densely populated area, with few resources and scarce rainfall, the population simply cannot satisfy its needs solely from the natural resources at its disposal. It must either think in terms of increasing the land it uses, or look towards diversifying its activities in the extra-agricultural sector.

At *operational level*, respecting this principle means rejecting approaches using only coercive methods ("It is forbidden to ... the maximum penalty is ...") that require the permanent mobilisation of repressive forces that are often costly and inefficient. In the same way, it is useless and harmful to try to make rural people feel guilty. This only slows down the thought process, the search for solutions, and can incite defensive reactions from the people who may turn away from further discussions.

4.2.2 A permanent mobilisation of local role-players

The second principle is about creating the conditions necessary for a permanent mobilisation of local role-players to define objectives and actions on a consensual basis. There is no point in driving people into action with galvanising attitudes and measures as this will only result in short-term mobilisation.

> "Popular mobilisation" accompanied by official speeches, mass meetings and food distribution certainly creates enthusiasm, but it is short-lived. For instance, it is not difficult to get together hundreds or thousands of people for one day to plant trees; it is, however, more difficult on the following days, when all the drums are quiet, to find people to water them.

Permanent mobilisation requires intensive work revolving around several aspects.

> Certain actions concerning the conservation of natural resources can be part of a priority programme. In this case actions result in rapid visible effects that the smallholders themselves have asked for: an anti-salt dam in a valley that has become totally or partially sterile, an anti-erosion system, and the like. Sometimes other actions can only be programmed for later for the following reasons:
> - At first they result in the diminution of available resources, even if the latter are degraded (this is sometimes the case when land is closed off in an area).
> - They require important changes to take place within the farm, the district or the village and therefore the actions are subject to certain conditions (changing management fertility practices for example, but also discussions among the various role-players).
> - The smallholders must regain confidence (through a series of successful actions) before embarking on larger-scale (material or social) actions.
> - They may cause conflict and tension between groups with varying interests who are not prepared to compromise. For example, it is preferable to avoid an action that will reinforce the land rights of immigrants living on the ancient land reserves of a village and whose rights are contested by the natives.

A. Smallholders' definition of multi-sectorial objectives for development

Opening a meeting with discussions on natural resources (their state, their development, the necessary changes) is not always very successful. It requires a lot of effort for medium and long-term results, while smallholders have more immediate priorities and constraints.

The general procedure for local planning (see Chapter I) makes it possible to integrate the analysis and the definition of short, medium and long-term objectives. It should be remembered that the procedure is a five-step permanent process:

- Consensual diagnosis (external diagnosis, feedback, auto-analysis and assembly)
- Defining a priority plan of action
- Implementation of a priority programme and definition of a mid-term local development plan
- Realisation of the plan
- Monitoring and evaluation

The consensual diagnosis must, of course, include the evaluation of local natural resources (inventory, evolutions observed, differences between the needs and resources, consequences) and *actions must be planned to improve them.*

B. Negotiation of objectives and multi-sectorial action programmes defined with other role-players who intervene at local level (see Chapter VII: Contracts between role-players)

> Crop and livestock farmers may sometimes come to agreements (grazing areas, access to water points) which can be made official after deliberation with local authorities (if they exist) or with the administration.
>
> Smallholders can obtain backing from technical services or financial support to realise their projects. For example the NGO Six S supports various types of local initiative projects, especially a number of development projects. The UN development programme has created a bureau that specialises in supporting local initiatives that deal with the environment (Africa 2000).

Firstly one must
- identify the other role-players, (direct or indirect users of the natural resources, administration, technical services, financial sources);
- characterise their strategies (objectives, interests, room for manoeuvreing);
- negotiate with them concrete means for carrying out one or more actions and set out limited but realistic objectives.

Some negotiations, however, are rendered difficult
- by strict regulations that do not always take into account the progressiveness needed to modify the practices of smallholders and other users of the natural resources;
- by some external role-players (charcoal makers, tradesmen and women specialised in harvesting products) who are sometimes interested only in short-term profits; in some cases alliances between smallholders and the administration, who have agreed on a monitoring procedure, can increase the efficiency of controls.

C. The importance of negotiating local plans and action programmes through dialogue with other role-players at regional and national level. A system of representation is needed. (See Chapter VII: Contracts between role-players)
The management of natural resources results from international, national, regional and local interventions. For it to be efficient, there must be a convergence of the different interventions. In fact, if a national policy is in contradiction with local practices and strategies, one can expect the effect to be globally inefficient or even negative.

4.2.3 Recognising rural people as central role-players
The recognition of the central role played by the rural population in the management of natural resources is the third principle.
This can be justified by the fact that the rural population has a unique knowledge of its own resources and that it remains the main user of them. At the same time resource management is carried out on a daily basis through practices and decisions. Lastly, only permanent residents can exert efficient control with regard to the decisions made concerning their environment.

On an operational level this requires four conditions to be fulfilled:

A. Rural people
Rural people should really be made *responsible* for the management of resources and this should be recognised by official texts and by those in charge of implementing them.

B. There should be precise identification of those in charge at the local level.
It is during the consensual diagnosis that it is important to identify with the smallholders the different spaces that exist and the decision-making centres that manage them: that means identifying which geographical area is concerned with the solution to which problem (farm, district, village, inter-village land). Who can make the decision to improve or change the current practices? Must he/she report to other authorities? Which ones?

This procedure results in operational zones being drawn up from agro-ecological and also socio-economic criteria (see Chapter II: Diagnosis).

C. Structures
Structures responsible for co-ordinating decisions and actions to be undertaken *must be set up*. These can be "traditional authorities" (if they have survived long enough), the risk being that they may perpetuate or even reinforce the socio-economic gap. It should be mentioned that if these structures do exist and are still active, it is impossible to ignore them as that will only result in conflict between the various parties.

They may be new structures (managerial

land committees) at village or inter-village level. Setting them up must be the result of a collective thought process and their tasks must be well defined (see Chapter VI: The farmers' organisation).

When decentralised local authorities really exist, they are often in the best position to orient, co-ordinate and control (see Chapter VII: Contracts between role-players).

D. Access to necessary information
However, rural people can only play a central role in managing their natural resources if they have access to the *necessary information* and if they acquire real expertise (see Chapter V: Training of farmers).

4.3 Methods and tools

An approach centred only on the management of natural resources has little chance of succeeding especially when the environmental degradation has been caused by forced overexploitation, itself resulting from a variety of causes (demography, unfavourable economic context, drought). The "vicious cycle" of degradation can only be broken, in fact, by taking into consideration all the causes as a whole.

This means that *an economic development strategy* must be defined in conjunction with the smallholders and the other role-players in general. It should take into account the different problems they face and therefore also the *management of natural resources*. In various cases (Sahel for instance) it is only through the implementation of consensual multi-sectorial development plans that the pressure on the natural resources will be alleviated and long-term, viable exploitation methods can be progressively defined (including, notably, processing activities, organisation of markets and processing networks, the creation of extra-agricultural activities and adapted credit systems).

The management of natural resources should therefore be perceived as a component of a wider strategy aimed at the recapitalisation of farms. Owing to this, the actions carried out in the economic field cannot be reduced to simple coercive measures; they are the essential component of the local development strategy.

The methods and tools used in the *general procedure* for local development are discussed in Chapter I: Local planning. We will limit ourselves here to the presentation of several *specific* methods and tools used in the management of natural resources that are efficient only if they are integrated in a global procedure.

4.3.1 Scale on which work is to be carried out

The level at which natural resources are managed is sometimes defined by the state (the village in Burkina Faso, the rural community in Senegal). Although this delimitation is a starting point, it is insufficient to understand the managerial rules in force and to define new rules. The different decision-making levels, their interaction and the spaces in question must be identified.

> The management of natural resources can be partially carried out at village level but some actions depend on higher (livestock management) or lower (prerogatives of areas, of farms) levels of decision making that must be taken into consideration.

Two tools make it possible to identify the different spaces and their delimitation:
- Field studies. The smallholders usually know exactly the boundaries of the lands "traditionally" under their control: physical signs such as trees or termite nests delimit them.
- The study of aerial photographs and maps.

In *agro-pastoral zones*, it is not advisable to begin work by physically delimiting the land:
- Livestock farmers sometimes see it as an attempt to confine them to a delimited area.

- From an operational point of view, the grazing area is defined less by its limits than by its centres. The spaces used are often defined by water points (boreholes, ponds, wells) and their nature (permanent or seasonal). The spaces therefore vary depending on rainfall intensity for example, and overlap partially the land used by other role-players. Identification of the different "centres" is therefore more useful than the identification of borders and limits that tend to move.

4.3.2 Characterising land types

Land can be identified by using two kinds of criteria:
- The nature of the soil, its position within the toposequence (plateau, hill, valley, lowland), the spontaneous type of vegetation
- Its use: cultivated areas (plot of land, big crop field, short fallow land), non-cultivated area (long fallow lands, grazing area, reserved land, forest)

The tools that can be used are the following:
- Thematic maps (of the soil, for example) and aerial photos
- Surveys (smallholders are often good at classification) completed by observations (transect)

It is important to determine the area of each piece of land identified (even approximately).

4.3.3 Evaluation of the state of the natural resources

This evaluation consists of assessing the state of the land (based on previously determined criteria) and the development of trends (over the previous 20 years for example) and, if possible, the speed of development.

The tools that can be used are the following:
- Observation (and analysis, if necessary) of the soil, cultivation practices, yields, grassing down
- Surveys at valley, village or farm level
- Comparisons with previous aerial photographs and discussions with village elders, and so on

4.3.4 Appropriation and access to resources

The above tools make it possible to identify resources and evaluate the global potential that they represent. The question is now to determine who has access and on what terms. This brings to the fore the land tenure system(s) in force and its (their) effects.

A survey is the best tool in this regard. However, because of the frequent complexity of land tenure systems, extensive prior knowledge should not be considered as a necessity before beginning an action, but should rather be seen as a permanent concern (see Chapter VII: Monitoring and evaluation).

Surveys on land tenure systems must be prepared (operators must be acquainted with previous studies and the land acts officially in force) They must
- highlight the *differences* regarding access to resources, that is
  - the inequality among families regarding land (area divisions within a village or district),
  - differentiated access depending on gender, age and social status,
  - levels of land security (or precariousness) depending on social categories as well as the reasons for this,
  - the means of transferring patrimony,
  - development noticed and its causes;
- make it possible to identify management rules and their development: Who decides? About what? Did the official regulations modify the management of resources? In what way? What effects did this have?

To be efficient, the survey must
- combine both quantitative and qualitative approaches;
- allow smallholders to express the perception they have about the land tenure system and how it affects their situation;
- cover the various social categories.

4.3.5 Quantification

Quantification is a tool used to support the planning procedure for development actions. It enables qualitative assessments to be transformed into digital data that do not claim to be a precise reflection of reality but rather an indication of size.

At the diagnosis stage this clarification enables one to
- estimate resources and productions (cultivable and cultivated surfaces, water quantity, the volumes of available organic manure and food crop);
- compare estimated current availability with the estimated needs of the population (food needs, water, animal feed, etc.);
- highlight in this way the existing equilibrium or differences, as well as their causes.

What can be quantified?
- Basic data: population, areas, livestock
- The estimated food needs of humans and animals
- Production potential
- The differences

On a prospective level (how to improve the situation), quantification enables one to
- identify priorities;
- analyse priorities: evaluation before conducting any action, evaluation of their effects (positive and negative);
- verify the "coherence" between the various planned actions.

Products	Potential production (360 ha cropped)	Current production (110 ha cropped)	Needs
Cereals	210 t	60 t	80 t
Wood	300 m^3	300 m^3	400 m^3
Forage	810 t	360 t	720 t
Water for livestock	0 m^3	0 m^3	780 m^3
Manure	–	24 ha fertilised	–

Example: Impact of an increasing number of livestock units in the community			
	Current situation	Scenario 360 TSU	Potential
Livestock	120 TSU	360 TSU	–
Forage requested	320 t (deficit)	960 t	810 t
Manure produced	48 t	144 t	–
Number ha fertilised	24 ha / 110 ha	72 ha / 110 ha	–

This scenario would result in an increasing deficit in forage and a continuing shortage of water (livestock currently migrates during the dry season). The scenario would require
- increasing cropped areas beyond the current 110 ha,
- building a dam or sinking boreholes,
- a policy of workforce management,
- changes in the relationship between crop farmers and livestock farmers and a better integration of crop/livestock farming.

Measuring gaps: an example

Initial data:

400 people, 100 cattle, 400 sheep (equivalent to 120 tropical stock units), 600 ha land :

	10% fertile lowlands	60 ha
	50% agricultural land (sandy loam)	300 ha
	10% bushveld	60 ha
	30% bushveld & dry savannah	180 ha

Estimation of the needs:

For humans:

	200 kg cereals/person/year	80 tons/cereals
	4 kg meat/person/year 16 tons/ meat =	80 sheep
	1 m³ wood	400 m³.wood
	3 m³ water	1200 m³.water

Needs of the animals :

Cattle	forage	440 tons
	water	500 m³
Sheep	forage	280 tons
	water	280 m³
Total	forage	720 tons
	water	780 m³

Potential production according to the full use of the natural resources (360 ha cropped, 600 ha used)

Cereals		60 ha x 1 t/ha = 60 tons
		300 ha x 0,5 t/ha = 150 tons
	Total	210 tons cereals
Wood		0.5 m³ x 600 ha = 300 m³
Forage		210 t x 3 = 630 tons of hay from cereal plots
		600 ha x 0.3 t dry matter/ha = 180 tons (dm) from grazing areas
	Total	810 tons (dm)

Current production (110 ha cropped, 600 ha used)

Cereals		10 ha lowlands: 10 x 1 ton = 10 tons
		100 ha land: 100 x 0,5 = 50 tons
	Total	60 tons
Wood		0,5 m³ x 600 ha = 300 m³ wood
Forage		60 t x 3 = 180 t of hay from cereal plots
		600 ha x 0.3 t dry matter /ha = 180 tons (dm) from grazing areas
	Total	360 tons (dm)
Manure		1 tropical stock unit produces 0.4 tons of available manure/year
		120 TSU x 0.4 = 48 tons manure
		At a 2 tons/ha ratio, this can fertilise 24 ha

Quantification requires calculation norms. The results of agricultural research can serve as a reference when choosing these norms. Often they have to be adapted to local realities.

At village level, quantification has pedagogic implications as it provides indications (even if they are approximate) of the magnitude of a problem that is familiar to smallholders and, therefore, stimulates review. All the same, the results must be reported back in terms that are accessible to smallholders (see sketches).

4.3.6 Contracts (see Chapter VII: Contracts between role-players)

Although contracts are important in all development-related fields, they are central in the matter of natural resources management for two main reasons:
- Long-term viable exploitation methods can only be promoted if all the role-players work together in a consensual manner.
- Local role-players have a determining role to play in this matter, but they need technical, material and financial support as well as collaborations that meet their requirements.

A. Type of contract

Contracts can be
- *internal contracts* between the members of the local community (crop farmers/livestock farmers, villagers/foresters and craftsmen/women) that should result from direct inter-professional negotiations (with the support of "mediators", if necessary) and should always be made official by the local administration.

This supposes that a professional organisation exists that represents the interests of the various producers and that it makes sure the decisions made by its members are adhered to.
- *contracts between the local community and external operators.* One village (or more) may be undertaking a development project that requires technical, material (means of transport) or financial (water and soil conservation for instance) support. It is often necessary for technical services or NGOs to contribute.

This support is more efficient if it is contractual (decisions discussed previously, work methods, consensual distribution of tasks), if it is part of a broader work programme and if there is a plan for the "after development" (change in farming practices, maintenance, etc.).
- *guidelines and development schemes* defined by the state. The state has an obligation towards the environment from which it cannot escape. It is, for instance, responsible for ensuring that there is a coherence between the different local and regional initiatives and also that the objective of maintaining production potential is respected. This responsibility is often translated into land planning programmes and development schemes that go beyond the local level. Local initiatives must fit into this framework.

In order for this to become effective, local communities (representatives) must be allowed to air their concerns and thus contribute to the definition of the general plans. In so doing the chances of these plans being respected are increased and the state will have more authority in this way.

> The construction of big dams on the Senegal River has made large areas in Mali, Mauritania and Senegal irrigable. However, in certain fragile areas prone to salinisation (delta area) general measures have had to be taken (structural measures that include notably a drainage network) to guarantee a sustainable exploitation. The local role-players must therefore take this into consideration in their installations and plan their own perimeters around the general layout (drainage).

It is imperative that these rules defined in a consensual manner be adhered to. The state and local communities must give themselves the necessary means.

5. Conclusion

The degradation of natural resources is usually the expression, at local level, of rural impoverishment and a weakening of the social rules of management that are no longer adapted to the present context. The main causes stem from the ever-increasing demographic pressure and widening gap between needs and available monetary and food resources. They are aggravated and accelerated by persistent phenomena such as drought and the migration of the youth, and sometimes by inadequate regulations.

All actions (repressive, incentive-based or participatory) centred only on the preservation of natural resources will have a limited effect because smallholders have short-term constraints to which they legitimately give priority.

To slow down the recent tendency, and if possible to reverse it, it is necessary to include the management of natural resources in a long-term strategy for economic development.

To achieve this it is necessary to
- give smallholders stimulating perspectives and the opportunity to capitalise collectively or individually (on the farms) through the creation of the right economic conditions;
- promote action programmes that include the short, medium and long term;
- seek a vertical coherence between national, regional and local policies as well as horizontal coherence. For example, it is unlikely that splitting the Department for Rural Development into two or three entities (agriculture, livestock and environment) will be conducive to promoting, at local level, the integrated approach that is necessary. Once again, the existence of strong local communities would be a good way of integrating both orientations and actions.

6. Recommended literature

Becker, J. (1998) Sustainable development assessment for local land uses. *International Journal of Sustainable Development and World Ecology*, 5 (1): 59–69.

Carney, D. & Farrington, J. (1998) *Natural resource management and institutional change.* ODI, Routledge Development Policy Study, London, UK.

Gentil, P., Waechter, P. & Yatchinowsky, A. (1992) *Environnement et developpement rural.* BDPA SCET/AGRI, Editions Frison Roche, Paris.

Green, J. & Thrupp, L.A. (1998) Gender, sustainable development, and improved resource management in Africa. In: *Africa's valuable assets: a reader in natural resource management.* (Veit, P., editor), World Resource Institute, Washington, USA.

Harper, D. & Brown, T. (1999) *The sustainable management of tropical catchments.* Wiley, Chichester, UK.

Valter, C. & Edigio, D. (1998) Sustainable development: global or local? *GeoJournal*, 45 (1/2): 33–45.

Volker, K. (1997) Local commitment for sustainable rural landscape development. *Agriculture, Ecosystems & Environment,* 63 (2/3): 107–120.

Chapter X: Product management
P. Deshayes, M.-R. Mercoiret, S. Perret

Chapter X links up with the following ones:

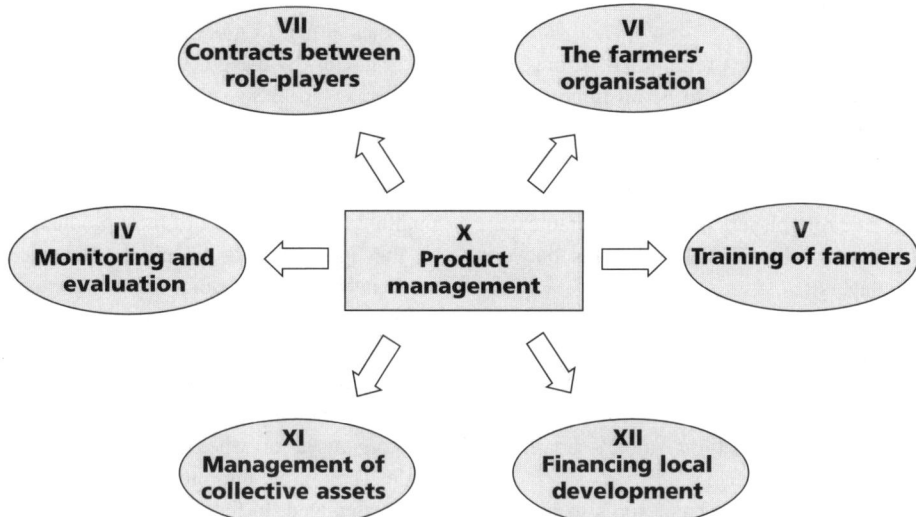

1. Introduction

Agricultural activities involve *product flows and transactions*. In modernising agriculture the number of products to manage has increased and product transactions have become more complex due to the diversification of their origins and destinations. Increasingly often financial operations accompany these transactions.

Many products have to be managed: seeds must often be bought, stored, treated, distributed, recovered; fertilisers and plant protection chemicals, agricultural equipment and spare parts, and so on. Downstream from the production level, harvested grains must be transported, stored individually or collectively (cereal banks); the marketing and sometimes the processing of the products must also be taken care of.

The *quality of product management* has an enormous impact on production. The inputs must arrive in time and at prices and conditions adapted to the farmers' possibilities; they must find a market and be sold at a price that enables acceptable profits to be made. Product management, therefore, is a *central issue* in agricultural development.

Long excluded from the management of certain agricultural products, developing farmers can take advantage of state withdrawal to increase their role in the supply, marketing and processing of products.

This requires them to set up economic organisations that are efficient (capable of living up to their members expectations), *profitable* and able to withstand competition from private economic role-players (industries, local businesses for example).

This movement is emerging in many countries. It is however faced with a number of problems owing to
- internal causes typical of emerging economic organisations (lack of experience in management for example);
- external constraints (statutes not being adapted, a not very stimulating economic and institutional environment that may even be limiting).

In coming years, the promotion of strong economic organisations is to be an important issue. It may enable productive conditions and farmers' incomes to be improved and also allow them to negotiate contracts with the other economic role-players in more favourable conditions (see Chapter VII: Contracts between role-players).

This chapter strives to
- highlight the constraints that should be taken into account when managing products in economic organisations run by smallholders;
- suggest, on the one hand, tools to improve management skills within organisations and, the support that may be given to them in this context on the other.

2. Smallholders and the management of agricultural products: elements for diagnosis

2.1 A role of varying importance in the management of products

Traditionally, smallholders managed their products (seeds, fertilisers, harvests, etc.) without any external support. They continue to do so
- for some products (organic manure, often seeds);
- in certain sectors neglected by development interventions: small-scale processing activities, small local businesses or trade at national or regional level (livestock), diversified agricultural production (vegetables, fruits, gathering).

Conversely, products connected to large-scale production (commodity crops) are partially out of their control.

When independence was declared in western Africa's countries, development programmes aimed to modernise the agricultural sector notably by using new inputs (high-yielding seeds, fertilisers, agricultural equipment). Networks for supplies and marketing, and credit systems were implemented and managed mainly by the governments. In many cases, smallholders lost control of their products. Often they ignored the origin and the real price of the inputs they were using, the way annual instalments and credits were calculated and the real costs of cultivation. The prices of their main products were imposed on farmers without them knowing how the commodity chain was organised or whether any profits were made. They rarely mastered the way discounts were calculated. In fact, they were asked simply to produce and to forget about the production up and downstream.

In the 1970s, large development projects enabled producers to once again play a role in the management of their products (supplies, primary marketing) but their responsibilities were still limited. The associations for development still controlled all networks above village level and were the middleman between smallholders and the other economic role-players (industries and tradespersons).

In the 1980s new agricultural policies advocated state withdrawal from its economic role in supplies, credits and marketing. In many cases this meant that the smallholders had to take responsibility for their inputs and productions. Not surprisingly, previous interventions had left them ill-prepared to assume these responsibilities.

2.2 Numerous management constraints

These constraints are often related to three main factors: the *regulations* governing management structure (statutes), its flexibility and level of autonomy, the *nature* of the products to manage and the *environment*.

2.2.1 Status of and statutes for the managing structure

(See Chapter VI, The farmers' organisation.)
For many years the regulations in force limited the status of smallholders' economic organisations. Three types prevailed:
- The *co-operatives* whose legal status theoretically enabled farmers to undertake financial and commercial transactions. In fact, smallholders did not make use of them owing to excessive supervision and the lack of flexibility in the running of them.
- *Farmers' groups*, often seen as "pre-co-operative structures" linked to development interventions and usually not having any recognised legal status. Their legitimacy derived only from the supervisory body. Without the latter's guarantee they could not undertake any financial and commercial transactions and sometimes could not enjoy direct access to agricultural credit.
- In some countries (Burkina Faso, Senegal), organisations obtained the status of *association* (similar to a non-profit local organisation) and sometimes aquired the status of a NGO. They therefore benefited from more favourable legal and fiscal conditions. This status was, however, difficult to obtain. Moreover, it did not really conform to the norms of a well run economic institution (the associations' non-lucrative goal prohibits any sharing of benefits).

During the eighties, a new statute appeared in certain countries. That was more favourable to the emergence of autonomous economic organisations run by smallholders (see below in the box). In general, however, excluding some exceptional cases, legislation is restrictive for economic initiatives.

Legislation must change so that economic agricultural organisations may implement, on a practical and legal level, the necessary official orientations that follow the withdrawal of the state and the transfer of responsibilities that ensues. Laws must be adapted to accompany new agricultural policies.

The law on GEIs (Groups of Economic Interest) in Senegal

Enacted in 1984, this statute allows two or more people (physical or moral) to constitute, without any initial input of capital, a group of economic interest with the view to "finding the means susceptible to facilitate and to develop its members' economic activities". The GEI in itself cannot make and share the benefits and as such it is not subject to tax; as for the members, they can of course make profits that are integrated into their incomes.

The terms for constitution are relatively simple: identification of the members, the aim of the GEI, the duration, and so on. It must be registered with the registrar of commerce and acquire a "contract" between members who are personally liable to the GEI for any debts incurred. The flexibility of this statute, which guarantees total managerial autonomy, as well as those it present (eligiblility for agricultural credit) explain the rapid growth of GEIs, which today number thousands.

Some groups, and or associations that are already autonomous or are supported by development projects or NGOs have acquired this statute. Although it presents numerous advantages, it sometimes allows "artificial" groups to appear whose grass-roots'- members are highly dependent on one investor/president.

> This is the case in Guinea where farmers' groups are recognised as pre-co-operative groups with limited possibilities of functioning: as the co-operative is perceived as a parastatal organisation it must, for example, have its budget approved by the public administration.

2.2.2 The nature of the products determines the particular constraints of management

A. Upstream from production

The following constraints can be mentioned:

- For smallholders' economic organisations specialised in the supply of inputs, the co-operation is critical as the progressive withdrawal of subsidies (for fertilisers for example) has resulted in the reduction, if not in the fluctuation of demand especially in the areas where production conditions are precarious.
- The management of inputs and of agricultural equipment is almost always linked to a loan system (see Chapter XII: Financing rural development) which complicates the implementation of any mechanism, especially when privatised agricultural banking operates according to its own rules (as is frequently the case for national agricultural and credit banks).
- The management of seeds poses specific problems: should one propagate the seeds on the spot? (This causes problems for the smallholders in terms of organisation, quality control and cash flow for the purchase of the production.) Which system should be adopted regarding the loan of seeds? (Interest rate? Reimbursement in kind? This raises questions about the quality of reimbursed seeds, their handling, their storage.) The recent increasing distribution of genetically modified crops (GMOs) adds to this problem and its complexity for smallholders, yet again dependant on external stakeholders (seed companies).

B. Downstream from production

Management constraints are linked to the products, to their perishability and their production period over the year, and to price stability.

- The management of products with fluctuating market value

> **Example: livestock for butchery meat**
>
> The seasonal price fluctuations of meat on the markets of the world's capitals are generally regular (although somewhat imprecise). Cattle fatteners and dealers therefore adapt their management and livestock systems (especially when to sell) to these relatively well-known fluctuations. Economic organisations must take into account price developments and take advantage of these fluctuations.

> Livestock dealers of East Senegal are true speculators: July is the time when livestock growers are in need of cash to buy cereals for the dry season. It is therefore a good period for buying (an increase in supply means low prices). The maintenance of animals during the rainy season is not costly. In early November, the animals can be resold with good profit margins on the big city market places. By this time cereals have been harvested and the livestock farmers do not need to sell their animals to buy feed.

> The livestock fatteners of Sine-Saloum (Senegal) benefit in another way from the fluctuations of market prices. Since they keep an important stock of groundnut haulms (from their fields) on the farm they can keep some cattle at low cost until the end of the dry season. They buy their cattle on the farm, from breeders at a time when prices are relatively low (at the beginning of the hot dry season before the animals body condition really starts to deteriorate). Then the fattened animals' are sold on the big cities' market places at the end of the dry season when animals in good condition are scarce and therefore expensive.

In most cases, in order to benefit from market fluctuations it is necessary to have adequate working capital and to follow market prices closely. Previsional management with this type of market remains difficult however competent the managers are. It requires both permanent readjustments and rapid decision-making. In order to secure market outlets, contracts must be signed with buyers. In return, certain regularity in supplies and quality is required, both of which smallholders' organisations are not always able to guarantee.

- The management of annual and seasonal rotation products (cereal banks for example). It requires significant working capital.

- The management of products with regulated prices (groundnuts, cotton)

Room for manoeuvring is restricted. However, prices are stable and the market is guaranteed, which makes previsions easier. The inability to gamble on the selling prices pushes farmers to reduce costs related to the harvest, conservation and sale of the products. It also limits investments and increases the volume produced without increasing structural expenses thus improving the margin of benefits.

C. Combining the management of both upstream and downstream levels of production

This facilitates credit management. For a long time external operators have understood the benefit of linking services rendered at the upstream level of production to marketing. Controlling distribution is a powerful tool when it comes to recovering loans: annual payments are automatically deducted from the products of the harvest.

An economic farmers' group can act in the same way. However, monopoly situations should be avoided. In fact, it is advisable that smallholders chose a structure for marketing that they control (rather than a private network), not because they are forced to do so but because it gives superior quality service.

2.2.3 Environmental influences on production

A. Institutional environment

The institutional environment is sometimes unfavourable towards smallholders independently managing their products. Here and there certain technicians still exert an influence by vigorously giving advice to smallholders' organisations that they are trying to orientate and control. A sort of authoritarian supervision is practised more or less openly.

B. Local social context

The local social context can interfere in the management of products by smallholders.

Farmers' needs are immense in all sectors, and their social need linked to the improvement of their living conditions are rightly a priority. Pressures are exerted so that economic activities finance social activities:

- Smallholders expect cheap supplies and profitable product sales from their economic structures.
- They would like their reimbursement problems to be taken into consideration.

- They would like transfers to be made from the productive sectors to the social sectors (financing a place of worship, a health care centre, etc.).

This interference sometimes results in payment for services that is below the real costs, and excessive transfers that jeopardise the economic activities.

C. Economic context

Currently, the *economic context* is generally not very favourable to the functioning of economic smallholders' groups. This is caused by a variety of factors: the increase in price of production factors, credit conditions that are not adapted to the means of many categories of farmers, and in some big industries, the low prices paid to farmers, not to mention open markets that are actually "closed" (livestock, for example in western Africa). Consequently, the achievements, of some organisations are uncertain and this risks slowing down their development.

Often, in addition to this, the communication network is poor. This leads to an increase in transport costs, irregular deliveries of supplies and poor distribution that in turn create a chain reaction: poor demand for inputs by remote isolated farmers and the abandonment of diversification activities (fresh vegetables for example).

The influence of tradespeople coming from town over farmers in isolated areas may be significant. Thus, in certain regions during the periods of fruit overproduction, it is the tradespeople coming from the town who fix the price to pay the farmer. The isolated farmers who have managed to organise transport are often confronted with the problem of rendering their new venture profitable, and also of penetrating the urban market controlled by traders who are already established. It seems to be that, as long as road infrastructures are not sufficient in areas with a small demographic concentration, only informal systems of supply will survive. Take the example of itinerant traders in western Africa, going from village to village on their bicycles. Their system subsists because their transportation costs are very low and because they get by on a very small income and they have an economical lifestyle.

Local competition should be mentioned as one of the constraints of the economic environment. The origins of the competition may be

- other smallholders' groups working in the same area and making the same product. It is usually easy to define their managerial and operational constraints and advantages, their techniques for gaining access to the market and developing customer loyalty;
- competitors who are often unknown, like traditional private businesspeople or those from the informal sector. They have a tremendous ability to survive where modern economic organisations cannot.

In certain areas in Mali, cereal banks are finding it difficult to develop as retailers exert a great deal of pressure on their clients. There are few retailers and they offer many services at once (or show solidarity). As a result, a client who no longer obtains his/her cereals from his usual retailers risks also losing his/her financial backing (credit) for his/her future purchase of basics such as sugar and oil. Although these retailers sell at a high price, the villagers cannot do without them in many isolated rural areas.

- lastly, certain government and non-governmental projects (charitable ones notably) are also sometimes at the source of upheaval in the local economy. Occasional financing through donations and subsidies are generous practices and respond to social and political objectives that can be justified. However, although occasional, such practices are sometimes all that is needed to break down initiatives. The arrival of a horticultural development project in a small area can cause a small local supplier of inputs to go bankrupt if the project is supplying inputs at unbeatable prices (which is possible since the products are tax free or subsidised). A question arises: What should be given precedence? Should it be a cheaper supply over one or more years and therefore the disappearance of existing or emerging companies with the risk that they may not reappear at the end of the project? Or should it be respect for normal economic functioning where the supply system is more expensive but more sustainable?

The constraints inherent in the management of agricultural products are many and smallholders have been ill-prepared to face them. At present, they must face up to them if they want to improve the mastery of their products and their benefit margins.

3. Work orientations

3.1 Management and accounting

3.1.1 What is management about?

To manage an individual enterprise or a collective economic organisation is to make *a number of decisions in order to achieve objectives* that have previously been determined. These decisions are of a financial or economic nature. They are also concerned with human resources (organisation, sharing tasks). This definition calls for two sets of observations about the objectives pursued: their diversity and their clarity.

A. The diversity of the objectives pursued

A business can be looking for maximum returns. It is therefore geared to obtaining maximum profits (benefits). However, there are limits that should not be exceeded if the enterprise does not want to lose its clientele and be condemned to disappear.

> Many examples show that it is not only in private businesses that these objectives are true. Thus, during a recent debate in one of the cereal banks about terms for selling rice, some members of the management committee supported the need to sell cash in order for the bank to avoid taking risks and make more profits. They were not very sensitive to the fact that such conditions would exclude the poorest consumers.

Economic smallholders' organisations frequently see themselves as collective service providers to individual farmers and therefore look to provide services at the best price. This often means selling at the lowest possible price. Many smallholders' organisations have discovered the limitations of such strategies when losses have resulted in the more or less rapid disappearance of the organisation and therefore of the services it was supposed to provide.

An economic organisation must therefore at the very least not record losses in order to ensure the future of the services it is providing.

> Recently a smallholder's enterprise specialising in tillage operations experienced the above. It calculated that the cost of one tractor ploughing was 30 000 FCFA/ha, but the managers increased it to 50 000 FCFA/ha in order to simplify the calculations and increase profits. As the smallholders protested, they went back to a price that was deemed reasonable: 15 000 FCFA/ha causing the enterprise to make a 100 000 francs CFA deficit during the crop year.

B. The clarity of the objectives pursued

Management decisions are more likely to be relevant if the objectives pursued by the organisation are clear to the concerned parties (management committee, general assembly) and in accordance with their aspirations. This underlines the limits of economic objectives set by outsiders and more or less strongly recommended to the smallholders.

Relevant decisions can only be made when managers have at their disposal accessible and accurate information that can explain their choices: financial situation, future prospects, different hypotheses available, and so on. For this to happen
- someone must keep the accounts;
- the accounts must be open to all those involved in the decision-making process and they must understand and be able to analyse them;
- a management committee must exist that helps the smallholders to interpret the accounts, to formulate hypotheses regarding decisions to be made and to evaluate the foreseeable effects of the different possible decisions.

3.1.2 Accounting

The purpose of bookkeeping and accounting is to record quantified information related to the different operations carried out by an enterprise. They enable the cost price of services and products to be calculated as well as the benefits and expenses incurred by the enterprise. Lastly they enable those concerned to know at regular intervals the results achieved.

Accounting is a management tool and not an end in itself. Bookkeeping is a hard and demanding task. Reducing it to a formal exercise, in which both managers and decision makers do not see its usefulness in improving the running of the enterprise, should therefore be avoided.

Bookkeeping has increased in many organisations, but it is generally only used to control money and product movements. Producing financial statements and analysing them is still quite rare.

In order for accounting to be helpful in the decision-making process, two conditions must be filled:
- Accounting systems must be adapted to the specific needs of the activities managed by the economic organisations and to the abilities of those responsible for the bookkeeping. The first requirement of an accounting system is that those who have to use it, understand it.
- When management is composed of a group of persons (management committee), a technique should be used which enables information to reach everybody so that it can be analysed and decisions made in a consensual manner. When a number of managers are illiterate, it is advisable to set up a means of communication that allows them to participate properly in the management of the organisation. When they are able to understand the calculations made, they usually show a close interest in the results of their organisation and actively participate in the search for solutions to improve them (see below paragraph 4).

This problem is an old and well-known one. Thus, in 1980 some researchers came to the following conclusion after having evaluated 33 cereal banks in Burkina Faso:

"The accounts (...) as presented today (...) are unusable (...) records are not systematically made (...) and data is not used. For instance, the total of purchases and sales is not calculated, neither are the annual benefits (...). In concrete terms, when a cereal bank decides to open a village boutique, its management committee does not know whether the money involved is taken from its benefits or from its capital (...). [Accounting] isn't used to monitor the running of the enterprise nor to draw the consequences resulting from it (...) [it] is never used in the decision-making process ..."

The accounting documents distributed by certain projects are useful to them, but incomprehensible to the villagers concerned who are supposed to fill them out (and sometimes use them).

For certain organisations it is generally appropriate to divide the accounting into two parts. The simple part does not provide the same possibilities but is generally better suited to the managers' abilities and will therefore be better understood and used. Some written proofs (or cash slips) are difficult to obtain. They may be replaced by the presence of witnesses.

3.2 Keeping the accounts

3.2.1 Tools for recording

Bookkeeping requires tools (notebooks, sheets) enabling all transactions carried out to be noted.

The tools to be used must be adapted to the organisation's activities, to the managers' abilities and to the future use of the accounts. In most cases the following documents can be found:

- The cash book allows all cash receipts and payments to be recorded. The following must be recorded in the cashbook:
  - the date of the transaction
  - the nature of the transaction (purchase, sale)
  - the amount of the transaction

- The banking book allows all receipts and payments made to or from the organisation's bank account to be noted. The headings in this book must be the same as those in the cashbook.
- The credit book, can take two possible forms:
  - A book (or file) in which loans made by the organisation to its members are recorded. For each smallholder who has borrowed money, the file must contain the date of the loan, the amount of each instalment and the dates for repayment (if it is a long-term loan);
  - A book (or file) in which loans granted to the organisation are recorded. The date of the loan, its origin (from whom was the money borrowed?), the amount borrowed, the sum to be reimbursed and the dates for repayment must be recorded in this document.

If the economic organisation has assets, possessions and property, it must also keep an inventory book in which infrastructure (buildings) and equipment (vehicle, tools, computers, etc.) are recorded.

Every time the organisation sells its products, it must keep a file and record in a sales' book the date of the sale, the quantity sold and, if necessary, the expenses related to the commercialisation (transport for example).

The *stock book* (or file) allows one to record per product the quantity retained by the organisation and the different inputs and outputs.

Example of groundnut seeds' management:

Date	Operation	Input (quantity kg)	Output (quantity kg)	Stock (quantity kg)
1998/06/15	Initial stock	–	–	500
1998/06/18	Purchase	1500	-	2000
1998/06/25	Distribution to farmers	–	2000	–
1999/01/15	Seeds recovery from farmers	2500	–	2500

The following example is an illustration of the process. It concerns a farmers' group that has to work out its' needs in inputs and in equipment, their storage and distribution, and the recovery of debts.

	Task to be performed	Document to be used
1	Organising the general assembly of the group Assessing the needs	Sheet for needs' assessment and listing
2	Listing demands. Organising ordering	Order forms
3	Controlling deliveries Informing the managers	Stock files
4	Taking stock of deliveries	Acknowledgements of receipt
5	Organising distribution among members. Keeping the books on distribution and costs recovery	Individual forms for debt recovery. Credit contracts
6	Calculating and issuing individual invoices	Invoices to members
7	Assessing the stock after distribution	Stock files
8	Direct cash sales of various inputs	Cashbox book/stock files
9	Organising the G.A. before the cropping season	Unpaid invoices and recovery/reimbursement files
10	Recovering unpaid invoices. Keeping the books of the operation	Unpaid invoices and recovery/reimbursement files
11	Calculating individual balance sheets	Unpaid invoices and recovery/reimbursement files
12	Organising the G.A. at the end of the cropping season	All documents + minutes

3.2.2 The use of recording tools

Mastering the tools is essential for those who have to make use of them, i.e. the person responsible for the accounts as well as those who will be examining them and making management decisions (management committee, general assembly). It is better to simplify the accounting tools rather than run the risk of having badly kept books that will be useless.

It is advisable to determine with smallholders the nature of the accounting documents needed by analysing with them the tasks to be carried out. Once the tasks are identified, one only has to answer the question: Which document for which task?

Mastering the use of accounting tools sometimes means resorting to local languages and always requires *training and close monitoring*. Regarding the translation of technical terms, it should be underlined that this is always a complex problem. Is it better to use a roundabout discription, an English word, or a local word that may result in ambiguities? Should one speak of a book or "remedy for forgetfulness" (as translated in Haoussa)? Should one use a decimal system when certain smallholders count money in multiples of five?

3.2.3 Financial statements

At the end of the year, it is necessary to take stock of the operations carried out during the financial year. This means establishing whether the enterprise has become richer (there should therefore be a benefit) or poorer (then there must be a loss).

To determine this, the accounts must be grouped together in end-of-year documents called the *balance sheet* and, if necessary, *profit and loss* or *income statements*.

The procedure in drawing up the balance sheet is as follows:

Assets (destination of money)	Liabilities (sources of money)
Infrastructures and equipment (values)	Capital of the group (contributions/subscriptions from the members)
Stocks (value)	Subsidies received
Accounts receivable (loans …)	Savings and reserve (money not redistributed to the members and reinvested)
Total cash (in the cashbox and in the bank account)	Deposits, retaining fees, interim payments … Financial statement (at the end of the fiscal year), debts

A. The balance sheet

If the organisation deals with a simple activity, the balance sheet enables it to know if the activity is showing a benefit or a deficit by comparing the situation of the company at the beginning of the year and that at the end (like comparing pictures).

Balance sheets can be drawn up each year from the following elements:
- Stock inventory
- Value of infrastructures and equipment (assets)
- Cash statement (at closure)
- Bank statement (balance)
- Credit situation (total to be recovered; total outstanding debt)
- Debt situation (total due)
- List of donations or subsidies received (capital injected)

B. Income statement (or profit and loss statement)

The balance sheet shows whether a profit or loss has been made, but does not explain why. In order to analyse the results and make decisions to improve them, it is necessary to draw up an income statement (profit and loss statement). This is possible if the bookkeeping has been thorough enough. It provides a detailed description of money flow, product transactions or any other values (e.g. depreciation of equipment).

The documents should allow the following information to be obtained for the whole year:
- Product sales
- Receipts originating from other sectors related to the activities (interest for example)
- Variations in stock value between the beginning and the end of the financial year
- The value of products consumed (petrol, inputs for example)
- Operating costs for the activity throughout the year
- Depreciation value of equipment and infrastructure

These documents may be
- the cash book,
- the bank account book,
- the stock sheets or files,
- the depreciation schedule,
- the inventory sheets or files.

Drawing up an income statement consists of summing up all the costs, by which we mean *expenses incurred* and all the products sold, meaning the *receipts received*.

Expenses can be purchases, personnel costs, transport costs, financial costs, management operating costs, depreciation (see subsection 3.3.3), et cetera. Receipts can be sales, and so forth. Initial and final stocks form part of the overall situation.

3.3 Management practice

There are two major phases: the analysis of the financial statements and the reflection on how the results may be improved.

3.3.1 Analysis of the financial statements

This consists of understanding the reasons behind the results obtained: Why was that much petrol used? Why were so few products sold? At such a low price? Why so much money to be recovered (outstanding debts)? Are there any possibilities of recovering it? and so on.

By analysing the causes of the results obtained, it is possible to see how future improvements may be made: reduce production costs, calculate cost prices accurately, explore the market, increase the volume of production if the market can cope with more, and so forth.

3.3.2 Evaluating the impact of the different options considered in order to increase the results

(See Chapter IV: Monitoring and evaluation.)
Several elements are to be taken into account:
- The feasibility of the decision envisaged (it is easier to talk about reducing charges than to do it effectively, to decide to increase the market than to succeed ...);
- The economic impact of the decision envisaged (What is the cost? What will the benefits be? What are the risks?).

Thanks to provisional income statements, a study can be made of probable results that depend on the options taken (increase or decrease of sale or purchase prices for example). Moreover, if it is planned to acquire new equipment, the provisional balance sheets and the provisional accounts will make it possible to study the profitability of such an investment, meaning the time it will take to recover the cost of the investment (payback delay).

The hypotheses envisaged to improve results also have financial consequences that must be evaluated: cash flow previsions are prospective simulation tools enabling the financial feasibility of different options to be tested. Will the necessary cash be available when it is needed?

3.3.3 Some frequent problems and ways to solve them

Among the shortcomings in the management of economic smallholder groups, two are usually underlined by personnel support: the lack of previsions (forecasting) and the lack of financial provisions.

Sometimes the lack of forecasting is simply caused by lack of experience and understanding of the economic stakes. Smallholders have been so used to being assisted that many of them still perceive the subsidies and donations to be renewable gifts. But whatever the interpretation, it is necessary to know above all else if those in charge of the organisation have sufficient information to be able to make estimates and forecasts. Some simulations may help in this respect. The problems encountered with forecasting are mainly due to the uncertainties characterising rural people's environment (climatic instability, attacks from predators or diseases, fluctuating prices, etc.).

Why is it usually so difficult for managers to integrate the concept of financial provision for renewal?

Making financial provisions raises, in the first place, the question of keeping money in a safe place. In order to be totally safe from all kinds of problems, the managers are not keen to keep too much money in the cash box. On another note, the banking system has shown many weaknesses in the past (reluctance to accept double signatures, unsuitable procedures). In certain areas the system has adapted itself but access to banks is still sometimes difficult for some economic organisations.

Sometimes there is confusion between depreciation and financial provisions. Although the word "provision" is understood by managers (including illiterate ones), the word "depreciation" rarely is. This term is often even wrongly used by agents not trained in accounting. Explaining this word to illiterate managers is difficult. It is therefore better to avoid it when possible.

> Depreciation is the fall in the value of an asset during the course of its working life. To amortise is to reduce the accounting value of this asset in order to record that it is depreciating.
>
> In other words, amortisation records this depreciation due to wear and tear or premature ageing (obsolescence). The amortisation rate is the depreciation rate. Accordingly, the firm is required to make financial provision for the depreciation of its assets, which will eventually have to be replaced.
>
> To facilitate communication with the uninitiated, it is preferable to use the words "wear and tear" and "ageing" rather than amortisation or depreciation.

Concerning the financial provisions for the renewal of out-of-order equipment, it would be simpler to show the value of the provision being equal to the depreciation cost. Thus, at the end of the year, after the yearly depreciation costs have been calculated and when it has been decided where the provisions are to be kept, a similar amount of money should be transferred from company funds or the bank to this special provision account (savings account or at least a special fund). It is, in fact, important for this provision to take place so that the transfer may be reflected in the yearly balance sheet (even if this is in fact only an exchange from one asset account to another one).

3.4 From records to decisions

When activities start to materialise it is nearly always necessary to get help from specialists (working on a part-time or full-time basis). Care must also be taken to avoid monopolising knowledge (others must be able to replace the specialists). It is very important to have regular audits (internal and external) and that chosen internal employees be trained to carry these out. Financial statements must be presented regularly in a comprehensive manner even to illiterate smallholders.

Not everybody has to become an expert in accounting, but everybody must be capable of understanding a financial statement (receipts and expenses) and be able to participate in the making of important decisions.

Who does what?

Operations	Operators	Tools
Data recording	The organisation's leaders and managers	Stock files / cash book
Data control	Manager / management committee. official auditor.	Cash control / stock control
Data processing	Manager / possibly external operator	Profit and loss account / balance sheet
Data presentation Report back	Committees, assemblies	Posters / boards
Sound decision making	Leaders / managers / committees, assemblies	Previsional profitandloss statement / simulations / discussions

4. Supporting management activities

4.1 Is literacy a prerequisite?

Many economic smallholders' organisations, boards of directors or management committees have only one or two literate members.

For a long time now there have been many literacy programmes attempting to improve this situation. Diverse experiences have been successfully conducted in this field (in Mali for example). The formula of intensive sessions that concentrate on accounting documents gives the best results (see Chapter V: Training farmers).

Literacy specialists have been proposing and implementing joint programmes in literacy and management for several years already. Moreover, in the case of some simple activities, records can be kept by using drawings to identify expense items. Therefore only a knowledge of reading and writing figures is required. This technique can be learned quickly (one week) and may be useful as a starting phase for example.

Although there are many interesting ways to increase the number of literate smallholders in management committees, it will undoubtedly take many more years before this is the case of all those in charge of economic organisations.

Moreover, literacy programmes usually target youths, and yet the latter are rarely elected to decision-making posts, but rather to positions where writing is considered essential (accountant, warehouseperson or secretary).

The decision makers are generally senior citizens of the communities for whom literacy is more difficult to envisage.

Thus, if it is impossible for illiterate people to keep the books, methods and techniques do exist which can allow those people to control transactions, analyse accounts, make economic and financial previsions on their own. In other words, such techniques enable mainly illiterate farmers to manage. The following points will try to summarise these techniques.

4.2 Advice for preparing a product management project

The adviser must make an external analysis of the economic project envisaged and provide support to the economic organisation in order for it to make its own analysis.

4.2.1 External analysis

During the external analysis, the adviser collects the necessary information in order to prepare his role in the next step. The analysis focuses on the strong points and the constraints associated with the nature of the products to manage, their place in the production process and the context in which the envisaged activities are situated.

4.2.2 Supporting the economic organisation for preparing its own project

This consists of helping the smallholders to consider all questions and issues, so that the planned project is realistic and feasible.

Once the future managers have debated the main objectives of their project, a detailed preparation can be carried out through a prospective simulation.

This consists of having the managers of the economic organisation sit around a table where they will attempt to simulate (act as in real life, through role-playing) the setting up and running of their enterprise. In this endeavour each one of them will play the role that he will be assuming in reality.

A cash box, money in the cash book, a bank account, buildings, etc. are presented, ready for the simulation to begin. However, before this can take place and simulated stock and equipment must be set up, there must be a discussion on the choice of investments, on the importance of the initial stock, etc. This is also an opportunity to discuss the market, financial sources and loan possibilities.

Once these elements are defined, money flow and product transactions performed by the organisation are simulated.

Then the running of the activities is simulated. This is the opportunity for the managers to clarify and distribute tasks, define the nature of the bookkeeping, discuss the contents of the internal regulations (if necessary), calculate cost prices and fix the prices (if this has not been done before).

The themes to be discussed during the sessions are the following:
- Internal organisation
- Internal regulations
- Establishing a bookkeeping system
- Drawing up a previsional profit and loss statement
- The cash flow schedule

The SIGESCO method (simulation, management and communication) set up by CIEPAC and GRDR is an example of a prospective simulation for the management of local economic organisations.

One must ensure that as many members as possible participate – especially those who are directly or indirectly responsible. The simulation (a role-playing activity where smallholders act out the projected situation) makes it possible for illiterate managers to participate and for smallholders to discover the questions that need to be thought about in advance as well as the constraints that the activity will be confronted with.

4.3 Management-oriented training

(See Chapter V: Training farmers.)
A training programme should combine:
- training linked to action (preparation, monitoring and evaluation);
- formal sessions demonstrating management mechanisms and how to develop the necessary skills to master them.

Simulation can also be of great use in this situation (if it is well prepared and based on adequate material) even for literate managers.

The themes to discuss are mostly
- internal organisation;
- the main orientations: Should certain social categories be privileged? Must the president definitley be a senior notable? How should temporary or permanent personnel be compensated?
- economic management (cost price calculations, investment choices, market studies, company profitability);
- accounting techniques (drawing up accounting documents and bookkeeping, analysis of end of the year results, etc.);
- communication with members of the organisation (How to report back the accounts? How to involve the majority of members in the decision-making process?)

4.4 Self-evaluation and external evaluation

(See Chapter IV: Monitoring and evaluation.)

Managers must regularly evaluate the economic and financial situation of their organisation and ensure that it is really adhering to the economic or socio-economic objectives for which it was created.

This must also guide their choices for improvements in the company's activities.

Unfortunately many organisations stop at a stage that should be purely preliminary to evaluation, namely at auditing accounts and noting money and product transactions. An external adviser to an organisation can get him/herself into a thankless situation if he/she limits his/her intervention simply to that of auditing. Would not it be better if he/she tried to get rid of his/her auditor image in favour of an image of management consultant/adviser to an independent organisation?

The role of a management consultant should therefore be
- to train managers to make periodical internal check-ups on both accounts and transactions. In this case, the consultant would only be present when the managers request his/her presence;
- to help managers at the end of the crop year to draw up and analyse the financial statements.

4.4.1 Auditing and control

In economic organisations, the management committee is not usually very literate. It is therefore useful to find appropriate and applicable methods that do not need any external support.

Three phases are to be remembered regarding auditing:
- An inventory of what is in stock, in the cash box and of receivables and debts. This must take place in the presence of designated people.
- Reporting back (presentation) on transactions having taken place since the previous report (depending on the number of transactions, the accountant will or will not regroup them);
- Finally, a comparison is made between cash sales and stock. Where is the stock sold? Is it in money in the cash box or in receivables?

In order to help the illiterate to understand the reporting-back procedure, communication aids may be used.

It should be noted that auditing is not a purely technical control. One should therefore be careful regarding the form it takes. "To control somebody" is sometimes perceived as doubting his/her honesty, touching his honour. It is therefore useful to talk about controls/audits as a "treatment against error" since "anybody can make mistakes".

4.4.2 The end of year economic evaluation

The end of year economic analysis requires competencies that are difficult for an ordinary accountant to master. Generally the technical support of an expert (management consultant) is useful to avoid mistakes, especially in the case of more complicated companies.

The two main financial statements that are necessary to make an evaluation are the profit and loss statement and the balance sheet. In order for both to be understood by the members of the management committee, it is important to find accessible formulas. Presenting these statements to the managers (decision makers) enables them to situate and measure the shortcomings of past management and thus formulate and adapt proposals for management changes.

4.4.3 Which method for an external overall diagnosis of management practices within an economic organisation?

An external financial and economic diagnosis can be made with the help of accounting documents, a stock inventory, physical investments' inventory, and financial statements (if they do not exist, they can be reconstituted).

The diagnosis of accounting procedures, work organisation and the management system practised is usually carried out through simple surveys. For a detailed diagnosis of the practices another method called retrospective simulation can be used. To simulate a past activity, it is played out during a meeting. The various people in charge of the organisation play their real life roles in order to simulate the running of the activity for one (or more) given period(s). The transactions are conveyed using accessories (images representing money or other objects).

Simulation enables one to avoid skipping certain steps that would not otherwise have been noticed during a simple survey, even though they may be essential to the management strategies or at the origin of problems. The simulation process is accompanied by questions that make it possible to have a better understanding of the motivations behind certain practices.

5. Conclusion

5.1 Reinforcing grass-roots economic organisations

Improving the performance of economic smallholders' organisations presupposes internal progress within the organisations, but also adequate measures that have been taken to accompany and support them.

In the *legislative and statutory field*, it would be appropriate to facilitate the procedures related to recognising economic organisations and to adapt the regulations.

In the *financial field,* banks, saving and loan co-operatives and other financial institutions (mutual agricultural banks, development funds) must be encouraged to think about their role. Thus, for example, some cereal banks are confronted with a lack of working capital. Loans for the agricultural year would be effective in enabling them to buy supplies at the best price.

In the *economic field*, the creation of a safer and motivating environment is the best way to stimulate smallholders' initiatives. This is essentially the responsibility of the state, either directly (favourable agriculture prices, stabilisation funds aimed at buffering certain variations) or indirectly (creation of an authority to initiate dialogue between role-players (see Chapter VII), tax incentive measures for economic organisations, etc.). Improved infrastructure is often essential in order to promote links between isolated areas and the market, and reduce transport costs. The creation of a market (internal or export) information network (or structures) may also be a useful service to the economic organisations that do not have their own means of carrying out research. In the same way, the creation of a new brand image for local products, through national campaigns (broadcasted advertisements) depends on political choices at national level, which have an important impact on local economic organisations.

As far as *technical support* is concerned, economic organisations need information and advice. Structures must be set up in order to respond to the demands for support in legal, administrative, economic, financial and marketing fields. They could take the form of professional chambers, management support centres or specialised departments managed by a federated farmers' organisation (see Chapter VI: The farmers' organisation).

5.2 Towards federations of economic organisations

Lasting economic organisations can only be born from groups of people who know each other and share common interests. They must therefore firstly be local. However, at a later stage, groups that are already organised may converge and become federative organisations. This is a transition from thinking of one's immediate interests (getting together in order to get an additional income), to that of thinking of negotiations (getting together to have enough influence to intervene in negotiations with other role-players). This may be at regional or national level, in the management of a commodity chain for example. It is already happening in southern Mali where one of the strong points of SYCOV (Cotton and Food Crops Growers' Union) has been to federate village associations that are already efficient at an economic level (see Chapter VI: The farmers' organisation). This is also happening in Guinea, in the potato chain, where a leader encouraged a rapid federation of groups with a common interest: selling. Further initiatives are currently emerging for cotton and coffee crops.

6. Recommended literature

Anandajayasekeram, P. et al. (2000) *Agricultural project planning and analysis*. University of Pretoria, FARMESA, 136p.

Belloncle, G. et al. (1982) *Alphabetisation et gestion des groupements villageois en Afrique sahelienne*. Karthala, Paris.

Gentil, D. (1984) *Les pratiques co-operatives en millieu rural africain*. L'Harmattan, Paris.

Chapter XI: Management of collective assets and facilities
D. Gentil, M.-R. Mercoiret, E. Arnou & S. Perret

The following chapters relate to Chapter XI:

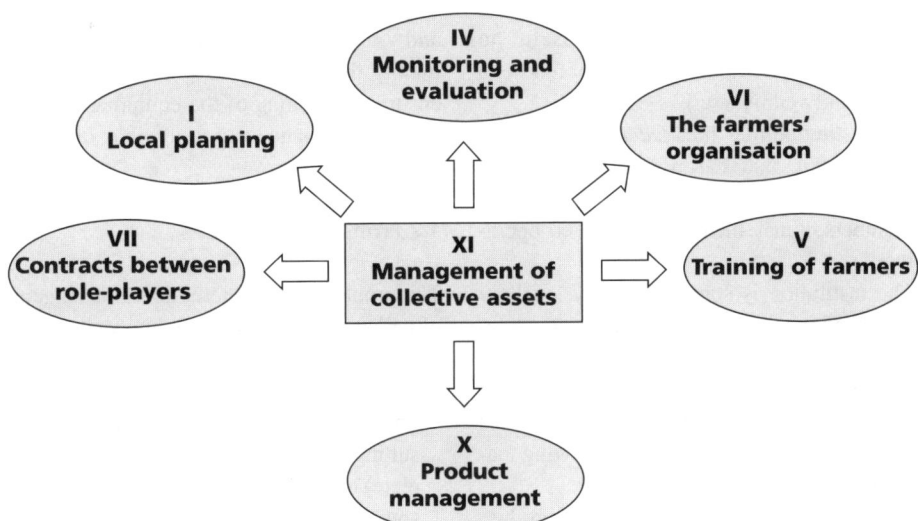

This chapter is composed of two distinct parts:
- The first part, after a few definitions, mentions some of the issues at stake in the debate on the management of collective goods, equipment, and facilities (sections 1 to 4).
- The second part is a case study on a communal water point and its management, with guidelines for action. In fact, owing to the great diversity of collective equipment, it has been decided to discuss one particular case to illustrate an approach, rather than to generalise. It should be noted that this particular case is common as it concerns about 100 000 rural communities in western Africa, that is to say 30 to 40 million rural people, and many more, if one takes into account the entire continent.

1. Definitions, orientations, issues for debate

Collective goods and facilities (generally equipment or buildings, infrastructures) belong to or serve an authority, a local group, or a community; they are different from privately or family owned goods and equipment:
- The authority may be a *state*: this is the case for roads, hospitals, big dams, and often school infrastructures. The goods belong to the country and are used by all the citizens.
- The authority may be the *local authorities* (a local district or a municipality) that have financial resources at their disposal and have been given the freedom to make decisions and manage their resources. This is the case,

for example, for youth clubs and maternity homes that have been built by rural communities in Senegal. The goods belong to the local community and serve all inhabitants living in the territory under its administration (see Chapter VII: Contracts between role-players).
- The local institution owning these communal facilities may also be a *ward,* a *village* or a *group of villages* that have decided, on their own or with an external partner, to acquire facilities that would be useful to everyone (a place of worship, a borehole, a community clinic, etc.).
- The institution may be a *federative farmers' organisation* that has acquired equipment (e.g. tractors, processing units) to help its members; it may then operate as a co-operative.
- The institution is sometimes very small: *a group* of men, women, smallholders, livestock farmers, or small-scale businesspeople or craftspersons who together purchase equipment, thus becoming both the owners and users of it (e.g. motor pump, sewing machine).
- More often, especially in the case of production facilities (e.g. small-scale irrigation schemes, garden schemes, shearing sheds), a group of people may access the facility and make use of it collectively, the initial investment in capital and infrastructure being often covered by the government or a development project.

1.1 Why collective goods?

Some facilities can only be collective owing to their nature: a borehole, a school, a community clinic and so on. These facilities are therefore of *public use* and are supposed to improve the living conditions of their users, reduce the differences between urban and rural areas, and "democratise" access to health and education.

Certain facilities are compulsorily collective as well, but they are used for production purposes by a limited number of persons (e.g. a community irrigation scheme, a shearing shed).

Other facilities or goods can, by their very nature be owned by individuals and families, or belong to and be managed by a private enterprise (mills, tractors, processing unit). However, owing to their cost and the difficulty of making them economically profitable, they are acquired collectively often through subsidies and sometimes by resorting to credit. The community that acquires or purchases them must ensure the running of the equipment or facility, its maintenance and if possible, its renewal.

1.2 Problems and issues

The problems encountered in the management of collective goods are well known:
- It is difficult to ensure they operate on a regular basis.
- There are shortcomings in maintenance resulting in the extended (and sometimes complete) breakdown of machinery; shortfalls in sound operation and maintenance result generally from a weak organisation and from the lack of rules regulating access, use and maintenance of the equipment.
- Usually there is little equipment capacity for renewing.
- Some influential persons or small groups may monopolise certain facilities or equipment.

Questions arise, notably:
- How can one ensure the management of equipment and goods, and cover operational costs? This refers back to questions of organisation, accounting and control (see Chapter X: Product management).
- Which conditions must be fulfilled in order to ensure the maintenance of equipment?

The example of a mill:
- How can both petrol and motor oil be paid for?
- How is the miller paid?
- How is the service paid for?
- How should the control be carried out?
- How can maintenance and repairs be carried out?
- How can one quickly find spareparts?
- How can the reparations be paid for?

2. Diversity of collective goods

Collective equipment, facilities and goods cannot all be managed in the same way. It is therefore important to ascertain what type of goods is to be managed, as this will determine the particular constraints and regulations. Collective goods are classified according to three criteria:

2.1 Nature of the goods: are they productive or a service?

Collective goods can be for production purposes (e.g. a motor pump in an irrigation scheme, a shelling machine) or may provide a service of a social nature (a community clinic, a water point, a school):
- In the first case, the goods are used to increase income or production; money and time are invested for the sole purpose of generating resources. The objective is therefore economic profitability: at the very least the cost of using the equipment cannot exceed the receipts.
- In the second case, the service provided (health care, drinking water, education, etc.) does not generate any direct income; it is supposed to generate welfare and to improve living standards and human capacity. Its costs have to be covered, either by the user (consumer of the goods) or by the whole society (through subsidies).

2.2 Management modes

User responsibility in managing facilities and goods is variable:
- Users may manage certain goods in an entirely autonomous manner; they must ensure the proper running of the facility or machinery, and must organise themselves as they wish.
- Other facilities and goods are managed by the government and are as such public services; the users may criticise the management but may not intervene.
- Certain facilities and goods are jointly managed (users intervene in the management, but through an administration or a local public authority). Often, in this last case, the users' responsibilities are defined without their being consulted.

2.3 Number of users concerned

The more users there are, the more difficult it is to manage directly. It is quite easy for a few women, members of the same group, to organise the running of a water pump or a sewing machine. However when the goods are used by a bigger, more heterogeneous group, (within an area, a community or several communities), it becomes more difficult to define the rules, to abide by them, and to get others to abide also. Often, many people, rightly or wrongly, feel cheated.

3. Some methodological indications

3.1 Guidelines for actions

3.1.1 Management capacity

The ability to manage collective goods increases when the goods correspond to real needs and are perceived as a priority by the users.

> Often, goods "given" as a response to a complaint, or proposed after external diagnosis and decision, are accepted. However, this acceptance, albeit enthusiastic, does not automatically mean that the community will look after the "gift". In fact, the acceptance may simply result from an opportunistic attitude, ("Let us take it anyway, we'll see later"), or from a desire to "modernise" (to be like the next village), or from a vague aspiration (to be able to get the same health care in the village as in town).

Before introducing facilities or equipment, it is necessary to *identify with the rural people* the problem the equipment or facility is supposed to address.

The participation of the future users (in work, in nature, in money) in the acquisition of the equipment or facility is a good criterion for appreciating the real interest the users have in the project.

3.1.2 Negotiating the technical decisions

The *technical choices* regarding the equipment or facility must be *negotiated* with the people.

For this negotiation to take place, it is necessary to make an objective presentation of the different technical possibilities, their *advantages* and *disadvantages*, as well as the constraints related to each alternative. And yet, in many cases, operators have made technical choices based on an external appreciation of the reliability of a technology, its economic possibilities, and the like. Sometimes these choices are dictated by options that are more ideological than technical. This was the case during the seventies and eighties when alternative technologies, that made use of renewable energy sources and so on, were used. Although commendable, they were not always found to be technically reliable.

> There have been some heated debates in certain villages between rural people requesting a motor pump and an external operator, "imposing" voluntarily or not, natural, solar or wind pumps which were still at an experimental stage.
>
> The same kind of debate has also occurred between operators, defenders of animal traction at family level and villagers wishing rather to acquire motorised agricultural equipment collectively.
>
> In the field of health, some operators have considered **village health care houses** as being inseparable from primary health care with the result that more money has been invested in them than in community pharmacies that often provide a similar service.

It is only possible to negotiate if external operators have not already made the technical choices and therefore there really is room for discussion.

Negotiations are efficient only if they are *concrete*:

- It is best to move beyond arguments such as "it is going to be very expensive" and establish a realistic *estimated income statement* for each of the possible technical choices (see Chapter X: Product management).
- It is also useful to announce the results obtained from different technical alternatives in comparable areas, the constraints related to their respective running and the necessary organisation (visits).

It is important to take one's time and not to make hasty technical decisions. This avoids rural people and external operators alike picking certain obvious choices dictated to them by previous models.

3.1.3 Defining management modes

It is absolutely necessary *to define concrete terms and conditions for managing* the approved facility before it is introduced. These terms and conditions concern

- the *site*: in order to avoid mistakes being made (e.g. a clinic or a community maternity facility being ill-situated) it is necessary to discuss at length where the facility or goods should be established. The discussions may also enable any internal divisions within the community to be identified (between different districts for example) at an early stage thus indicating problems that could occur later in the management of the facility or equipment.
- the *everyday running:* the operating charges (mostly financial), maintenance, repairs and supplies (if inputs or spare parts are required) must be taken care of. These should all be discussed as well as the necessity to provide for the renewal of a facility or equipment that is subject to deterioration. The analysis of the costs related to the goods must go hand in hand with identifying the means to meet the costs. This requires making realistic and accurate calculations as well as defining clearly how the costs will be covered (see Chapter X: Product management).
- It is necessary to support the internal organisation of the users by identifying precisely the tasks to be carried out (see Chapter VI: The farmers' organisation). Even if certain people's ideologies must suffer, one has to admit that direct collective management, even with management committees usually made up of influential local persons, is not always the most efficient way. Contracts between the community and a private operator or a service provider are usually preferable.
- Internal organisation cannot be separated from environmental organisation; spare parts, fuel, inputs, medication, and so on, must be available. Networks must be organised, connections created and contracts signed with the private sector or administration (see Chapter VII: Contracts between role-players).
- External and internal organisation is efficient only if the rules of the game remain the same, with modifications taking place only once an agreement between the different parties has been made, and if each one fulfils his commitments and respects the decisions (see Chapter VII: Contracts between role-players).

Three observations:

- Voluntary (unpaid) working systems last only a short time (see Chapter V: Training of farmers) and cannot ensure the stability that is necessary for community facilities and goods to be used and managed correctly.
- Users' fees, especially when they are indirect (collective crop field), are usually disappointing and unreliable. When compatible with the nature of the facility or goods (millet mill, health care house, tilling operations), payment for the provision of services is more successful.
- Rural people tend to overlook forward expenditures, while external operators tend to overestimate them. Undoubtedly it is a good idea to start off with an average estimate and readjust every year according to the results obtained.

> It is for instance very difficult to get smallholders to accept an increase in the charges they have to pay especially if the principle was not discussed at the beginning and the decision was made unilaterally.

3.1.4 Training the users

For any goods or facility to be durable, farmers must be given the necessary technical and managerial skills. These can only be acquired through an adapted training programme (see Chapter V: Training of farmers, and Chapter X: Product management).

3.2 Different management methods for different collective goods

3.2.1 Productive goods

Although there are many different types, we will only discuss two cases:

A. First case: the equipment is acquired and used collectively
(e.g. a motor pump for a garden scheme)
This type of equipment is usually managed collectively:
- A *management committee*, nominated by the group of users, is required to make all the economic, technical and financial decisions necessary to operate the equipment correctly. The management committee is efficient only if its members really have the authority and necessary abilities to make relevant decisions.
- The management committee must, in agreement with the users, have the means to work properly. This means may be qualified personnel (a pump attendant for example) who will be directly responsible to the management committee (for instructions, salary, possible sanctions), or contracts with external partners that only the management committee will be entitled to negotiate.
- The system can only work in the long term if the users who nominated the management committee submit it to periodic controls. For this to happen, accounts, decisions, and the existence of sanctions and the like, must be publicised and made clear (see Chapter X: Product management).

If inefficiencies and shortcomings appear, they may be due to three main reasons:
- The activity for which the equipment is used is *not economically viable*: the production from within the irrigated area does not cover the charges and make a big enough profit for the farmers. This may be a result of the technical choices made, the lack of market outlets or the prices paid by the farmers. It is no longer, therefore, a problem of managing the equipment but rather a problem related to *managing the activity*, its organisation, networks, and so on. In this case, a general review must be carried out. How can the economic results be increased? How can production costs be reduced? How can market outlets be increased? How can one negotiate better prices?
- Perhaps *the managerial and technical skills* of those in charge are *insufficient*: this is more often due to unsuitable training or to training carried out too quickly. It can also, however, be due to unsuitable management tools, or a poor relationship with the technical environment (Chapter X: Product management).
- *Social divisions* between members of the community can paralyse the management of collective goods, the latter merely being a pretext to express their differences of opinion. If goods really correspond to farmers' needs, solutions may be found. Otherwise, it is unlikely that solutions will work in the long term.

B. Second case: Goods or facilities are acquired collectively, often with external aid. However, utilisation is not collective and may be optional.
(e.g. a tractor for a group or various groups, a forge, a means of transport)
The management of this type of equipment is subject to specific constraints, as often members of the community feel they are not benefiting as much as they would like to from the equipment. Since management committees are often slow moving and inefficient in dealing with these matters, more and more communities choose to privatise or outsource the management of these facilities by creating joint ventures. This consists of the following:
- The community receives the goods and entrusts an individual or small group (a group with economic interests or a private small-scale company) with the goods, in order to manage it as a business.
- The running of the equipment or facility becomes the company's responsibility (the equipment becomes its working tool). It sees to its maintenance and renewal.
- The company is paid to supply services to the users, be they members or not of the community that received the equipment. Priority is, however, given to members of the community, and sometimes at preferential rates (e.g. on a marginal cost recovery basis); the search for new clients is essential in such a situation, in order to cover operating (marginal costs), wage, maintenance and replacement costs properly.
- The profits from the use of the facility or equipment go to the managing company which in turn will pay a certain amount (fixed or proportional to the benefits) to the community from which it received the equipment or facility.

If matters do not run smoothly, they may be related to four factors:
- The community and its leaders do not really entrust the company with the goods, and continually interfere in its management. Consequently, the manager loses responsibility, entrepreneurship, and ultimately does not carry out his/her functions properly.
- The payments made back to the community are so high (or the service fees paid so low) that the manager is unable to cover his/her costs and make an acceptable income. The manager becomes unmotivated and stops taking any initiatives; the goods may deteriorate with time, and/or the company collapse.
- The manager is incompetent: he/she does not have technical or managerial skills. He/she, therefore, cannot fulfil his/her contract with the community (bad quality services, confused management) and moreover cannot find new clients.
- The company created is not economically profitable: the demand is not enough for instance, or the prices charged are too low or too high. One has, therefore, to think about the activity (find new clients, establish acceptable prices for clients) with regard to a financial equilibrium.

Generally, the privatised (outsourced) management of collective goods and facilities is an efficient alternative to inefficient and slow-moving management committees.
- It requires a clear contract to be established between the community and the company, such a contract guaranteeing real managerial autonomy for the company and mutual benefits for both parties.
- It does not exclude regular control by the community, according to modalities defined from the outset in the contract.

3.2.2 Managing social facilities
This concerns facilities that aim to improve the lifestyle of rural people: the provision of drinking water (wells, boreholes, reticulations and water points), clinics, literacy centres, and the like.

Productive goods are not considered here.

For example, mills, shearing sheds, boreholes for market gardening or livestock watering, and so on, are not social facilities. Although they do improve living and working conditions, they are firstly productive equipment linked to economic functions and must be managed as such.

Social facilities are often located at village or community level (less often at district level). Users are numerous and diverse. The services provided by certain social facilities contribute to improving production: better quality drinking water and better health care limit the outbreak of disease and, therefore, have an indirect effect upon production. Literacy increases access to information and management skills and therefore also contributes to increased production. However, these services do not directly generate resources but rather require financial means for remaining operational.

Two questions arise:
- How can one ensure that the facilities are run efficiently (measured by the quality and regularity of the services rendered)?
- How can the necessary resources be generated?

The answers to these questions are inseparable as seen in the example below.

The community health care house

The people's contributions to the maintenance and running of a health care house are easier to obtain if the first-aid worker is competent, available, and has medication on hand. However, for this service to be provided, financial resources are needed to purchase medication, to pay the first-aid worker, to maintain the house, and so on.

Some answers:

It is necessary to discuss what is needed for the health care house to run properly:

- Ensure that it addresses a real demand, a priority.
- Clarify its function and avoid creating expectations that cannot be fulfilled (interests and limits).
- Negotiate the installation site with the users.
- Collectively designate the management committee and the first-aid worker, and define their roles (contract) as precisely as possible.
- Train the management committee as well as the first-aid worker in their respective functions.
- Organise medical supplies.
- Organise monitoring and evaluation.

It is also necessary to define the financial support for the health care house:

- Evaluate the costs involved (investment, yearly running).
- Analyse the various possible means of covering them (e.g. payment for treatments, individual fees, deductions made directly from productive activities, etc.).
- By consensus select an appropriate way to finance the centre (see Chapter X: Product management).
- Evaluate the results in economic terms (Are the charges covered?) as well as in social terms (Are the people satisfied? Are some people excluded ? Why? How can this be corrected? etc.)

Two additional observations:
- In many cases, village people prioritise social equipment because, although not directly productive, it is meant to address serious problems (e.g. diseases, water shortage). This is normal, but it is important to highlight the fact that the sustainability of social facilities depends on income from activities producing or creating resources. It is not, therefore, realistic to think of social facilities as being independent of production: *new expenses require new resources.*
- Village people tend to transfer, as much as possible, resources from their productive activities to those of collective interest; this is done through a multitude of fees or different funds, and mainly concerns activities that are totally or partially collective. One must be cautious and make sure that these transfers are not made by the capital accumulation within economic units (farming households or organisations) or, even worse, to the detriment of mere reproduction of the productive activities.

If this happens, the productive activities may stagnate or even regress (for example, if the productive equipment cannot be repaired or replaced); then the social services will in their turn decline.

4. Issues for debate

4.1 Should collective management always be preferred?

One may ask whether it is better for a mill to be managed collectively but to work intermittently, or for a mill to work continuously under the management of one responsible individual who benefits from a share of the profit.

Realism must prevail over idealism. Collective management should not be imposed; it is often inefficient, cumbersome, and often implies social interference. Once the users have debated the advantages and disadvantages of both types of management, they must choose between collective management and outsourcing (privatisation), which does not mean that social controls are excluded (a ceiling price for a service may be fixed, accounts should be audited, etc.).

4.2 Inter-community goods and facilities

Sometimes it is not justifiable for a village to have its own facility (community clinic, school, social centre). It does happen that two neighbouring communities both have schools but lack teachers and have only a few pupils, or have clinics without enough patients, medication or nurses, and so on. How can such a situation be avoided?

Villages often compete to obtain social facilities not only for reasons of prestige, but also for the quality of service they afford. The formation of new villages has aggravated the problem as external operators and governments alike have tended to accumulate all the facilities and goods in the same village. This has resulted in those who are disadvantaged feeling increasingly bitter.

There are two ways forward:
- The "traditional way", which consists of leaving the conciliatory role to the village elders and wise men. This is not a possibility, however, if long-standing problems are already disrupting relationships between villages, as these may flare up in quarrels based on modern political sympathies.
- The "modern way" through local authorities or government, if they exist: the locally elected members should mediate between villages, which should mean that they do not always privilege certain communities (those that are bigger or more central).

4.3 Type of contract between the users and local authorities

The users and local authorities (district, government) could jointly manage the public facilities (school, community clinic) providing the responsibilities are clearly shared and defined in a consensual manner (see Chapter VII: Contracts between role-players). In fact, it is frustrating for the users to be called upon only when problems arise (e.g. to repair a leaking roof at the school or at the community clinic).

4.4 Creation of an amortisation fund

The first objective of such a fund should be to operate and maintain facilities and goods. Under certain conditions the question of making provision and of renewal arises (a depreciation allowance should be set aside for equipment renewal purposes) (see Chapter X: Product management):
- It is useful to calculate how much should be put aside each year for making provision (the cost of the equipment or facility divided by the number of years it is predicted to last) and to attempt to save that sum. The way in which this is achieved (totally, partially, not at all) can be indicative of the efficiency of the management and activity. It should, however, be mentioned that it is very difficult to reimburse a loan for a facility or equipment and make provisions for replacing it. It is easier when the facility/equipment has been obtained through a subsidy (in this case, the annual provision can be equal to that of repaying a loan).
- When provisions are made, it is a good idea to transfer the money to an interest-bearing bank account. Money saved in this way will contribute to the replacement of the equipment or facility that will in the future be more expensive (inflation, technology changes, etc.). It can also contribute towards acquiring, by means of a loan, another piece of equipment or facility.

5. Succeeding in the management transfer of collective goods and facilities: lessons from experiences

5.1 An essential new component: the users

5.1.1 Changes
An era has come to an end. It was dominated by the *principle* of acquiring or using a number of collective goods or facilities *for free*. Despite such conditions, the needs of the people were not necessarily satisfied. For collective facilities at local level (water point, village pharmacy, community clinic, school, a small-scale irrigation scheme) changes have resulted in
- users making a physical and/or financial contribution to the investment;
- maintenance charges being transferred to the users;
- decentralised and/or privatised (outsourced) management.

5.1.2 Constraints
Most of the facilities have only recently been introduced to rural society and require what is largely seen as alien technology. Their introduction has been the subject of many errors and failures.

Each programme or project is therefore subject to the following constraints:
- Promoting changes in financial and managerial methods that require more effort from the users.
- Introducing innovations that rural people have directly or indirectly learned to distrust: such innovations do not refer only to technological aspects or new equipment, but also to the new way of managing them.

5.1.3 What users want
To overcome user-related constraints, previous practices must be modified. In the past, social facilities were allocated according to objec-

tives set by *national* (or *regional*) *planning* or by the populations' political representatives. As soon as populations become role-players in the management and payment of maintenance, a third component arises: the *will of the rural communities who are the beneficiaries of the facilities*. This will becomes the key to success. Either it is present or it is not. In cases of failure to arouse such a will, years of extension services will never be able to repair the damage.

5.1.4 Innovation put to the test

This can be summarised in the following way: either the facility and management method (including maintenance) proposed demonstrate their superiority over what existed before, or they will be rejected. There is, however, a possibility that the new facility may be partially utilised if it is complementary.

> The manual pump is a good example. If it is located in a village where it reduces the time it takes women to get water considerably, then the latter will stop using the traditional water points as long as the pump works regularly and is not too costly. Then there will be **substitution**. If, during the rainy season, the more numerous and closer traditional water points appear, there will be competition and the pump will be momentarily abandoned. There will be **complementarity**.

5.1.5 Negotiation is essential

To make a new facility or a new way of managing a facility look attractive, in-depth knowledge of the situation of each group of users is required. Their situation depends on a multitude of variables (natural, economic, social, political, religious, etc.), the complexity of which is far beyond the control of a project promoter. *The participation of users is therefore indispensable before and during the operation. The users are the only ones to know their* own constraints in detail just as the project manager knows his. Permanent negotiation between both sides will therefore be necessary.

5.1.6 Past experiences often prevail

During negotiations, the rural people's viewpoints do not simply depend on criteria directly related to their needs (e.g. water shortage), but also on the *perception they have regarding the innovation proposed*. In Togo, certain villages refused boreholes because of the problems they had previously experienced with the type of pump proposed.

5.1.7 Spiral effect

There is a *spiral effect in failure or success*. Through negotiations a multitude of communities

- an in-depth knowledge of the users' constraints, perceptions and positions can be obtained, which in turn improves the initial methodology which obviously could not predict everything;
- the elements of the debate concerning the propositions can be widely diffused in rural communities.

The two sides learn to know each other better. As a consequence, the failures will be *more resounding* and the successes *more promising*.

5.1.8 Possible confusion

Owing to the precarious economic situation facing rural communities, it would be *catastrophic to underestimate* the communities' obligations with respect to the *collective facility or goods proposed*. A typical and frequent example is the proposition made by humanitarian organisations or charities to build a borehole at little or no cost to the community despite the fact that a number of measures must be undertaken prior to the installation of a modern water point (see the following case study in Togo). Unfortunate misunderstandings may occur, leading to *confusion as to the level of engagement to be undertaken by each party*.

5.1.9 Importance of having a national methodology

To avoid such misunderstandings or at least to reduce the damaging effects, it is essential to compile a *national methodology* for reference purposes. It will be more successful if it has been drawn up taking into consideration the lessons from the field and if it is the result of *constructive dialogue between the various partners*.

5.2 From theory to practice: some recommendations

From these premises, it is possible to formulate some practical recommendations. They are grouped into four themes, none of which can be applied on its own.

5.2.1 Adopt a negotiation approach with beneficiary communities

See Chapter VII: Contracts between role-players.

5.2.2 Provide the necessary conditions for adopting the proposed innovation

It is necessary to
- *answer the populations' needs* with a technological solution adapted to their financial means. For example: the transition from free water (from a river or a dam) to payable municipal water is not difficult to overcome. On the other hand, the installation of a supply scheme and service for drinking water presupposes a higher water price that not everyone may be able to afford.
- *introduce reliable equipment*: defects in the concept and/or materials used to make manual pumps are also one of the reasons that they have been difficult to establish and are responsible for numerous costly and lengthy breakdowns. Much other equipment has known the same fate.
- *promote the technological environment* by identifying, supporting and training the professionals who will be in charge of maintaining equipment (craftspersons or local mechanics, spare parts suppliers).
- *standardise equipment*: too many different types and brands of the same sort of equipment only multiply maintenance problems as
  - repairers are then required to attend complex training courses;
  - it restricts the market for each part resulting in reduced profitability which in turn leads to that activity being abandoned.
- *identify the beneficiaries' motivations* and take them into account. For instance, whether the water is initially drinkable will be a determining factor in the appropriation of a modern water point.
- *negotiate the location for the installation*, taking into account the technical constraints (generally expressed by the project) as well as the social constraints (revealed during the rural community's internal debate). For in-

---

Communities, villages and wards

Beyond the strictly administrative meaning that may exist in certain countries, **the term "community" often implies very different realities.** For example the following have very little in common: a village of 300 inhabitants belonging to the same ethnic group whose houses are close together, and a community of 5 000 inhabitants with many wards scattered and established on the basis of ethnic origin, or extending along 8 km of national road.

In the first case, the social cohesion is generally strong enough to enable them to manage collective facilities independently; in the second, centrifugal forces make autonomous management very difficult.

Community members and committees

Owing to pressure from projects and the multiplication of collective facilities in rural areas, tens of thousands of different management committees have been generated. These committees are supposed to be dynamic and efficient as they theoretically represent the users whose main interest is a correct and permanent running of the facility, equipment or goods.

In reality, the socio-political pressure exercised by local influential people interferes in the institutional procedure proposed or imposed by the project. When the notables are too influential, the committee is condemned to being ineffectual, or it is used as a screen behind which the facility is misappropriated to the detriment of the majority of users. When it is too weak, the committee becomes illegitimate; for example, it is difficult to collect the contributions. Generally there is some form of constructive or destructive tension between the two tendencies. The older the community, the more acute the problem. In contrast, in recently created villages (in a new agricultural area) there is usually a more egalitarian climate, especially since these are frequently small communities. Regarding the community water supply, they use the concept of a consumer unit (see above, "Communities, villages and wards") and the socio-technical implementation stage has contributed to the integration of this organisational system management committee or water point committee in the village structure.

---

stance, the best location for a water point is not necessarily the one that would give the best water output, but rather that which will shorten the time it takes to fetch water.
- *implement proper training programme* that enables everyone to become familiar with the utilisation, maintenance and management of the facility.

If there are *difficulties in using a facility*, try to get the beneficiaries to formulate some users' rules (who can use what, when and how?).

5.2.3 Describe precisely the nature of successes and failures, and instigate the necessary methodological changes

It is useful to organise a data collection and processing system tailored to the project. Data collection not only allows blockages, regressions and successes to be identified, but also enables those concerned to react rapidly, to amplify successes or minimise failures. This requires
- a limited number of relevant indicators to be defined: the related data will be easier to collect and will therefore be more accurate;
- frequent assessments between field agents and management to be organised;
- data processing to be limited to a strict minimum.

5.2.4 Favour the emergence of a national methodology based on lessons from the field

Contradictory speeches, promises, and differences in treatment can only encourage rural populations and field agents to think they are being treated unfairly and make them avoid their responsibilities.

Conversely, a common methodology applied to all programmes enables
- the responsibilities of each party to be clearly set out;
- a better distribution of information;
- developments to be mastered (there is a possibility of comparing the relevance of measures proposed in different situations).

This methodology should develop at a regular pace according to the lessons learnt in the field, keeping in mind the fact that people's attitudes to the facilities and goods proposed, as well as to management methods, also change over time.

Lastly, a national methodology must not be a constraint but a guide. In this respect and provided that it does not encourage regressions, a methodology should
- of necessity adapt to particular local situations;
- allow for experimentation on a smaller scale that should foresee possible future developments.

5.3 Some pending questions

5.3.1 Dispersion of human resources

Implementing a transition procedure as described above requires both important investments in personnel training and the temporary recruiting of contractual agents. Two problems arise:
- Government agents are constantly *transferred from one area to another* and are under pressure to renew their training. It is therefore difficult to create a close, integrated team that feels really involved in the project.
- At the end of the intensive realisation period contractual agents should be dismissed (as the philosophy of community water supply programmes is to privatise/outsource maintenance, and therefore to leave in the field only a minimal administrative presence). Contrary to what was practised in the past, it is rarely possible to relocate contractual agents in the public or parastatal sector. Consequently, there is both a human drama and loss of highly valued expertise.

If the FORMENT project (see the following case study in Togo) has managed to avoid the first problem (personnel movements), the second has been treated only empirically. In as far as community water supply projects have succeeded one another, contractual agents have moved from one project to another (although regardless of linguistic barriers).

In the future, without pretending to be exhaustive, two solutions can be envisaged:
- Line administrations (e.g. Department of Hydraulics and Energy in the case study) could recruit a limited number of agents who would each be responsible for monitoring water points in a given administrative area.
- Other agents could be redeployed to supervise the maintenance of water supply equipment in the private sector. Theoretically this would be made possible by a tremendous increase in the number of existing pumps and through the skills of the concerned agents (pump mechanics, and development and training agents).

5.3.2 Replacing certain equipment

The problem of renewing facilities and goods cannot be treated in a generalised manner. Specific answers must be found depending on several factors: the nature of the material involved, the service provided, the group of people concerned, financial sources, the technological and socio-economic environment.

Thus for the same material (e.g. a manual pump), several solutions coexist:
- If the model used is still manufactured and is satisfactory, it will simply be necessary to change or repair certain spare parts or sometimes to adapt the part (from an old type to a new one).
- If, on the contrary, a certain model is no longer manufactured or is a technological failure, another pump must be bought.

The costs for either operation will obviously not be the same.

However, the key to renewal remains the motivation of the users. The more convinced they are that the facility to be replaced, at a reasonable cost, corresponds to the problem it is supposed to resolve, the more motivated they will be to pay for the operation themselves.

In certain areas (e.g. in Mali, Aqua-Viva)

where the maintenance of community water supply systems has been successful for many years, the villagers change the pump at their own expense. In other geographical areas where the opposite situation has been observed, the same model of pump has been rejected and the users have turned away from the modern water point in favour of wells and backwaters.

However, in both cases, if it was necessary to fix or re-drill the borehole itself, the operation, because of the high cost involved, would be financed by *external operators* (project).

If the principle of partial or total *self-financing* is accepted, *the problem of acquiring the necessary funds remains*. Saving money from water sales or fixed contributions is often recommended. This is absolutely essential when the time comes to renew important equipment (certain spare parts of a solar pump cost about US$ 3 000). However, there are three main difficulties:
- A certain resistance from users to immobilise money when there are so many needs in the village, some of which may be given a higher priority
- The deficiencies, even bankruptcy, of banking systems in rural areas and the fear of money being misappropriated
- The poor return on capital regarding usurious rates and the price increase of the equipment to be replaced

In the future, the renewal of small to medium-sized collective facilities and goods could be provided for by a variable *combination of self-financing (one-off payment or savings), loans at a low interest rate or donations*.

6. Recommended literature

Cusworth, J.W. & Franks, T.R. (editors) (1993) *Managing projects in developing countries*. Longman, London, 236p.

Esman, M.J. & Uphoff, N.T. (1982) *Local organisation and rural development: the state of the art*. Ithaca, Cornell University, USA.

Moench, M., Caspari, E. & Dixit, A. (eds.) (1999) *Rethinking the mosaic: investigations into local water management*. NWCF & Institute for Social and Environmental Transition (Publ.), Boulder, Co., USA.

Whiteside, M. (1998) *Living farms. Encouraging sustainable smallholders in Southern Africa*. Earthscan Publications, London, 217p.

Managing collective equipment: the example of community water points in Togo

1. Context and issues

1.1 Collective equipment providing a service and managed by a village

The case described in this chapter (the FORMENT project) concerns the *implementation of a new maintenance and management policy* for domestic water points that were installed in rural areas in northern Togo from 1980. The project concerns collective goods (water point) providing a service (drinking water) to a small community (100 to 2 000 inhabitants) where the responsibility for managing the point has been transferred from the government to the users (living in the community or village).

1.2 Tens of thousands of water points

This example does not intend to review all the problems pertaining to the management of collective goods, but rather to focus on the case of the *many tens of thousands* of boreholes equipped with manual pumps that have been installed in western Africa. Almost as many again are planned or are being installed. In many villages, this is the *first piece of collective equipment*. Therefore a study of the conditions, which will decide on the success of both the pumps' appropriation and their management by the rural people can be very revealing.

1.3 Regional context: a deteriorated situation

Western Africa went through two decades (1970s and 1980s) of drought combined with a high demographic growth rate. To address this situation as quickly as possible, western African countries and a number of financial sources (bilateral and international aid, NGOs, and other credit organisations) were mobilised to equip rural areas with tens of thousands of water points.

Generally, for both financial and technical reasons, *boreholes activated by manual energy* were chosen rather than wells that were preferred by the rural community who had used them for generations.

Furthermore, the villagers who were to benefit from the installations *were seldom even consulted* as to the principle of the equipment, its implantation or the chosen technology. As for maintenance (repairs and servicing), the question was often considered only long after the installation of the pumps.

This situation often resulted in the beneficiaries either totally rejecting the collective equipment or being incapable of maintaining and repairing it. As a consequence, in 1985–86, 40 to 60% of the water points were out of service.

2. A story of management transfer: the FORMENT project in Togo

2.1 More than 50% of pumps broken in Togo (Central and Kara regions)

2.1.1 Unsuitable maintenance service

The first boreholes equipped with manual pumps were installed in 1979 and sponsored by the European Community (European Development Fund).

In 1985 there were approximately 1 600 boreholes in place. From the beginning, the Department of Hydraulics and Energy (DHE) was aware of the maintenance problems encountered by neighbouring countries and therefore created a pump maintenance service. This consisted of 11 mechanics sent to different parts of the country to carry out pump repairs. In certain cases, a villager in charge of a water point was able to make small repairs after a short training.

Despite this arrangement, the surveys conducted in 1987 in the Central and Kara regions, regarding the pumps installed in 1980, showed that 53% of pumps had broken down, with a maximum of a 75% breakdown in certain regions.

2.1.2. Users not consulted

Further surveys revealed that
- the boreholes were drilled without prior consultation between the administration, the engineers and the beneficiaries, which resulted in many rejections;
- the pumps were better maintained when they were the only available source of water;
- an improvement in health resulting from the use of borehole water was rarely perceived by the beneficiaries (except in the case of outbreaks of Guinea worm infections);
- the villagers had three major problems in maintaining their pumps:
  - Lack of financial reserves for serious breakdowns;
  - Unavailability of the pump maintenance service mechanics to make repairs when pumps broke down owing to the wide areas they had to cover;
  - Difficulty in obtaining spare-parts;
- the villagers found it difficult to accept having to make a financial contribution towards maintenance following years of a free maintenance policy (1980s).

2.2 Defining the conditions for the users to undertake management responsibilities

2.2.1 Users

The FORMENT project, stemming from the French words for training and maintenance (*formation* and *entretien*), began in the last quarter of 1986. It was applied to two kinds of drilling works:
- Those carried out and equipped with manual pumps from 1980 (53% of which were broken)
- Those carried out in 1984–1985 and not yet equipped with pumps, that is 350 boreholes in 250 villages

Both types can coexist in the same village, with two types of users: those who have become suspicious after experiencing failure with one system, and others who have not yet been exposed to the difficulties of maintaining the pumps.

2.2.2 Two principles

The project is based on two complementary principles:
- *The recognition of rural communities as fully-fledged partners*, and as such, negotiations should be organised with them. The aim of such negotiations between the water administration and the rural communities should be that the latter entirely take responsibility for the management of the water points given to them in the context of the Eu-

Three levels of maintenance:

Level	1	2	3
Type of maintenance	Routine servicing Replacements of some parts in the upper structure	Almost all repairs	Repairs needing tools or skills higher than level two
Maintenance operator	Community mechanics Maintenance manager	Contracted mechanics Private workshops	Administration workshop or a contractor

ropean Development Fund projects.
- A coherent, decentralised and privatised maintenance service must be implemented, which will allow villagers to exercise their new responsibilities. In fact, it is not possible to try to make villagers responsible if the concrete conditions for exercising this responsibility do not exist.

2.2.3 New maintenance structure
A. The pump: a fragile element
Maintaining a water point is not just about maintaining the pump since the latter is made up of a coping, an anti-mud system, a fence, and a protected area (as the basic components of a "modern" water point). However, the pump remains *the most sensitive element* because firstly, it is a *prerequisite to the use* of the entire unit, and secondly, it is the only part that contains moving pieces. When one talks of maintenance, *one therefore means the maintenance and servicing of the pump.*

B. Three levels of technical maintenance (cf. figure below)
- *Level one*: minor maintenance carried out by a community member (a local craftsman, or a mechanic, or a caretaker).
- *Level two*: most repairs carried out by a specialised rural craftsperson or a private mechanic.
- *Level three*: repairs that need specific tools and higher skills than in level two (e.g. oxyacetylene welding). This type of operation is very rare and can be performed either by the administration's workshop or by a qualified and equipped private workshop.

> As equipment becomes obsolete and worn out, an increasing number of pumps might need level three interventions. Up till now, when such cases have occurred, restoration programmes have been financed by the government and a sponsor during which operation the pump supplier has intervened with specialised technical teams. In the long run, it will be necessary to identify and train private sector entrepreneurs who will be able to intervene in individual cases.

C. Three maintenance pillars
For maintenance to be carried out efficiently, the three inseparable and interdependent pillars on which it depends must work together. If one is inefficient the entire structure no longer works and the situation deteriorates rapidly (see figure above).

These three pillars are
- management committees for water points (water committee, water point committee) handling financial reserves;
- private repairers, craftspersons or mechanics, trained, registered and available;
- a supply network for spare parts with (generally private) shops in the main communities.

Administration should intervene in the setting

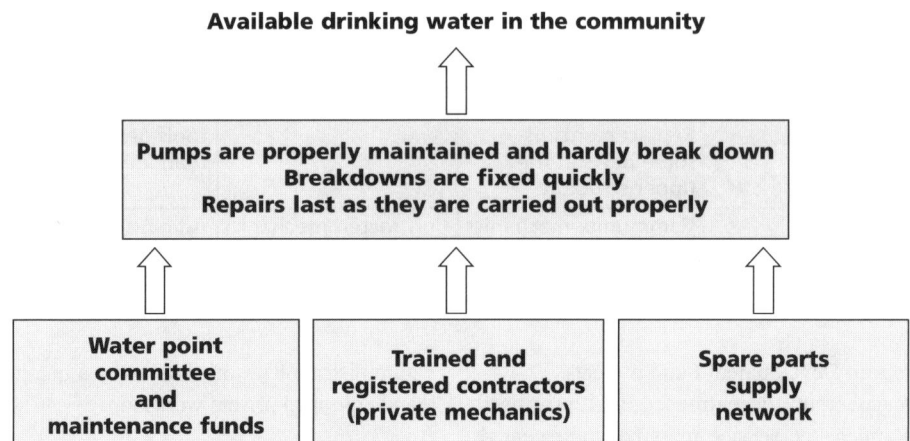

up and monitoring of the structures. The more autonomous these structures become, the more the rural development agent must gradually reduce his/her interventions.

2.2.4 Negotiations between administration and communities

A. The right to accept or to refuse a water point

The task of the rural developmental agent (called FORMENT agent) is to propose to the villages and communities, in his/her area, a contract in which it is stated that

- the community commits itself to maintaining its water point in a clean and good working condition;
- the administration undertakes to rehabilitate an old pump or to install a new one, to set up a maintenance structure and provide the necessary training to the local managers of the water point.

The community may refuse to sign such a contract, with the result that,

- if the equipment is old and broken down, the pump will not be fixed, until further decisions are taken;
- if the pump is new and working, vital parts will be removed until further decisions are taken.

Most of the time, the community will take the initiative and ask for the pump to be restored or opened.

The contract obliges the FORMENT agent to be perfectly clear regarding the nature of the engagement between the partners, thus avoiding a situation where the community's responsibilities are obscure in order to gain their acceptance more easily. The contract can only be finalised once the water committee and acceptance funds have been set up.

The contract is read and signed during a ceremony at which regional authorities are represented.

B. The contract as a communication medium

The government's responsibilities, represented by the Department of Hydraulics and Energy, are

- to look for an exploitable source of water (groundwater);
- to install a manual pump with coping and an anti-mud concrete platform;
- to post an agent in the community, in order for him/her
  - to help set up a water committee and to train its members;
  - to advise on how to collect fees for the maintenance fund;
  - to provide information to the community

regarding maintenance and hygiene norms for the equipment;
- to train some people to fix the pump (e.g. local mechanics, craftspersons);
- to ensure of the availability of spare parts within the supply network;
- to monitor the water resource (quality and quantity).

The village's responsibilities are
- to complete the construction of the water point (fence, walls, doors, drainage system and sink for waste water);
- to form a water committee in charge of the utilisation, the maintenance and the repairing of the water point and pump, and whose members are committed to undertaking training;
- to organise an equitable system whereby the water point users make financial contributions to cover maintenance costs: spare parts, repair works, transport;
- to maintain the structure permanently according to the conditions of use and hygiene established by administration and norms;
- to call a contractor in case of a serious breakdown and pay for the service provided.

In case of a violation of the commitments made by the rural community, the Department of Hydraulics and Energy reserves the right to stop the water pump from operating.

C. Each community has its own method of making financial contributions

The community members are only requested to make one contribution to the initial investment (coping, anti-mud concrete platform, fence).

In 1993, the initial financial input from the village amounted about US$140, prior to construction. From time to time the fund is replenished with a deposit equal to the expenses made. This constant level fund no longer requires annual fees.

The contribution for the maintenance fund amounts to US$70 per pump annually. The committee retains this amount either in cash or it can be deposited with the National Agricultural Bank. Community members are free to make their contribution as they see fit. The FORMENT agent only describes the pros and cons of each practice: collective plots which benefit from the contribution; direct financial contribution per family, per capita, per month, per year, water sold per bucket, and so on. In many cases, several practices coexist. Payment per bucket is however seldom used.

2.3 Motivating relations between the various role-players

2.3.1 Clarifying the personnel's responsibilities

The project depends on the Department of Hydraulics and Energy, and is under the authority of the managers of the hydraulics subdivision of both regions concerned and of a principal co-ordinator who is a director at DHE. Technical assistance by a socio-economist completes the set-up. Each agent is responsible for an area consisting of approximately 30 villages. He/she lives in the area and masters the local language perfectly. Except in unforeseeable cases, the same agent remains in that area during the entire project.

2.3.2. Organising the procedure

The project runs in five consecutive phases:

Phase 1: Contact–information–contract
During this phase, the FORMENT agent contacts the communities and informs them of the new maintenance strategies established by DHE. He/she presents the contract, answers questions related to it and organises for those communities that agree to the project to sign it.

Phase 2: Organisation–installation–equipment
The FORMENT agent participates in the rehabilitation of old pumps, or in the installation of

> Let us take two examples; the first is almost statutory, the second is more common sense:
> - It is out of the question to install a pump if the village has not signed a contract and set up a maintenance fund.
> - In phase three, the agent must train the water committee members. To do that, he/she must know the residential addresses and names of local repair persons, then, he/she present them to the communities. Otherwise, the training process will remain abstract and without real operational effects.

new pumps in the communities that have signed the contract. At the same time, he/she identifies, with the help of the villagers, those rural craftspersons or mechanics who may be capable of becoming pump repairing operators. The latter are then given theoretical and practical training, as well as a set of tools. Once the pumps have been put in place or restored, the FORMENT agent discusses with the community the reconstruction or installation of the water point. Staff members of the project would normally be responsible for approaching spare parts suppliers; however, the agent must monitor the availability of spare parts in the shops in his/her areas.

Phase 3: Training the committee members
The FORMENT agent trains the different members of the water committee in their respective roles (president, treasurer, secretary, maintenance and hygiene keeper).

Phase 4: Monitoring-support and training
This is the phase during which the FORMENT agent consolidates the system. To do that, he/she oversees the way things are running, and intervenes only when weaknesses appear.

Phase 5: Monitoring-evaluation
The FORMENT agent intervenes less and less but is still able, through a number of indicators, to detect progress and regression, and delivers warnings or alert calls if necessary. His/her plan of action will take into account individual communities.

The actions are therefore organised in five consecutive phases. The agent cannot begin a new action if the objectives of the previous one have not been fulfilled.

2.3.3 Active training as a steering tool for operation and development

The success of the project lies mainly in the agent's ability to negotiate with the communities, and the capacity of the latter to take care of their water point. For this to happen, the agents must train and inform the people. As their job requires them to accomplish many complex and precise tasks, one single quick training programme is not sufficient to give them the necessary skills. This would result in *errors, omissions and a lack of precision that would lead to and sustain throughout the project zone a climate of confusion regarding the*

> When a relatively high percentage of communities in each zone have signed the contract, or are about to do so, the agents go back to training before beginning phase two. In the meantime, periodic assessments take place. They are useful, not just for record keeping, but to **bring about the necessary corrections to certain points of training** that may have been misinterpreted by the agents, and especially **to encourage exchanges** between the latter on the problems encountered and the way to face them.

rights and responsibilities of each partner (administration and community) throughout the project zone. Equally, between the beginning and the end of a project, developments occur that can confirm or invalidate certain principles of the project. They must therefore be taken into account in the training.

The agents are therefore trained in cycles, alternating training with implementing phases in the field.

Training and informing entails
- knowing exactly what is to be transmitted (knowledge);
- being able to transmit (know-how);
- adopting an attitude that allows transmission (presentation).

The project has therefore adopted a method of training that focuses on both theory and practice alternatively; classroom work and practical work require maximum individual and collective participation from the agents.

2.4 Accompanying and supporting the process

2.4.1 A rapid and significant improvement in the situation

The result of actions being implemented has been
- a spectacular decrease in broken pumps: 55% in February 1987 to less than 10% at the end of the same year;
- shorter repair times: only 5% of repairs took less than a week before February 1987, compared with 41% between the end of August 1988 and April 1989.

2.4.2 Defining indicators

The pump breakdown rate is an important criterion. The delay in repairing pumps provides even more interesting information for the agent, as the pump breakdown rate depends on the time it takes to have them repaired: the longer it takes, the more broken pumps there will be at a given moment. Short repair times indicate that
- communities feel the need to get their pumps fixed quickly, and to pool the necessary means for that;
- spare parts are available;
- local repair operators are efficient.

Conversely, long repair delays must make the agent think about the reasons for this.

In practice one can identify
- communities where the pump is considered as vital equipment, and which have solved their organisational problems and benefit from a favourable technological environment (repairing operators, shops for spare parts), and where repairs are carried out increasingly quickly;
- communities where one or more conditions necessary for sound maintenance are lacking and which show long delays before pumps are fixed.

2.4.3 Understanding the local reality

To have a better understanding of the factors that can extend the time it takes to repair a pump considerably, it is possible to make a diagram (see table on next page) demonstrating the procedure.

2.4.4 Identifying the obstacles

All the problems and delays are related to one of the three (inseparable and interdependent) pillars that form the maintenance structure (see above 2.2.3).

Sometimes, owing to an accumulation of obstacles in getting the pump repaired, community members feel discouraged, and may eventually abandon the water point. It is therefore important for the FORMENT agent to identify these obstacles.

From a faulty pump to a working pump: flow of actions and possible causes for repair delays

Actions	Causes for delay
1. Water committee meeting and decision to call a repairer	Lack of agreement in the community, conflict Members absent from the water committee Insufficient maintenance fund
2. Contacting the repairer	Poor accessibility and/or communication facilities Transportation costs Community members' unavailability (time constraints)
3. The repairer comes and diagnoses the problem	Poor accessibility Repairer not available Repairer not competent or without proper tools
4. Water committee meeting and decision to purchase the part	Lack of agreement in the community, conflict Members absent from the water committee Insufficient maintenance fund
5. Getting to the shop to purchase the part	Poor accessibility Transportation costs Community members' unavailability (time constraints)
6. Purchasing the part	Shop closed or part unavailable Costly part, requiring cash withdrawal or a cheque from a bank account (signatures) Price has changed
7. Warning the repairer	Poor accessibility and/or communication facilities
8. Bringing back the part to the site	Poor accessibility
9. The repairer comes and fixes the pump	Poor accessibility Repairer not available Repairer not competent or without proper tools

A. Obstacles emanating from community organisation

The origin of most problems can be found in *the absence of negotiation* prior to drilling for the borehole, meaning prior to the initiation of the FORMENT project.

Although the following list of problems is not exhaustive, those most frequently encountered are that
- a modern, fully equipped waterpoint may well not be a priority for the community;
- the location of the borehole may not satisfy the users, thus causing conflict regarding water use or management, or power conflicts (especially in the case of communities divided into many wards sharing one water point), or it may simply be impossible to use (a borehole situated in a lowland area may flood during the rainy season);
- the technical solution proposed (drilled borehole with a manual pump) may not be considered reliable because of past experiences. Often the result of such a situation is that insufficient contributions are made to the maintenance fund for repairs, and more generally that the atmosphere is not conducive to making decisions.

B. Obstacles originating with repairers

Generally repairers themselves do not give rise to problems except for *their transport*. Many repairers working in the FORMENT project *make use of public transport*, the frequency and reliability of which changes with the destination, to the point of sometimes being non-existent.

A repairer is, on average, in charge of 15 pumps. In many places, a pump represents the first if not the second water supply project in the community. As a consequence, in 1987, the pumps were very far apart obliging the repairers often to travel long distances (85 villages out of 245 were more than 20 km from a repairer).

Through the repairer's skills and equipment the majority of breakdowns can be successfully managed with the exception of those at level three that are rare (see above 2.2.3) but can become more frequent when the pumps these start to become old.

C. Obstacles related to the spare parts network

This pillar of the system is by far the most difficult to organise. The decision to sell or not to sell spare parts in a particular place is not made by the FORMENT agent, nor the communities, nor indeed the repairers, but by external decision makers (pump suppliers, commercial networks) situated far from the sites. The only people that can influence such decisions are the administration, the regional authorities, or sponsors involved at national level.

This pillar of the system, owing to its deficiencies, can stop repairs from being carried out for many months. Out of 26 broken pumps on 31 July 1989, at least 17 (roughly 65%) were not repaired *because spare parts were not available*. Similarly, the absence of spare parts is the main reason that more than half of all repairs take more than ten days.

Easy or difficult access to the shops of the supply network impacts on the rapidity of repairs.

2.5 Evaluating the elements of success

2.5.1 From a project to a national methodology

Owing to its success, further developments have occurred in the FORMENT project:

- Firstly, the project has expanded to all water points within the two regions covered by its activity.
- It has also been applied and adapted to the newly created structures in northern Togo (new European Development Fund project).
- Moreover, its principles have gradually shown their worth in other geographically close projects carried out by the United Nations (new infrastructures) or the French Cooperation (rehabilitation).
- Lastly, it has enabled a national policy to be developed based on
  - preliminary and continual discussions between partners involved in the setting up of water points (administration and project management on one side, and rural communities on the other);
  - the promotion of decentralised and privatised maintenance structures.

In 1991, broken pumps accounted for around 5% of a total of 1 000 pumps (in the Central and Kara regions).

2.5.2 A relevant concept, flexible administrative practices and an enabling climatic, socio-economic and institutional context

If the relevance of the principles leading to the realisation of a project are not to be undermined, the conditions in which they are applied are nonetheless important. It is therefore important, before concluding, to recall *the elements of success*. There are three types:

- Those that neither the project, nor the administration can influence, not rapidly in any case; that is the general context, the environment or national policy.
- Those about which the project can make demands, the satisfaction of which depends mainly on the sponsor's and administration's positions and policies. We will call them administrative elements.
- Those depending directly on the project's conception.

A. Elements referring to the general and institutional environment

The following elements must be implemented or considered before a project can begin:
- A national policy clearly stating the principle of the decentralised maintenance of water points, under the responsibility of the users;
- Severe water shortages affecting many communities or villages during the dry season, which result in an increasing demand for new water points;
- A communications network extending across the whole socio-economic area, which enables rapid contacts to be made;
- Enough local motorised equipment to favour the installation and activities of small-scale mechanics;
- A supply network of shops for spare parts able to supply communities with half the total number of pumps;
- The absence (or the existence) of other community water supply schemes in the same territory. This obviates the making of different, sometimes contradictory, speeches on the nature of both the rights and the responsibilities of the different role-players.

B. Administrative elements

Several key conditions must be mentioned:
- There should be one dedicated administration responsible for the project (Department of Hydraulics and Energy), with, as a consequence, a certain flexibility at operational level in the field, and increased simplicity in the decision-making process at a higher level.
- The recruited contractual agents should be selected according to criteria relevant to the projects' activities and should reinforce the administrative team in place.
- The FORMENT agents should remain in their positions throughout the project. They are attached to the Department of Hydraulics and Energy. Each is sent to live in an area he/she knows well and where he/she masters the language. It has been noticed that, generally, the majority of people do stay in their positions. *There is therefore little loss of expertise or experience.*
- The FORMENT agents have benefited from intensive training.
- Their (the agents') intervention makes use of a methodology based on negotiation and concrete arguments.
- The lessons from the field are taken into consideration and enable the methodology to be readjusted (by this we mean modifications to the way contributions are made, creation of a spare parts shop in the village, innovations leading to more adapted tools for local repairers, etc.)
- Generally, great care is taken *to set up sound implementation.*

2.5.3 Conclusion: motivated role-players

The Togolese case is not isolated.

For all the development operators and promoters of water supply programmes, the supply of drinking water in sufficient quantities is one of the main priorities as far as rural African communities are concerned.

This assumption is more easily made for regions in the Sahelian strip than for those situated further south. Although *easy access* to water is a permanent or temporary concern for many villages, the fact that it is drinkable is not a motivating factor.

Despite this physical need experienced by millions of rural people (and more especially by women), many village hydraulics projects initiated between 1975 and 1985 in western African countries have resulted in massive underutilisation of equipment owing to maintenance problems.

Maintenance firstly depends on socio-economic actions before being a technical problem. In order for collective goods or a facility to be maintained, the community set to benefit from it should have desired it. That is to say that it is one of their priorities. If this condition is fulfilled, then the problem concerning the technical environment of the equipment can be raised. Although the implementation of a facility or equipment may depend on political motivations, its durability depends on the decisions taken by each of the role-players, and such decisions depend on the interest people have in the project (see figure below).

If the technological environment is reliable, the technical solution proposed can be accepted, provided the equipment concerned is one of the village priorities. The villagers then frequently call on both local craftspersons and spare parts suppliers. The market is expanding for both the latter and their performances and services are improving. Innovations are therefore becoming more compatible and reliable and consequently are better accepted.

Chapter XII: Financing local development
Y. Fournier, D. Gentil, S. Perret

The following chapters are connected to Chapter XII:

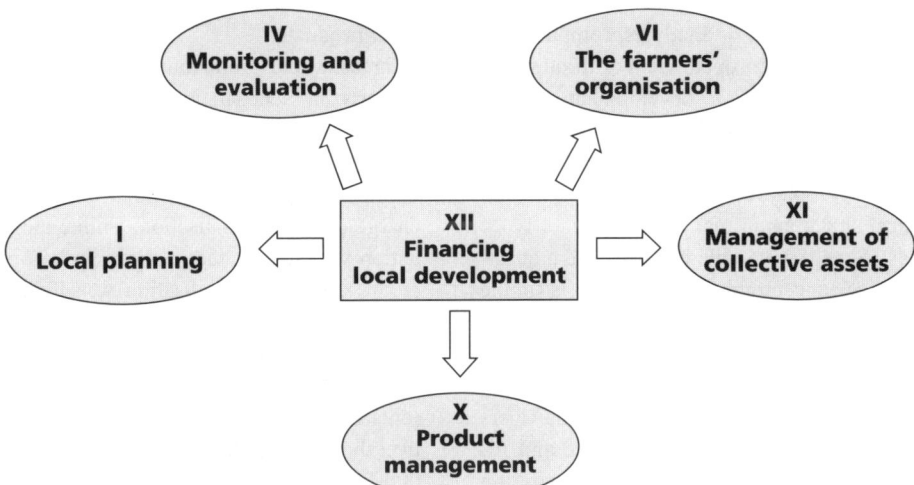

Any local development process needs financial support. The nature and means of this support are numerous, as well as controversial. This chapter strives to present the different experiences, pinpoint the major issues and draw some methodological orientations from them that seem applicable to most situations.

1. Typology of rural credit systems and history in western Africa

1.1 Agricultural credit through national banks and credit funds

Attempts to set up "modern" agricultural credit are actually as ancient as the colonial era. These attempts consisted of duplicating French models of agricultural credit funds in the colonies. These systems took the form of massive and easy credit provided by local companies during the 1910s, or of mutual funds (solidarity collateral) as of 1956 in Cameroon.

> The main problem of credit in Africa has always been collateral, since the usual guarantees existing in Europe (land, residential or industrial buildings, livestock, stocks of goods, heavy equipment and machinery, etc.) are hardly applicable. This has accounted for most of the failures of classical banking schemes. Therefore, other forms of guarantee (moral guarantee, solidarity) have been resorted to.

With independence, development banks were created though they were not interested in agricultural credit (despite offering traditional services supporting some commercialisation campaigns).

National agricultural credit banks and funds were then created: CNCA in Niger (1967), CNCA in Togo (1967), BNDA in Ivory Coast (1968), CNCA in Benin (1975), CNCA in Burkina Faso (1980), BNDA in Mali (1981), CNCA in Senegal (1984).

These banks granted loans either directly to the beneficiaries (individuals, companies or farmers' groups, co-operatives) or through development projects or development companies. A review of these different experiences in West Africa in the early 1990s highlighted a variety of problems: bankruptcy (in Niger) and critical situations in many banks (Togo, Ivory Coast, Senegal), volatile balances and a quasi-exclusive concentration on cotton areas (in the case of Burkina Faso and Mali).

The causes of these situations have been numerous: loans granted according to political rather than technical criteria, project managers lacking professionalism, burdensome and ill-adapted procedures, exaggerated operation costs and salaries in the banks, over-centralisation, inadequate strategies in the event of low-yielding crops and certain farmers attempting to dodge repaying loans.

> Farmers show solidarity (as groups) and for several reasons farmers often see the government and not banking structures as the predator. Thus, it becomes socially acceptable not to pay back loans and therefore difficult for banks to recover their money.

Two situations where recovery rates have been high are worth mentioning:
- In cotton areas, the loan recovery rates have reached nearly 100% over the past 30 years. This exceptional situation can be explained by the organised and monopolistic nature of the commodity chain (from seed supply to processing and sales on the world market) and by a sound knowledge of the producers and their needs (either directly or through village associations). A direct levy at the commercialisation level is very efficient since the debtor can sell only on a controlled basis through the cotton company. Any attempt to bypass the system (for example selling elsewhere) results in the cessation of further supply and credit for the farmer from then on. Such conditions do not exist for other products.
- In areas where groundnuts are grown, particularly in Senegal, integrated credit systems worked well on a large scale for more than ten years up until the 1980s. Credit was granted to co-operatives, which were also in charge of commercialisation. Unlike the situation with cotton, the relationship between credit and commercialisation was indirect. At the end of each planting season, profit and loss accounts would be issued by the co-operatives, showing benefits and arrears (outstanding payments). Should a co-operative show a 300 000 FCFA profit and 100 000 FCFA unpaid, the bank where both accounts were held would consider only 200 000 FCFA as profit. Shareholders of the co-operative would then be liable and share together the loss incurred owing to outstanding payments (principle of joint and several guarantees). This guarantee was not merely a moral guarantee as the profit resulted from commercial activities. Unfortunately, this system was abused and fdeteriorated. Local influential people and board members systematically covered the loss with ever-decreasing benefits to the shareholders. For the system to work, profit would have had to have been much higher than the amount of outstanding payments. If the latter was high (owing to a bad yielding season, or the farmers' lack of willingness to pay back loans), an overall non-payment movement was quickly triggered along with a series of rebates and moratoriums, which ultimately resulted in bankruptcy (Senegal in the 1980s).

> **Transparency and strictness may help:**
> In Niger during the early 1970s it was clearly demonstrated that abuse of the system is avoidable if there is transparency and strictness in the accounts. Equally, the level of solidarity (joint and several guarantees) must be socially manageable (one village or one ward, and not several distant communities involved in one co-operative). A total blockade (and not simply a reduction) of the shareholders benefits must be implemented until repayments are fully recovered.

> **Solutions that can make the situation worse:**
> Rebate is a suppression of part of the debts; for example, it is a consequence of a bad yielding season. Usually, if the season which follows is mediocre, the farmers will demand further rebates again and will stop repaying (payments strike).
>
> Payments may also be postponed to the following years (moratorium). Often, farmers facing regular payments plus moratorium payments are inclined to think that they will never be able to pay the entire debt. Often they will not pay at all, as they may be facing other problems. There is always the hope that the government will get tired and decide on a general write-off. Psychological and political issues as well as the concept of threshold are important in the field of credit.

1.2 Agricultural credit by the projects

Projects or regional development structures often play an important role in rural credit, either as intermediaries for the national banks and credit funds, or as independents (with the aid of external funds). The problems discussed earlier do occur, and are often aggravated by a number of factors:

- Lack of competence on the part of project managers (who are usually agricultural operators interested in credit distribution rather than in loan recovery).
- Lack of independence on the part of the credit offices vis-à-vis extension and supply.
- Lack of concern about the long-term sustainability of the system. The objective is to distribute inputs and equipment and then exhaust a fund rather than to find a financial equilibrium. With such a perspective low loan recovery rates are not a concern.

1.3 Savings and credit co-operatives (COOPECs and credit unions)

These institutions have different names: "*caisses populaires*" (Burkina Faso, Cameroon, Zaire), "*caisses rurales d'épargne et de prêts*" (Côte d'Ivoire/Ivory Coast), "*banques populaires*" (Rwanda), and so on. In western Africa COOPECs started to appear during the 1970s. They were introduced earlier in eastern Africa (known as credit unions).

New local co-operatives were created during the late 1980s (Mali, Burundi, Senegal, Congo, Guinea) in response to the deterioration of centralised systems for agricultural credit (CNCA, BNDA). International sponsors also became increasingly interested in alternative ways of organising savings and credit.

The use of such alternative systems have demonstrated the following:

- Promoting savings has been successful: despite past scepticism concerning the possibility of promoting monetary savings by smallholders with limited revenues, all experiences have proved that savings activities have been significant.

- There is a diversity of needs:
  - The need to set some money aside to protect oneself firstly against hazards and losses (theft, fire, termites, etc.), and secondly against excessive personal expenses or external demands (parents, friends, neighbours), which are is easier to resist when no money immediately available
  - The need for available liquidity to face life's necessities (disease, death, ceremonies, etc.)
  - The need for small loans (house improvement, social needs, inputs for production, etc.)
  - The need to earn interest on this saved money, even though such remuneration is often modest
- The link between savings and credit varies according to experience: credit can be quasi-simultaneous with savings, or it can come only after a long preparatory phase (often 1 or 2 years). It can represent only a small part of savings (10 to 20%) or conversely more than 50%.
- Loan recovery rates have been satisfactory (often higher than 90%), although there have been increasing delays and outstanding payments (as well as embezzlement, as has been the case in certain co-operatives in southern Cameroon) even when it concerns the farmers' savings.
- Requests for credit often prioritise social needs (housing, health care, education), rather than productive investments.
- Co-operatives with a non-homogeneous social base, where there is a tendency to use the farmers' savings to generate credit for local businesses, often result in outstanding payments. Farmers generally ask for small loans.
- A large part of savings can ultimately be injected into classical banking systems, which

Popular banks of Rwanda

These are a typical example of a successful savings function. About US$ 30 million have been saved, and about 150 popular banks created in 25 years and more than 33% of all Rwandese families are involved.

This success can be explained by

- an enthusiastic welcome by the population to the creation of popular banks in 1975 (people being very interested in a secure saving system);
- a favourable institutional environment: the state, the central bank and the administration supported the action and the approach used without interfering, which guaranteed independence and management autonomy.

In addition, some simple methodological principles were applied:

- The first banks were created in a region with a history of co-operative development (church initiatives, local savings co-operatives, etc.).
- Some simple principles of management were applied, which promoted the dissemination of information, and the participation of the people in the elections by the general assembly of an administrative body (management board), the control board and a local manager (from the area).
- The practice of co-operative democracy was implemented with the support of the central bureau of popular banks (promotion) and the union of popular banks.
- There also was close system of coaching and control by regional representatives and decentralised development units as well as strict management (monthly and quarterly financial statements to be submitted to the central bank).

favour urban areas. The money is rarely injected back into the rural economy through credit.
- This type of organisation is often proposed from the outside. Base members tend to lose their control, while only economic criteria are applied along with an increasing gap between the central organisation and local groups.
- Training often favours a transfer of models rather than support for the invention of new types of organisations.
- It is difficult for COOPECs to achieve financial equilibrium, even after 10 or 15 years of operation. Although commendable in that there is more autonomy vis-à-vis external funds, the search for such balance may have undesirable effects such as limitations to both training and control costs, and a search for security with non-risk investments and loans.

The COOPEC model also raises some concerns:
- National organisations (unions) are costly and heavy.
- There is a tendency to favour development at top levels, for example, through the creation of a federative bank which is actually a commercial bank.
- It seems difficult to recycle local savings into credit which would benefit local projects.

1.4 Informal finance in rural areas

In the absence of organised credit systems, rural people resort to two main types of other systems:
- Informal interpersonal credit systems through family and friends (low or no interest rates, based on kinship or neighbourhood solidarity and reciprocity), or through moneylenders or loan brokers (quick, short-term credit, high interest rates, up to 300% per annum), credit merchants and traders (goods supplied on a credit basis), pawnbrokers (material collateral basis) or landlords (credit for production inputs to a sharecropper).
- Various systems of rotating savings and credit associations (ROSCAs), savings groups, "tontines". Such systems have been widely developed all over the continent with many variations.

ROSCAs (known as "tontines" in western Africa, "stokvels" in South Africa) are especially popular among women. Various forms and levels of contribution exist (from US$ 0.2 per week, up to more than US$ 1000 per month, as seen among Bamileke merchants in Cameroon). The system relies on reciprocity: each member contributes on a weekly or monthly basis, at a fixed date, and each member receives alternately (fixed, or after a toss or a draw) all the contributions. Other systems consist of a money keeper collecting and giving back money on an individual basis, according to his needs.

Despite a certain willingness to incorporate ROSCAs into the COOPECs, it seems preferable to let both structures function separately. According to the farmers, they provide different services and are subject to different operational rules.

Savings groups are also very popular in developing areas. A group will be formed with a common objective (be it ultimately for individual or communal purposes). Funds are deposited with a money keeper or directly in a bank. Communal purposes include building a clinic, a school, water services, and so on. Some groups may even invest collectively in a business run by the group or by an appointed manager. Savings groups also provide the mutual motivation to save, something which may be difficult to do as an individual.

Burial societies are very popular in rural South Africa (Coetzee, 1993). In certain communities, almost every household belongs to a burial society, which uses the money to cover burial costs of members who die.

1.5 Example of the Grameen Bank (in Bangladesh) and its application to Africa

The Grameen Bank experience in Bangladesh is well known. Its basic idea is to support rural populations, especially the poorest of the poor with an emphasis on credit rather than on savings.

The Grameen Bank system has managed to develop
- in a context of precarious livings conditions (low employment, rural small-scale trading and craftwork activities, meagre revenues, great dependence vis-à-vis local usurers, rural exodus, urban unemployment and poverty, marginalisation of women and regular natural calamities).
- through the initiative of a scientist working on action research based on local solidarity and organisation; the first objective was to create self-employment. Subsequently it developed into a banking institution for the poor, especially women (85% of the members); the system relies on the so-called "strength of the poor": solidarity, a vivid experience of the value of money, tenacity and the struggle for survival.
- with an approach which clearly departs from classical banking systems: no forms are filled in, no collateral is requested, there are no tills or tellers but a staff of well-trained agents work with the villages.
- based on a set-up of joint and several guarantee groups made up of five people who are socially and economically homogenous, who know and trust one another, and generally with men and women in separate groups.
- based on the credit activity at the outset and later on savings; the credit includes
  - individual productive activities (one year loan, at 16%) redeemed in 52 weekly payments; the interest is paid at the end (51st and 52nd payments).
  - collective productive activities (under the same conditions); the credit ceiling is multiplied by the number of members (but granted only to previous beneficiaries who have paid back their loans).
  - housing loans, for 12 to 18 years, granted only to good members.
  - emergency aid, in the case of disaster (floods) payable according to the same conditions as the productive activities.
  - a rescue fund (in the case of death, disability, unexpected problems), fed by a contribution of 25% of interest paid on credits.

Loans are first granted to two members, then to two more and finally to the fifth member (if the first ones have been paying back). The whole group examines the demands for loans and decisions are made collectively.

The economic impact of the Grameen Bank system is first measured by the magnitude of the credit activity:
- There are about 1 million beneficiaries (of which about 85% are women) from about 20 000 villages.
- There is about US$ 170 million in credit.
- About US$ 17 million has been transferred to the rescue fund.
- There is a high loan recovery rate (over 98%).
- The Grameen Bank system employs 6 000 persons.

At the beneficiaries' level, the impact is measured through
- the accumulation of productive capital by the households,
- increased income (20% to 50%),
- job creation,
- a decline in usurious rates and an increase in agricultural salaries owing to pressure for higher wages.

This success story should not lead to this system being merely copied in Africa. One of the reasons for its success is that the Grameen Bank system has developed its own type of organisation adapted to the very conditions of the country after a relatively long phase of experimentation. Beneficiaries form heterogeneous socio-economic groups, who experience rapid individual turnover rates (small-scale entrepreneurs) and pursue diverse activities, thus generating different financial needs. Unlike farmers, who are all bound to the same planting schedule in a given area, Grameen Bank group members do not have credit needs simultaneously, and do not all pay back at the same time. Thus risk is diversified across different sectors.

Recent experiments in Burkina Faso and Guinea have drawn their inspiration from this example. In Guinea, the following principles have been applied:
- A joint and several guarantee group is formed with five members: these members choose one another freely; they are from the same village and share a common socio-economic situation; women and men form separate groups; traditional authorities (village wisemen) have a controlling and supervisory function in these groups.
- Credit is based on the needs expressed and takes account of the multiple nature of rural activities (agricultural inputs, labour, small-scale businesses, etc.); it gives priority to economic criteria (e.g. profitability of the activities supported by the credit); each member presents and justifies his/her credit application before the group and the credit agents.
- It is based on rotating credit: two members are served, then two others and then the fifth member (who is the head of the group).
- There is regular and close monitoring of the beneficiaries both at their homes and workplaces;
- Management procedures are simplified and flexible (the credit ceiling may be increased if the previous loan was reimbursed without problems; medium-term collective credit may be granted after a successful short-term experience); monitoring-evaluation and dialogue with the beneficiaries determine adaptations.
- An interest rate higher than the inflation rate is fixed to preserve the value of both the invested capital and the credit.
- A 100% loan recovery rate is required, with strict adherence to due payment dates, otherwise no further credit will be granted.

1.6 Besides COOPECs and ROSCAs, what else?

COOPECs, ROSCAs and informal credit systems are not the sole instruments for credit and savings in rural areas. Other systems do exist as briefly described below.

1.6.1 Seed stocks as savings and credit systems at community levels

For the past 30 years in Madagascar, Niger and Mali, a successful collective savings system has existed through seed storehouses. The norm for an individual loan is the reimbursement of two bags of seed harvested for each

---

Given an initial stock of 100, and annual storage fees of 10, the following progression occurs:

First year: Seeds borrowed = 100; Seeds repaid = 150; Available final stock = 140

Second year: Seeds borrowed = 140; Seeds repaid = 210; Available final stock = 200

Third year: Seeds borrowed = 200; Seeds repaid = 300; Available final stock = 290.

bag of seed borrowed during the sowing period. However, farmers have collectively adopted a 50% interest rate (1,5 bags paid back for 1 borrowed). Even if a small part of this interest is used for storage fees (seed treatment, possible loss), the initial stock is tripled in three years.

In practice, part of the stock is used by the community for sowing, the rest can be sold inside or outside the community and the money is then invested for community purposes. This operation reduces the costs for borrowers and makes it difficult for money to be stolen (money is handled and stock usually located right in the middle of the village). This allows for collective investments and enhances the quality of the seeds (which are stored and treated under good conditions, with the possibility of stored seeds being renewed with improved seeds).

1.6.2 Savings and credit from grain banks

There is a wide variety of grain banks with diverse purposes (e.g. food security at community level), and with as many successes as failures. A grain bank is a community or intercommunity organisation that manages a stock in kind. This stock may be pooled through purchases among members, purchases from other areas with excess production or from external inputs. The stock is either resold or lent to members during harsh periods (e.g. dry season, low-yielding years). It can also be kept in total or in part to form a safeguard.

In all cases, an important factor of success is the adoption of a differentiated policy, which depends on the results of the growing season (surplus or deficit). To generate revenue, it is essential to buy the products at the beginning of the harvesting period so that the bank may benefit from the difference between the price at that time and the increased prices during the "hungry gap" before the next harvest. Then loans in the form of bags of grain can be considered throughout the year, with the option of having different interest rates depending on whether it is for a member or a non-member.

1.6.3 Initial donations of capital to farmers' organisation

In situations where it is impossible to initiate savings or where loan repayments are uncertain, the solution may be an initial donation (initial capital), in the hope that this will reproduce itself in the long term.

> An example in the Gao area (Mali) after the harsh 1973 drought that dramatically affected fishermen's activities:
>
> A consortium of NGOs released capital in successive packages to fishermen's co-operatives in order for them to re-establish their means of production (pirogues and nets). All modes of credit as well as borrowers' choices (nature of credit, period of repayment) were carefully discussed and decided on in a general assembly. The first package was earmarked for about 20% of the fishermen. The second package was granted only after the first group had paid back 100% of their loan. There was therefore social pressure from the group of future borrowers, which in turn resulted in a high loan recovery rate. When all the fishermen interested had been served, the money from the repayments was used in different ways depending on the co-operative (e.g. to set up workshops for the manufacture of pirogues; to buy stocks of grains; to set up a shop for the commercialisation of salt fish; to buy a motorised pirogue for faster transport of fresh fish to the nearest towns; etc.).

The World Bank has financed small-scale rural projects (e.g. FONADEC in Senegal) through initial donations (e.g. a motor pump for a small-scale irrigation scheme). In this case a bank account is opened and managed by the farmers. Farmers' payments are calculated taking account of the equipment's maintenance and renewal. The basic assumption, based on past experiences especially in Senegal, is that farmers are not very motivated to pay back any parastatal organisation or development project. Conversely, they are eager to contribute to their own accounts, especially when it comes to protecting their own activities and making them sustainable in the long run.

When inter-community solidarity exists, an alternative solution may be a rotating system whereby the repayment of the first investment realised in the community serves to finance a new investment in a neighbouring community. This principle has been adopted with the mills in Yatenga (Burkina Faso).

In Niger, agricultural credit to co-operatives has been initiated by USAID (USA cooperation and aid agency) and other sponsors. Their approach combines credit, and training and support to farmers' organisations through low initial capital. Other organisations combine subsidies on credit for initial investments with compulsory savings. These approaches have proved relevant and efficient in certain situations.

1.6.4 Development funds

The objective of such funds is to promote savings at the village level and to make use of the savings in productive community investments. The community members themselves decide on the modalities of the loans. Another objective is to get the community acquainted with local financial systems and then to promote further self-managed systems for credit, savings and investment.

> One village development fund is located in Segou (Mali) and supported by IFAD (International Agency for Food and Agriculture Development). Each village was requested to pool and to manage a fund to be used for the community's needs and to serve as a guarantee of the debt's recovery.

Village people can use the credit for diverse activities, for example, for cash crop fertilisation, improvement of small stock herds, domestic industries, rural small-scale industries such as forges, handicraft activities, woodworks, and the like. The high percentage of loan recovery (104% in Segou, Mali, which means a total loan repayment plus additional contributions) shows the extent to which the villagers feel responsible for the village funds.

In many cases, small shops have been established to generate additional funds. IFAD's long-term objective is to make these shops into real development centres on the community level, where farmers can receive production inputs, credit and technical advice.

These are but a few examples showing that, over and above the official structures for agricultural credit, there are many modalities for financing rural development; savings; credit; subsidies; contributions in cash, kind or labour; investment; ROSCAs and the various combinations of all these modalities. Secondly, experience shows that in each situation the farmers and rural people should lead the discussions and make the decisions themselves.

2. Some issues for debate

Section one has clearly shown that there has been more failure than success in the field of rural credit. There are of course some success stories but they are either linked to very specific situations (cotton) or they are too recent to allow conclusive lessons to be drawn. Many points remain controversial, and different practices with different philosophies can still be observed. The views expressed here are not unanimously shared.

2.1 Agricultural credit or rural credit?

Up to now, official programmes which are linked to different operations undertaken by various governments have been preoccupied with agricultural credit in its narrowest meaning, that is with supporting access to inputs, seeds, agricultural equipment, and so on. When a free choice is given to the farmers, it becomes obvious very quickly that their credit priorities concern a wider range of activities: small stock production; fattening and feed-lotting of pigs, sheep, goats; domestic food processing; local trading; handicraft; storage; labour hiring; spare part supply; and the like.

If one wants to respond to the farmers' needs, thereby increasing the diversification of activities by rural people, financial support should undoubtedly not be limited to agricultural credit. Free choice and direct involvement by the farmers helps to prevent misappropriation, complicated procedures, low recovery rates, and mistrust of credit monitoring.

2.2 "Hot money" vs. "cold money"? Is the mixture possible?

"Hot money" is money which comes from the farmers who control it and which deserves both attention and social control. The rural people look negatively upon and even apply social pressure or ban people who fail to pay back a fellow farmer or the local farmers' cooperative.

"Cold money" comes from the government and external sponsors. "Cold money" is often stolen, misappropriated, or not paid back. The communities do not necessarily see this in a negative light.

The social value granted to both types is obviously different, as are the ways to manage them. These distinctions explain the significant differences of loan recovery rates between the COOPECs or ROSCAs and the official credit institutions.

The mere idea of farmers putting their savings back into the rural areas may result in improvements. However, this is not sufficient to respond to all the needs, especially those of the most disadvantaged with low savings potential (resource-poor farmers, women and youngsters). Financial input from outside is almost always necessary. This input can have an initiating and multiplier effect on savings. The most important thing is that this input should not be perceived as easy or cold money. The farmers' involvement in the setting up of the system (selection of the borrowers, modalities of repayment, types of guarantees, etc.) is critical to the sustainability of the system (through high loan recovery rate).

2.3 Savings and credit: where to start?

This is another way of asking the question of "cold" vs. "hot" money. It is undoubtedly better to start with the farmers' savings, insuring that farmers learn about and get involved in rural finance, that they rely on their own strengths and become autonomous vis-à-vis external intervention. However, this must not become a dogmatic rule.

In some instances (i.e. a climate of mistrust, high inflation, etc.) and for certain social categories (women, resource-poor farmers, youngsters), credit services (or a combination of credit and savings) may have to be provided. The most important points still remain. First, the farmers must be included in the discussion of the objectives and the decisions concerning the modalities of credit with potential beneficiaries. Second, they must participate in the decisions about loans, so that they fully accept the system and its rules.

2.4 Individual or collective credit?

Besides the specific cases of collective investments (e.g. mills, grain banks), credit can also be granted to individuals, although in most cases a group will guarantee the loan. The bank, the project or the NGO grants credit to a group which usually corresponds to a village, a ward, a community or a farmers' association, which is in charge and controls the granting and recovery of individual loans.

Funding provided in this manner relies on mutual trust between the farmers and the local democracy. Problems may occur such as the local, influential persons and their relatives or friends having first access to loans; the misappropriation of repayments; difficulties in exerting social pressure successfully which may lead to the risk of local conflicts; and low loan recovery rates. In such situations, external financial institutions often impose a joint and several guarantee system, which makes the work easier for them. However, such systems generally do not correspond to the desires or to the social and managerial capacities of the farmers, particularly when the groups are big. "The dog that swallowed the egg is the one that must be beaten" say the Haoussas people in Senegal.

Again a sound approach should avoid imposing pre-established, uniform schemes. Several formulas should be proposed and their advantages and disadvantages considered. There can be guarantees within a COOPEC through individual collaterals (e.g. the possibility of freezing savings as seen in Cameroon). Alternatively, a borrower can offer as a guarantee his/her turn in a rotating savings and credit association (as seen in Benin). In other experiences, small groups of five members who are in turn savers and borrowers have proven to be sustainable (i.e. the Grameen Bank in Bangladesh, and similar groups in Guinea).

2.5 Productive vs. unproductive credit

Often the tendency is to prioritise productive credit and to distrust unproductive credit (e.g. hunger-gap aid, social credit and credit that supports commercialisation). In fact this tendency is not always relevant, for several reasons:
- The boundaries between unproductive and productive credit are not clear and the farmers express both economic and non-economic needs (credit for production and credit for social needs).
- Most attempts at giving agricultural credit based on offers predetermined by development operators without accounting for the farmers' choices and strategies have failed.
- The implementation of unproductive credit (for social needs or consumption) can be considered acceptable if it is known that the beneficiary will repay it with the revenue derived from profitable agricultural production. Such credit may actually stimulate agricultural production.
- If unproductive credit is discarded the farmers might resort to usury to face their social obligations. The farmers have to pay higher interest rates which makes them dependant (the whole harvest may have to be given as a guarantee to the lender or trader).
- Loans do not always support a sole investment by a farmer; they often cover two or three activities, thus allowing the farmers to minimise risk through diversification (animal husbandry, crop production, off-farm activities, etc.).

2.6 Credit to the rich vs. credit to the poor

Contrary to common belief, rich clients are not always the best payers and low loan recovery rates are found among civil servants, traders and local notables, as well as the poor people.

Can one target only the poor (as in the Grameen Bank experience)? The social and economic structures in Africa differ from those in Bangladesh. Discrimination among potential rural borrowers is not acceptable. Moreover, the poverty criterion (or threshold) is not easy to determine when farmers have access to land and own animals. Discrimination would deprive many farmers who have no credit alternatives.

It seems preferable to fix an upper limit for credit (loan ceiling) (e.g. between US$ 100 and US$ 200). Such credit is not likely to interest big, potential rural borrowers. Categories of people that are often excluded from credit (especially women, resource-poor farmers, small-scale entrepreneurs, etc.) would qualify for this type of loan.

2.7 Interest rates, the cost of credit, and the search for a financial equilibrium

The majority of development operators favour low interest rates for the agricultural sector and higher rates for savings. This viewpoint has been reinforced by the regulations of the central bank. Unfortunately, such an approach seldom allows for the system to be sustainable, and it is the reason for the many failures which occur soon after external funding stops (subsidies). Sound management of credit and saving systems for smallholders remain costly, even when the procedures are simplified and when farmers contribute through unpaid labour. Transaction costs are numerous and high owing to the small but numerous operations being treated, the numerous trips, the risks, the time spent on negotiations, the discussions, the information, the training, and so on.

Observations and past experiences show the following:

- The necessity for very low interest rates has never been confirmed during discussions with the farmers. When involved in setting up a credit system, farmers compare the offers with the usurers' rates (100% to 300%) and consider as acceptable any interest rate lower than what they are used to.
- Two credit systems may coexist, different but complementary, as illustrated in eastern Senegal where a cotton growers' association benefited from
  - a formal credit offer, initiated by the cotton processing company SODEFITEX (credit for input supply and animal traction), which was challenged by the farmers owing to ever-increasing input prices and interest rates for medium-term credit for animal traction being considered too high (15%);
  - an informal credit offer initiated and organised by the growers' association (financed by the cotton tax, pooled and collectively managed by the association) which supported social and consumption needs at the annual rate of 25%.

In addition, most credit granted is planting season credit (less than one year) which mobilises small nominal amounts for a limited time (usually from 6 to 8 months). Hence, payments are small, with few variations whatever the interest rate. The major concern of farmers is to access

---

Towards sustainable local financing structures

The profitability of a local credit and savings structure is determined as follows:

Gross bank profit = income – expenses

Bank income = interests on the loans + interests on investments + commissions or fees charged to clients

Bank expenses = clients' remuneration

Net profit = gross benefit – standing operation costs – variable costs (risks)

The standing costs are very high owing to the high transaction costs they include (numerous operations with an average deposit/credit operation amount of US$ 40). The variable costs may remain low and constant as long as the loan recovery rates are high.

From the above and from past experiences, the following factors should be considered:

1. In terms of management costs:

- Relying on benevolence, unpaid labour
- Simplifying management to the extreme
- Remunerating clients through savings rather than through quasi-free credit, and granting credit at acceptably high rates
- Maximising profit from available cash and funds (through investments)
- Limiting the role of external organisations with regard to cash management
- Introducing cross-control procedures among the operators

2. In terms of risks:

- Maximising the solidarity of the community associations
- Letting the farmers' groups decide on credits
- Avoiding granting credit to civil servants and local big traders
- Training farmers in active management and risk evaluation

To solve the tricky challenge of long-term equilibrium and sustainability, one should have a banking spirit, and always bear in mind that every cost incurred, every risk taken will eventually impact on the beneficiaries who might then discard the system if they deem these charges to be unexpectedly burdensome.

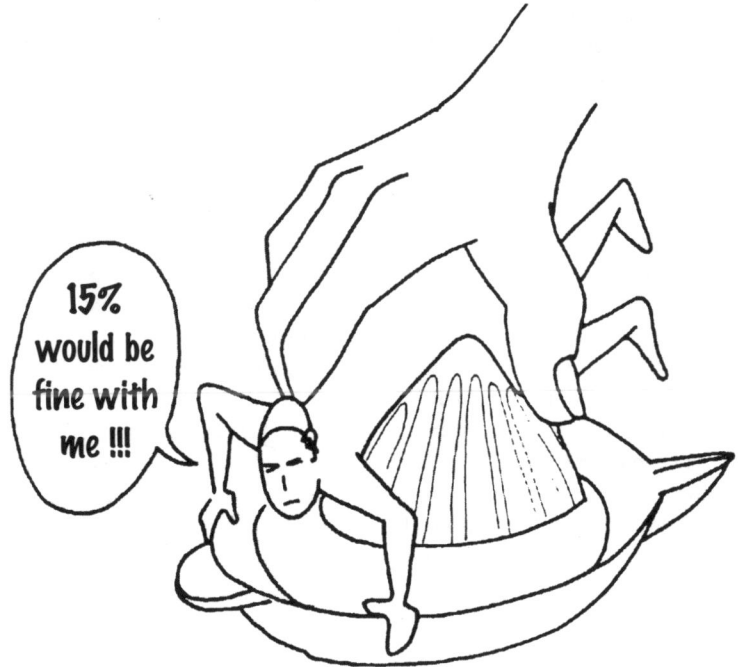

Rural people are used to high annual interest rates (beyond 100%).

credit under conditions that leave them room for choice in its utilisation.

It is necessary to set up a balanced system with the farmers (with a five-year perspective for instance). One can simulate with them the consequences of a variety of different factors (amount of savings and credit, recovery rates, management and transaction costs), using a simplified income statement for instance. The analysis can focus more on the differences between savings and credit rates than on each rate separately.

A reasonable differential is about 10%, which often supposes that the credit rate is higher than 15%. For farmers, such rates (ranging from 2% to 3% per month) are acceptable since they are used to monthly rates of 10% to 20%.

2.8 Centralisation vs. decentralisation

Many problems facing credit systems are linked to excessive centralisation, that is long procedures, expensive overheads, gaps between the decision level and the implementation level, gaps between the COOPECs and their local groups, decentralisation of decision making of both local agents and farmers.

One must therefore get the grass-roots structures more involved, and have them take more responsibility for decisions on granting credit and recovery, and for personnel and financial equilibrium. Structures must be geographically and socially close to the beneficiaries (both savers and borrowers) who will in turn consider them as their own concerns. However, these grass-roots structures must also benefit from regional and national support, especially in the fields of training, control and monitoring, in representation vis-à-vis the authorities and in negotiations with the official banking structures.

Part of the support costs must be paid by the base structures, while external financial support (e.g. on training) may temporarily be important. In any case, these services should be implemented by light structures (e.g. a tenth of staff members at national level) and must truly serve the grass-roots structures, and not become a steering and burdensome hierarchy.

3. Methodological orientations

The diversity of experience and debates shows that there is no universal formula relevant for all situations or one model that takes priority. However, there are some constant elements.

3.1 Several basic principles about rural credit

- Rural credit must respond to the real needs expressed by the farmers (the credit offer must coincide with the demand), and should match the realities and constraints of the rural environment and of the potential beneficiaries.
- Credit should be linked to local savings; experience has demonstrated that credit is better recovered when it mobilises the beneficiaries' savings.
- Credit does have a cost. This is reflected in the interest rate which must include the real costs (e.g. cost of the resource, management costs, cost of the risks incurred), and which is necessary for the system to be reproducible, that is sustainable in the long run.
- Specific agents must carry out credit monitoring and management. They must be trained and be able to make a proper diagnosis. Such a diagnosis should rely on observations and regular dialogues with the beneficiaries and result in procedures being adapted if necessary.
- The structure in charge of credit operations must be autonomous from other development operators (agricultural extension, input supply, etc.). It must also be psychologically and physically close to the beneficiaries (without intermediaries). It must seek the

active participation of the people concerned.
- The state and its administration should support the initiation of actions and contribute to their evaluation. But it should not interfere with the operations per se, since the farmers' trust in the system is critical and is more easily gained if the rural people do not feel that the government is intervening.

3.2 Elements of an approach

3.2.1 Starting from a situation diagnosis
Special attention must be paid to defining and ranking the financial needs of the farmers. This process should consider the local diversity (types of farmers) and include any other diagnosis related to other relevant functions. Discussion can lead to a critical analysis of past credit experiences by the farmers and of the current credit conditions (official or informal, interest rates, constraints in getting loans granted, recovery systems, guarantees, etc.). One should also try to analyse past and current ROSCAs' experiences.

3.2.2 Defining a strategy
This stage follows the diagnosis. Several cases may be considered:
- There is no existing credit system or institution in place. In such a case, a model should be progressively designed through experimentation and dialogue with farmers.
- There is a system or a rural institution of credit in place (e.g. national agricultural bank, other bank). In such a case, two strategies may be envisaged depending on the situation:
  – Setting up a complementary system of rural mutual savings and credit close to the farmers, which takes into account the needs which have not been satisfied (e.g. adapted Grameen Bank system);

  – Buying out the rural credit institution with or without mutualisation (such situations currently occur in Africa and Latin America) by farmers' organisations or by a decentralised savings and credit system; this supposes that the buyer has the necessary financial means or that he/she receives external support (e.g. a sponsor may hold the social shares, which will be progressively repaid).

The objective is to adapt the rural credit institution to the needs of the farmers. One must however keep in mind that it is often difficult to change the logic of action. Actually, such experiences have not proved successful. Even when farmers' associations become shareholders (e.g. FONGS in Senegal being represented at the National Bank for Agricultural Credit), such a strategy has been risky and uncertain since it is not easy to induce the changes that are expected by the farmers.

3.2.3 Setting up sound systems
Difficulties arise when a completely new system has to be developed, when a system is new in the area but exists somewhere else, or when a system must be altered or adaptations have to be made. Points for review and decisions are not that numerous. One of the most acute problems is that of guarantees (collateral). The following examples present a rehabilitation process in a co-operative for savings and credit in Benin and the setting up process of a new system inspired by the Grameen Bank in Guinea-Conakry.

Close attention must be paid to the procedures, the forms, the accounting system, and so on, in order to achieve simplicity and transparency in the system and to ensure the involvement of the local people. Finally, one should avoid reproducing the burden of heavy national structures while setting up vertical structures.

In Benin: an experimental credit programme in the regional and local agricultural credit institutions (CLCAM and CRCAM)

Financing credit:

At the farmer level, it has been clearly stated from the outset that money for credit should come directly from the farmers' savings and not from external sources. This creates complications when arranging the conditions for credit granting, the modalities for recovery and the nature of the guarantees.

A proposal to use external funds (same amount as the loan in demand) has been made. This external financing actually corresponds to a de-freezing of a part of the debts of the regional banks owed to the national bank (Borgou, Atlantique and Zou regions) and to the reconstitution of the equity capital (Mono and Zou regions).

As far as cash management is concerned, there is a need to grant the first loans in March (for the planting season). These can be covered either by new savings in the Borgou region or by recovering previous loans in the Zou region.

It is therefore necessary that external funds are made available to the different regions in a timely manner.

Modalities of credit:

- The loans granted at the local level are planting season loans (less than one year). However, in the Borgou area, the local board can be consulted by the regional branch about loans for animal traction and equipment.
- For the first year, loans are used for agriculture (labour, seeds and inputs), livestock (fattening piglets, poultry and kids), and the women's activities (small-scale trade, products processing, etc.).
- Interest rates are slightly below those usually practised in rural areas (reimbursement of two bags for one borrowed, 10% monthly interest rate, etc.). They must however cover the different charges of the local financial system (manager's salary, running expenses, savings recovery, etc.). Interest rates have been set at 15% yearly for the Borgou area, and at 2% monthly for other regions. These rates cover the interest rate per se, as well as a contribution to the guarantee funds and diverse fees for setting up and monitoring the file.
- Loans are granted exclusively to members, including individuals and co-operative structures.
- The different management levels have established a number of conditions for granting credit. Although these conditions differ according to the variety of structures at the local level (this pinpoints the importance of local boards), they generally have the following characteristics:
  - The application must be accepted by the board.
  - The beneficiary must be a member of the co-operative, with no outstanding debts.
  - The beneficiary must have saved a minimum of 10% of the value of the loan in demand (usually about US$ 15); savings are frozen until the credit is totally repaid.
  - A credit ceiling is set and strictly respected (generally between US$ 150 and US$ 600).
  - The beneficiary must produce sufficient guarantees, moral guarantees (seriousness, honesty, hard-working behaviour, etc.) as well as physical or financial collateral (guarantee by a cotton growers' association, personal guarantee of fellow members of the co-operative, turn of ROSCA, agricultural products, house rent, vehicle, etc.).

- The schedule and modalities for loan granting, monitoring and recovery have been established for each local structure (e.g. in cotton areas, direct levying is carried out by local managers when the cotton company CARDER pays the cotton growers' association). The manager and the board play an active role in each of these different areas. In the case of outstanding payments, the board should be involved (applying moral pressure and implementing the guarantees stated in the contract). Resorting to police or judiciary authorities must remain absolutely exceptional.

Implementation:

Procedures, formalities and paper work must be simplified as much as possible. Information about the conditions of granting credit must be distributed as widely as possible (i.e. general meetings). Public notices about these conditions must be displayed at the structure's offices and should include a version in the local language. Members' demands should be collected and submitted by the manager to the board (or to its credit committee).

Monitoring and evaluation:

The sound operation of the programmes is supervised by staff members at regional level, with the part-time support of a consultant. Regular reports (in June and March) should be written, summarising and then evaluating the results.

In Guinea-Conakry: an experience inspired by the Grameen Bank principles

Objectives, elements of procedure	Planned	Achieved
Nature of the credit	Profitable, diversified, agricultural / non-agricultural loans	Yes
Beneficiaries	Resource-poor farmers, men and women, crop and livestock farmers, craftspersons, anyone who wishes to develop current or new activities	Small and medium-scale farmers formed the majority of those interested. Commercial farmers, civil servants and big traders were not interested owing to the low ceiling
Ceiling	US$ 35 to 45 (depending on the local structures)	US$ 55 for all situations
Duration	Less than one year at the beginning	Yes (12 months)
Interest rate	3% per month on the remaining debit balance (the overall interest rate representing about 20% of the loan)	Yes (+ 5% for file management)
Repayment modalities	10 monthly payments, an 11th one for interest, a 12th one for guarantees (death, accident)	12 equal monthly payments, including capital, interest and guarantees (called a solidarity fund, representing 20% of the interest)

Objectives, elements of procedure	Planned	Achieved
Guarantee	Self co-opted groups of 5 to 10, with rotating presidency Local chiefs check the groups' quality informally The guarantee itself is the social pressure among group members in the event of outstanding monthly payments	5 members in all groups Yes, with complementary control by development operators Yes, there are 2 first beneficiaries, then 2 others after 2 months, and finally the president
Partnership share	Low amount, provided when submitting the application or at loan granting	US$ 0.5, levied from the loan granted
Monitoring and evaluation	Loans are distributed and recovered on each market day (once a week) Regular and frequent visits to the groups to check loan utilisation, and to take stock of activities and results Yearly assessment of loan impact and of the level of satisfaction	Loans were distributed for the first time in July 1989. Further distribution and repayments are to take place at markets or at certain convenient locations chosen by farmers Visits once to twice a month (at market places, farms, plots) Yet to be organised
Training programme	Explanations and instructions provided on the conditions for credit Distribution of documents in written Poular, in Arabic, in French	Yes, generally in 2-3 controlled training sessions

3.2.4 Mastering the system's expansion

Once the system is set up, one of the biggest difficulties is to resist pressure from the farmers, the government or the financial organisations to expand too rapidly, particularly when the first experiences have been successful. First, they must recognise that a successful start does not necessarily imply long-term viability, since many new systems tend to perform well in their first stages. Second, one should make sure that the conditions for long-term success are fulfilled (i.e. complete information about the rural people, training, quality of the staff, control capacity, etc.). It is advisable to establish a realistic plan for further expansion after two or three years of experimentation. This plan will be re-examined periodically depending on the results.

3.2.5 Monitoring and evaluation

A sound accounting system and a thorough analysis of the credit statistics and figures remain the basics for efficient monitoring and evaluation. The most important indicator of success is the loan recovery rate. Another critical viability indicator is the financial equilibrium of the local structures and of the system as a whole. There are other indicators which should be gathered, presented and discussed with the farmers.

As an example, it was proposed in 1984 (during a conference in Lomé) that several simple indicators be systematically collected from COOPECs, as a starting point for national and international reviews. The following indicators still remain valid (data must always be calculated in the same manner to allow for comparisons from one year to the next):
- Number of local credit fund structures and their development
- Number of members, their development (in absolute figures, and compared with the number of potential members) and distinction between shareholders and beneficiaries (users)
- Total savings, savings development, and contributions according to socio-professional categories, the amounts and the numbers of accounts
- Savings per member (in current and inflation-adjusted currency) according to socio-professional categories and to status (shareholders or beneficiaries)
- Savings compared with monetary revenues or revenues of the main commercialised products (in broad terms)
- Proportion of credit operations compared with savings operations (amounts and number of operations per category)
- Overall outstanding payment rate per loan type and per socio-professional category of the beneficiary
- Credit utilisation (type of activity) per category
- Self-financing of the local financial structures, of the whole system and the management regulations to reach it

Collecting simple indicators is but a starting point before the more thorough evaluation. It is important to
- identify and know the promoters of any rural finance experience (the members and the external operators);
- identify the promoters' real objectives (savings security, financial equilibrium, credit distribution, financing local development, etc.);
- acquire an accurate knowledge of the respective roles of the social groups involved in COOPECs or ROSCAs;
- measure the actual autonomy of the members.

These elements allow a real evaluation to be conducted with the members with whom possible new orientations can be considered and perhaps undertaken.

4. Conclusion

Financing the rural sector is becoming increasingly important. It makes it possible to reinforce the farmers' organisations and their autonomy, as well as to promote agricultural and non-agricultural productive activities, including processing, craftworks, and services.

Following the failures and poor results of both development banks and specialised banks (national banks of agricultural credit), two alternative formulas seem particularly promising:
- COOPECs (co-operatives for savings and credit) in which credit is generated only from savings previously collected by the members
- ROSCAs (informal or formal experiences of solidarity and rotating credit, which have existed for a long time in Africa (tontines, stokvels and also the Grameen Bank of Bangladesh) in which money may also come from external sources and where credit may precede savings

Other formulas also offer interesting possibilities (seed stocks, grain banks and village development funds).

Such diversity clearly shows that there is no universal panacea. In each situation, a system should be developed with the farmers after a sound analysis of the situation.

Two golden rules may be set:
- The system must be socially appropriate and adapted to the farmers.
- Financial equilibrium must be targeted, which often implies relatively high interest rates in order to achieve long-term viability.

5. Recommended literature

Compiled with the support of Prof. G.K. Coetzee, from the University of Pretoria, South Africa.

Adams, D.W. & Fitchett, D.A. (1992) *Informal finance in low-income countries*. Westview Press, Boulder, Colorado, USA.

Adams, D.W. (1992) Building durable rural finance markets in Africa. *African Review of Money, Finance & Banking*, 1 (1): 1–15.

Bathrick, D.D. (1981) *Agricultural credit for small farmers' development*. Westview Press, Boulder, Colorado, USA.

Bryceson, D. (2000) *Rural Africa at the crossroads: livelihood practices and policies*. ODI Natural Resources Perspectives, April 2000 (52).

Coetzee, G.K., Mbongwa, M.M. & Nhlapo, K. (1996) Restructuring rural finance and land reform financing mechanisms. In: Van Zyl, J., Kirsten, J.F. & Binswanger, H.P. (editors) *Agricultural land reform in South Africa*, Oxford University Press, Cape Town, SA.

Coetzee, G.K. (1993) The credit component of the Farmer Support Programme. In: Singini, R. & Van Rooyen, C.J. (editors) *Serving small-scale farmers in South Africa*. Development Bank of SA, Midrand, SA.

Dale, W., Adams, D.W. & Von Pischke, J.D. (eds.) (1984) *Undermining rural development with cheap credit*. Westview Press, Boulder, Colorado, USA.

Donald, G. (1976) *Credit for small farmers in developing countries*. Westview Press, Boulder, Colorado, USA.

Epargne sans frontiere (2000) *La microfinance en Afrique. Evolutions et strategies des acteurs*. Techniques Financiers & Developpement, 59-60 (July–October 2000), 148p.

Fenwick, L.J. & Lyne, M.C. (1999) The relative importance of liquidity and other constraints inhibiting the growth of small-scale farming in KwaZulu-Natal. *Development Southern Africa*, 16 (1): 141–155.

Huppi, M. & Feder, G. (1990) The role of groups and credit co-operatives in rural lending. *The World Bank Research Observer*, 5 (2): 187–204.

Miracle, M.P., Miracle, D.S. & Cohen, L. (1980) Informal savings mobilization in Africa. *Economic Development and Cultural Change*, 28 (4): 701–723.

Seibel, H.D. (1986) Rural finance in Africa: the role of informal and formal financial institutions. *Development and Co-operation*, vol. 6: 12–14.

Schoombee, A. (2000) Getting South African banks to serve micro-entrepreneurs: an analysis of policy options. *Development Southern Africa*, vol. 17 (5): 751–767.

Von Pischke, J.D., Adams, D.W. & Donald, G. (editors) (1983) *Rural financial markets in developing countries*. EDI Development studies, Johns Hopkins University Press, Baltimore, USA.

Chapter XIII : Women and development
A. Corrèze

Women and development is a general theme found in all other chapters.

1. Why a specific approach for rural women?

Although marginalised or ignored for so long by development policies or intervention schemes, over the past two decades women have had an important role to play in the regeneration of social groups and sometimes, because of an economic crisis, their role has been essential for the survival of the family and the society in which they live.

In many African societies, their status, based on their role as mother and spouse, has for a long time masked the role they have been playing in the economic field through their involvement in food production, transport, storage, the conservation and processing of agricultural products, marketing, small stock farming, artisan work, small and sometimes large-scale trade and to a lesser extent, except in urban areas, service and salaried activities.

Because women were seen almost exclusively in terms of their maternal and domestic duties, the action programmes earmarked for them were essentially based on education, child health care, nutrition, health, literacy, and so on.

Nowadays, important changes have taken place in rural society: there has been decentralisation and the empowerment of communities; accelerated monetarisation, and the expression of individual strategies (departure of dependants, of heads of the family, of school youngsters leaving behind social constraints; female-derived strategies, the increasing inability of urbanites and civil servants to face up to family solidarity, etc.).

Women, especially in rural areas, are strongly attached to the values on which their social status and identity are founded: this attachment is often noticeable in their anguished daily attempts to fulfil their children's needs. From this concern and the numerous activities undertaken to address these needs, an awareness of the importance of the women's economic input has emerged, firstly in the context of the family, but also within a general context. As a consequence of this development, important changes have taken place. Women have developed a taste for economic independence, and with this, feelings of inequality and unfairness regarding access to means of production, to training, to organisation of community life, and the willpower to get organised and have a say in matters concerning themselves.

These changes as a whole have led to a certain recognition of the many roles women play, both from the community itself and from external operators: national officials, sponsors, international organisations, NGOs, and the like. In the same way, in many cases these changes have enabled women to express their own views on the problems and constraints they face and the wishes they have for the future.

Women's opinions and the organisations which favour their active participation are recognised in both economic and social fields as being the necessary basis for a specific approach.

2. Some methodological prerequisites

If a specific approach to women's problems is necessary, it cannot, however, be dissociated from the general context of the society in which they live, the place they occupy and the role they play in it. Their power to innovate is determined by the social organisation of the communities, their history and the changes that have taken place. Women, just like men, are subjected to the same heavy constraints associated with the political and economic environment; these constraints may, however, have a different effect on women.

> In Burkina Faso, in the province of Yatenga, the exodus of men, accentuated by drought, means that many women find themselves acting as farm managers; they are therefore obliged to divide their time between the food crop, the small stock and many other smaller activities (artisan, trade).

Women are never a homogenous group. Discovering each woman's status and the differences resulting from this (regarding access to natural resources, to equipment and to work, to the availability of time for specific, individual or communal activities) is a prerequisite to any review of the contents, means of intervention and evaluation of a programme intended for women.

The starting point for a specific action regarding women must also be well identified:
- It can be a request for support emanating from the women themselves.
- It can be an initiative coming from local authorities (department for promoting women), from sponsors or from men wishing to see their wives and daughters "move on/ broaden their horizons" (training in hygiene, health care for children, etc.).

The method of approach and the organisation of work will be different in each case.

Lastly, one should be careful with regard to women expressing a need for assistance/support. The existence of previous development projects often give women "an image" of development. The tendency is for them then to request support to recreate the previous projects. Therefore, when women request sewing or crochet training, it is difficult to tell at the beginning whether this is a genuine need or the result of previous projects.

3. Procedures for giving support to rural women

There is no universal formula but rather a five-step approach, which, of course, needs to be adapted to each situation.

3.1 A twofold diagnosis

It is indispensable to make a diagnosis before action is taken. The methods of diagnosis discussed in Chapter II are twofold and must focus on the following:
- *The general situation of the community* in which the women are intending to work, and are living; the internal situation and the external constraints. Four points need to be taken into account:
  - The allocation of resources and their utilisation by the different role-players: people (household heads, young men, women, girls) or groups (livestock farmers, small-holders, associations, co-operatives); it may be a question of access to land, to water (irrigation) or to equipment
  - The changes that have taken place (under pressure from socio-economic changes in the environment)
  - The origin of the idea to undertake a specific project with women and the reac-

tions this has aroused within the community
- Identification of the structures and institutions which at various levels (local, regional, national and international) have a direct or indirect impact on the lives of the population
• *The specific situation of a given group of women* by identifying in each field the various categories of women, their rights, their constraints and their power. It should be noted that the demand[1] determines the methods used to diagnose a particular situation. The initial diagnosis will depend on whether the request is formulated by an administration in charge of promoting women wishing to establish/target programmes with that specific goal, or by an operation of rural development interested "in women", or by a group of women concerned with a particular problem, or alternatively still, by a village assembly of women called to express their "needs" (some examples are to be found on this subject in the second part of this chapter).

3.1.1 Contents of diagnosis

The main aim is gradually to develop a general diagnosis, which includes all aspects of the socio-economic reality accompanying the action. A diagnosis must focus on the following points:

• Access of different categories of women to the means of production: land (different types of soils), water, seeds, labour, credit, and the like.
• Changes that have taken place or are still currently taking place: these changes reveal either strategies developed by the community or social groups, or personal strategies developed by certain women (for example, the acquisition of land for security reasons or the use of paid labour; in the urban or semi-urban environment: access to credit, to services, to professional training, to collective/public organisation).
• Existing or potential services: technology, infrastructure (roads, water availability and installations, sanitation), supplies (of staple food products, inputs, medication) credit (formal and informal structures). For each category a study must be carried out showing how it functions and to what extent each category of women has access to each category of service and for what reasons some may not.
• The level of training and the women's access to the various training programmes existing in the area or nearby: schooling, literacy, technical and professional training, activities and responsibilities possibly undertaken by trained women.
• Organisations designed especially for women and their representation in existing or-

Women's access to production means must be part of the diagnosis.

---

1. In this context the word "demand" will mean "the origin of the action": who had the initiative of supporting the women?

ganisations; the characteristics of women who enjoy a particular social status or exert some responsibilities.
- The demographic situation: fertility, death rate (especially maternal mortality), gender ratio (men/women), women exodus (women and girls).
- The role of the state and its intervention policies relating to women household taxation, and the fiscal system on goods and activities; legal situation, institutions responsible for women or intervening on their behalf
- A woman's place in a cultural context, in common law, and any development observed.

3.1.2 Realisation of the diagnosis
Information should not be collected as a prerequisite to starting a project, but should be gathered from various sources as the project unfolds, from the women themselves, the community at large, and from available documentation.

Even if it only deals with one particular aspect of the women's lives (health, production, training), the report back on the diagnosis and the discussion which ensues (not only with the women, but also with the community as a whole, either at a meeting or preferably through existing channels: co-operatives, communities, specialised associations, village council) is an essential methodological element. The inevitable debates will highlight not only the feasibility of the programme being considered, but also any resistance to it, the constraints, the support needed and any precautions to be taken.

3.2 Finding partners

Analysing the demand
The procedure to follow will depend on who has made the demand. There are two possible situations:

Analysing the demand is important.

1st case: The women concerned have already been identified: the application comes from a group, a designated category – the customers of a social centre or of a SMME, the spouses of producers involved in a development operation. It is therefore necessary to go over the history of the demand with the group (is the group or the institution to which it belongs behind the demand?) and to gain an impression of the group concerned in order to evaluate its homogeneity and reliability, its composition according to age, matrimonial status or activities (and therefore the availability of resources and free time), the level of training, and the motivation for the project: what is expected from it individually and collectively? (It has been proven that most community projects can only last if the participants have an individual or collective interest in it, be it material or not.)

In short, one should verify that the people concerned share a similar commitment, a common interest and similar abilities thereby avoiding incompatibilities (in funding, investing time and ideas). It is clear that it is easier to find these criteria in smaller groups rather than in bigger ones, and that smaller groups are often preferable for some given activities than larger ones. If necessary, they can be combined to form a bigger group and thus increase their power (negotiating, bargaining.).

In Bangladesh, the Grameen Bank, which gives credit to poor women, has established a means of organisation based on small groups (less than ten women) whose members have many things in common including a strong feeling of solidarity. This does not prevent meetings between the various groups taking place including important gatherings on certain occasions.

2nd case: The women affected by the action are to be identified: this is the case when the action is initiated by a national campaign to help "women", or a project is launched by a NGO, a sponsor, or so on. In this case, firstly, information either on the possibilities being offered to women (training, credit) or simply on the wish to work with those who are interested must be circulated. This should be demonstrated by the presence of agents or of certain means of work. Then one can set out the criteria for a specific action and the means of identifying them. This information concerns the community as a whole, and then more specifically the women themselves. At this stage the inventory of the existing formal or informal organisations can be made in order to study what keeps them together, how they work, what constraints they are subjected to and eventually the demands for support that have already been formulated. It is often deceptive to begin working with all the women at once. They quickly become disappointed with the vague programme and the lack of tangible results; it is better to generate motivation based on one or two small groups sharing a similar interest, as in the first case.

3.3 Pinpointing priorities for actions (and their content)

3.3.1 The demand

The demand directly emanating from the women, or formulated in connection with information previously supplied to them, must be discussed with them:
- Does the planned action involve all the women?
- Which resources are being mobilised?
- How secure are these resources and who can guarantee them? (For example, there was a group of women who invested in a market gardening project on a plot of land which was taken away from them the following year.)
- What investment is necessary? How to find it?
- How will the action work and what organisation is necessary? What is the time frame? For how many people? At what time, and so on.

> As stated regarding the diagnosis, there are many examples of bias in women's requests concerning their needs or their wishes as far as development actions are concerned. To illustrate this point here is an anecdote: Some years ago, in south Mali a formal meeting between women from one village and the woman representing the National Union of Women at that time resulted in a demand for a social centre for the village. Surprised, the representative wanted to understand the request, and the explanations given by the women (learning to sew, to read and write) did not convince her because all the previous discussions had dealt with problems related to agricultural production. This was, however, the last word at the women's assembly. With time moving on, the women invited their guest to have lunch and, in the informal discussion that followed, the women stressed what their priority was: the layout of the lowland where they were planting rice and where part of the infrastructure was collapsing. The astonishment of the representative with regard to this obvious contradiction was answered in the following way: the neighbouring or rival village X had obtained a social centre …

- What sort of training is necessary, in which fields: technical, management, literacy?

By "dismantling" every aspect of the planned action one can help the women to take a deeper look at the demand, encourage them to think and attract the attention of the women concerned, encourage members to join or to leave. Comparisons may also be made with more useful, easier or more instructive actions. Priorities may also be established.

3.3.2 Priority actions

The priority actions identified by the women must be discussed with the community leader and with the community as a whole, especially if they need to use resources belonging to the community or family (land, working time, income). This is an opportunity to inform, and to carry out a collective study of the feasibility of the project and share responsibilities. Through this procedure the specific action targeting women becomes integrated into the community dynamics. This seems indispensable to the process and should be seen less as a necessary means to obtaining the men's agreement, or at least their neutrality, than as a methodological and pedagogic tool to help communities reflect on themselves and the place women occupy within the community.

Women's priorities must also be placed in the context of local, national and regional development, thus allowing the women to benefit from existing structural support systems and from both information and training.

The definition of the contents and priority actions is not an opportunity to isolate or to marginalise women, but rather an occasion to scrutinise the socio-economic functioning of the community and to highlight its external and internal constraints. It is the constant proposals and discussions between men and women which allow this continual analysis throughout the action.

3.4 Support for implementing the action

Having identified the action to be implemented, the support agents (development agents, leaders of the rural organisations, etc.) need not have an active role to play in the action. As the responsibilities have already been defined and assigned to members of the community, the development operator should only intervene when asked to do so by the group. He may be asked to help in the following fields:
- Giving information.
- Training, or acting as an intermediary to find ad hoc trainers.

- Making a first contact, if necessary with certain structures: banks or credit organisations, administrative or technical services. At the beginning of an action it may be necessary for the development agent to take an active role in guiding a non-experiment group, but generally this is not advisable. His role should be more that of helping the women to express their needs.
- Leading a group discussion aimed at analysing the possible difficulties encountered, or the results obtained (monitoring-evaluation).

The most important thing is neither to take the place of the women nor to render them dependent on an external development agent to manage the action. This may go against the wishes of the women especially if they are less familiar with collective actions and public relations. It is about helping the women to maintain the review dynamics. It is important to insist on this point because the study of existing female working groups reveals that, even for those who are considered models of dynamism and "success" in certain fields, a high level of dependence is recorded vis-à-vis the external elements of the group. However, through action, women discover their own abilities and thereby gradually gain confidence in themselves.

Example: A group of women from Office du Niger were running a mill. They had so much trouble with their successive millers that they decided to train two of their female colleagues for this job (reputed to be a man's job). The performance of these women was satisfying to both men and women.

3.5 Evaluating the action

There is no particular method for evaluating women's actions: all the methodological and pedagogical thinking (not to mention the other dimensions!) are applicable to both men and women. It is more with regards to the effects that specific questions need to be asked:
- To what extent have the activities carried out led to changes and of what kind in the allocation of resources: access to land, to water, to equipment, to family workforces?
- Were any particular constraints discovered owing to the status of women and to their domestic duties; how have these constraints been addressed?
- What is the impact on the women's time: has the activity been a benefit or an aggravation to the workload?
- What has been acquired in terms of knowledge and skills?
- What changes in the level of training, information and organisation?
- What changes have there been in their representation within the community and as a group. Has their point of view been taken into account when a collective decision has been made? Are they represented in community structures? Have the development organisations and institutions taken into account the actions undertaken and the problems (wishes, strategies) these have revealed?

Not equal, though not different ...

As a conclusion to the "support procedures for rural women" one can say that because "they are neither exactly similar nor extremely different" women need a specific approach, but these approaches cannot isolate them from the community in which they are living.

Talking about specific approaches for women does not mean that women are any less concerned than men, heads of families or youngsters by what their country is going through: the degradation of the environment, the disintegration of rural society, the rural exodus, the dramatic liberalisation of the economy, and economic and often political crises. "Women and development" simply draws attention to the particular constraints that confront women. From diagnosing the situation to measuring the effects of the actions undertaken, the method of work should always take account of this. The essential differences are linked to the social status of women: how they use their time, the level of training, access to training and the level of fatigue (the joys of literacy classes in certain training centres after a whole morning of pounding the earth/corn with a child on one's back, or breastfeeding or taking care of a crying child!), the level of training, and access to information.

Women are now emerging in the process of development. This means that they are becoming visible, and society owes them much over a long time. They are organised and dynamic in many fields, but there is, however, still much that needs to be done and much progress to be made in the way of doing things.

4. Methods and tools

4.1 Methods

It seems useful to present some examples to illustrate the previously mentioned procedures. The examples described are not meant to be "models"; they attempt only to show how the main steps of the procedures proposed can be carried out. The questions raised are not exhaustive as every agent has to think about what information is important to him. However, they mean to indicate a reflective and methodological procedure which does not necessarily lead to success, but does lead to an action based on reflection and the participation of the various role-players.

4.1.1 Example 1: Demand for support for market gardening production at village level

This demand, emanating from the women, is transmitted to a development agent through the mediation of the person in charge.

Step 1: Analysis and diagnosis of the demand
• Analysing the demand
Is the person in charge talking on behalf of all the women? Who of the women are already productive? Or who of a group of women would like to take up the activity?

Are the village leaders informed about this demand? Have they already expressed an opinion? What is their opinion?

In this example the hypothesis is that a demand has been formulated by certain women wishing to undertake an activity and that the position of the village leaders and the men in general is not known.
• Diagnosis
A general diagnosis on market gardening in the village and its environment:
– *History:* Do they already practise market gardening? Since when? What changes have occurred: expansion or reduction (in both cases, find out the reasons)? Resources used: land allocation, water resources. It is important to let the people themselves state what they mean by market gardening. The introduction of plants of European origin must not result in the abandonment of the ancient practices of crop vegetables cultivated for stews, or the recent attempt at diver-

sification that combines cereals, fruits and medicinal herbs, and the like.
- *Role-players:* Who is practising market gardening and why (subsistence farming, sale)? Who is working? Who provides the seeds? Who is in charge of marketing and possibly processing? Is this a secondary, a main, a seasonal, or an occasional activity?
- *Technical aspects:* The calendar and cultural practices: ploughing, sowing, nursery techniques, care, protection and management of water, and so on; conservation and processing techniques.
- *Economic aspects:* production costs (seeds, perhaps salaried work, pesticides, fertilisers); market conditions and markets (variations, competition, transport); supply system used for the seeds and the products.
- *Reaction* to the women's demands: resource sharing, support or fear of competition, doubts about their technical abilities.

Who should do the diagnosis? (see also 4.2 Tools)
- The village leaders
- The existing market gardening producers
- The women making the demands
- The development operator from the technical services concerned
- Specific categories: tradesmen, carriers
- The role-players who have a broader view of the problem (market controllers, regional services.)

The following should be considered when conducting a diagnosis of a demand proposed by women:
- The reasons behind the demand for this type of activity.
- The assessment of the general diagnosis (in which some women were involved and which should be reported back to the group).
- Ideas on the resources available: Which land is being considered? Where will the water come from? How secure is it? Suggestions for guaranteeing the security.
- Room for the activity in the rest of the women's occupations (domestic and productive).
- Ideas on the possible means of organisation.

Step 2: Finding partners
In the above example, the starting point is the demand for support emanating from several women wishing to undertake a new activity or at least a more organised and systematic one. The participation of these women during the two stages of diagnosis should have clarified the problems arising from the activity of market gardening, the main constraints and the necessary conditions. It is possible that certain women may withdraw at this stage.

The second step is to clarify which women can really commit themselves to the activity and/or what kind of support would be necessary for those who still persist in spite of seemingly unfavourable conditions (for instance an urgent need for a cash income).

A discussion with the group of women still wanting to continue should be based on several themes:
- The necessary *means of production* such as land and water: Do all women have collective access to these? Under what conditions? From which organisation?
- The *capacity of investment* in the activity: Monetary resources (and/or credit); implementation of the different steps: production, processing, marketing, ideas on individual (or group) work, the use of salaried workers.
- The *objective sought:* Is it the same for all the women or are there differences in their motivations? For example, young mothers may be looking to improve their babies' diets, while other women with more time on their hands and more autonomy may firstly be seeking a better financial income.

This task should enable the women or groups of women interested in the activity to be identified and their differences (such as objectives, means) to be taken into account. The work can then be organised accordingly: collective or individual work, intensification or back-up work.

Step 3: Establishing priorities
The women should decide which possible areas of support are priorities: access to resources, technical support, supply, credit, aid to organising transport, marketing and processing.

It is useful at this stage to come back to the entire community (through the people associated with the diagnosis) to integrate the women's dynamics into a social context.

Step 4: Support for implementing the action
The kind of support must be discussed with all the women concerned and with the villagers. Depending on the priorities defined, resource people or existing structures may be called upon without the direct involvement of the development agent/extension officer.

To get the operation underway, it may be useful to introduce the women concerned to possible suppliers if there is a problem in that field (see Chapter VII: Contracts between role-players), or to the appropriate technical services if technical support is needed (this may also be the local market garden producers if they accept this role on a basis to be discussed). Visits to other groups can also be organised, or help may be offered in getting the necessary documents together to apply for credit, and so on.

It is also useful, in case of difficulties (damage caused by animals, land taken away, shortage of water), to encourage discussion between the women and the rest of the community to solve the problems in a concerted manner.

It is important for the extension officer promoting the action to be neither the initiator nor the person in charge of the action; he should only provide support for the demand without replacing the women and the other role-players.

Step 5: Evaluating the action
The evaluation (for instance at the end of a crop year) must be assessed in terms of the objectives sought by the group(s) of women. If the objectives were financial, priority must be given to the economic indicators: production costs, yield, losses, market flow, prices. If the objectives were to improve the food supply, of course the indicators will be different.

Whatever the immediate objectives are, more general criteria of evaluation must be taken into consideration:
- In what way has the action undertaken contributed to increasing the women's ability to organise, manage, or increase their knowledge (techniques, structures, networks)?
- In what way has it gained recognition for the women by the community: new access to resources, to freedom of expression, to decision making?

4.1.2 Example 2: A group of organised women wishing to acquire a paddy husking machine

Step 1: Analysis and diagnosis of the demand
Are all the women in the group (community, association, co-operative) informed of the demand? (Sometimes the demand emanates from a woman in charge who is more in touch with organisations, or from the extensions officers, the head office or more influential women.) What are their *motivations*?
- Economic (supplying money to the community's bank account)
- Social (providing a service to the women so that they can save time and reduce their work load), strategic or political (proving their competence, their economic ability, their social existence as a group)

Is the demand known to the village authorities? What has been their reaction? And that of the other role-players in the village (rice producers, tradesmen already in possession of a husking machine)?

- General diagnosis
  - What is the situation with the production of paddy? Are there any estimates of the production? Is it on the increase? What are the prospects (in the village, in the zone)?
  - Husking: Is it a bottleneck in agricultural tasks? Are there any existing husking machines in the village, in the neighbouring villages (any recent developments)? What is known about their functioning: volume of activity, profitability, technical conditions of their maintenance, and so on?
  - Are there any organisations (NGOs, co-operatives, development operations) intervening in the village? Do these organisations have any projects regarding husking?

Who does the diagnosis?
- The village leaders
- The rice producers
- The owners of the husking machines
- The agricultural services
- The women interested in the project
- The mechanics

> A group of women in Burkina Faso managed to get a husking machine just before the (male) co-operative gave its members a family husking machine made in China, at a much lower price!

A diagnosis particularly for women:
- Study the general diagnosis.
- Conduct their own analysis of the problem and the justifications for their choices.
- Study the expertise required: technical, managerial.
- Study the constraints: waiting time, the necessary organisation, the expertise to be acquired.
- Study the market and costs in relation to the seasonal character of the activity, and so forth.

This diagnosis must be compared with the objectives the women have prioritised, with particular attention, if the objective is economic (paying money into the community account), to its probable profitability, of course, but also to what the women are hoping to get from the increased earnings – individual advantages (mutual aid fund or any other aid) and/or collective advantages, such as a communal project whether it is defined or not – fixed in time. There are unfortunately many examples of community activities which deteriorate because the women involved become tired of "accumulating" (in the best of cases) without knowing why and whom the fruit of their investment (in time, work, money) will serve!

Step 2: Finding partners
In this example, the women involved have been previously identified: it is the group of women themselves who have made the demand. However, some will be more involved than others and it is necessary to find out (having identified the necessary expertise and the responsibilities to be taken during the implementation of the action) which women are the best prepared for these tasks and the most motivated to acquire the necessary expertise.

It is also necessary to think about the means by which the women in charge will report back, firstly to all the members of the group, but also to the community regarding the progress and the results of the activity.

Lastly, the partners are also those who contribute to the success of the operation: mechanics, energy and spare parts suppliers as well the clients who need to be loyal in what is often a really competitive situation.

Step 3: Establishing priorities
In this example, the most important thing is that all the components of the action envisaged should be identified and the various steps and responsibilities as well as the necessary means of organisation clarified.

It is as well to remember some useful questions (assuming that the diagnosis has established the viability of the choice made):

The machine:
- Where is it to be bought?
- Which one to choose (capacity, robustness)?
- At what price (it is necessary to compare suppliers)?
- Under what conditions (deferred payment, application for credit, etc.)?
- Terms and conditions of delivery.
- Supply of spare-parts and maintenance.
- Its operating procedure: location?
- Petrol supply
- Days and hours it runs
- Organisation of husking, and so forth.
- Its management: cost to the users?
- Who receives the money and who keeps it?
- Who does the bookkeeping?
- Who makes the decisions about necessary expenditure and who records it?
- Who does the balance sheet and how often?
- How is the information given to the interested parties?

It may not be possible to give straightforward and precise answers to all the questions raised, but asking them starts the thinking and decision-making process. It encourages the group to think in real terms of the feasibility of the action and in so doing trains them.

It is also very instructive, at this stage, to put the group in contact with individuals or well-equipped groups who can share their experiences.

Step 4: Providing support
The support of the development agent, or the person in charge of the rural organisation, is mostly situated upstream of the husking activity. His job is to help the group make an in-depth analysis of all the different aspects of the action: situate it in its environment, anticipate the often complex organisation which will be necessary and which must not fail – a broken husking machine opens the gate for competitors, which is not profitable!

During the course of the activity, the development agent should not intervene, unless it is to respond to specific demands and help the group resolve problems it cannot solve alone. This can be done, for example, by putting it in contact with services or useful people or by encouraging the community as a whole to think about the problem (if, for example, the women's action conflicts or is in competition with the activities of other groups).

The agent can also intervene or bring in outside support for training in bookkeeping or financial management.

Step 5: Evaluating
The evaluation process, in which the agent's support role is once again important, must cover many aspects:
- *Economic results*, which, strictly, can only be evaluated through sound bookkeeping. This can in turn help to determine what if any training is still required in this field.
- *Technical aspects*, in particular concerning the maintenance and the reliability of the material.
- *Social aspects:* Is the action providing a real service to the women, or could it be improved, through a more relevant type of action, and the like?

Last, as for all our examples, two fundamental questions must be asked: In what way has the action improved the women's organisation; has it enabled them to acquire new knowledge and skills? In what way has it reinforced their presence in the community and given them the ability to express themselves and make decisions?

4.1.3. Example 3: Setting up a regional literacy programme for women

Step 1: Analysis and diagnosis of the demand
- Is this the regional repercussion of a decision made at national level? Which authority? What are the objectives? Based on what analysis?

- Is it a regional decision? Which authority? What are the objectives and what was the result of the analysis?
- Who is being called upon? Just a few services? All the development agents? All the organisations working with the population?
- In the example studied it is supposed that this is a national decision from the Ministry of Education, but that the person in charge of the women (female promotion, social action "female branch" or a specialised department in charge of development projects) is free to organise the programme in her region of intervention.

The general diagnosis collects the existing data:
- The number of women and young girls who have been to school
- An inventory and evaluation of the literacy actions conducted over the past 10 to 20 years (by whom, how, with what results: quantity and quality?)
- An inventory of the literacy documents distributed in the region and the methods used by different operators
- A lightweight survey among literate women to understand in what way reading and writing have made a real change in their lives (access to responsibilities, reading of documents) or if they have simply gone back to being illiterate
- What the men think about the women learning to read and write

Depending on how big the area is, it may only be necessary to conduct one survey on a well-chosen sample group of people.

Who does the diagnosis?
- The agents in charge of the services concerned
- The community leaders
- The women in charge of the women's organisations

The diagnosis with the women:

- By usingsing a carefully chosen sample group (representative age groupings, socio-professional categories, literate and illiterate women attending school or not), an attempt must be made to establish what the women really expect from learning to read and write:
  - Do they have a fundamental need to enter the "modern" world and be recognised?
  - Do they have a practical need to carry out the duties and responsibilities they have undertaken (bookkeeping, writing reports, etc.)?
  - Are they imitating others. There is a literacy centre for men in the village, so the women want one too?
- Study with them the constraints: the necessary time, the place, the time of day (most heads of families are reluctant participants if the classes are in the evening), the gender of the teacher.
- Clarify with them the use they hope to make of the new skills.
- An important issue to be discussed with the women is the language in which the literacy classes should be held. The common language usually selected is often less used by the women in certain regions than the men. Thus classes in this language means both learning to speak the language and learning to read and write it.

One must be sure to report the women's diagnosis back to the community for open discussion.

Step 2: Finding partners
The diagnosis should enable the development agent to establish the categories of women interested in the literacy programme and their motivation. It can be assumed that the groups will comprise women with different motivations and this will entail the necessity for different formulas: pacing, duration, practicalities, material requested, etc.

It should be noted that using the women's motivations as a starting point means abandoning the idea of a general programme which concerns all women equally.

It is also necessary to find people or associations (role-players) who will be able to play an active role in the forthcoming training process: women or men, young people, NGOs or development projects willing to provide buildings, supplies and human resources.

Step 3: Establishing priorities
This type of intervention cannot hope to reach all women at the same time as this would result in a high level of loss even from the very willing women. Some groups may be considered as priority cases based on certain criteria: motivation, need, matters of urgency.

If the means allow it, different types of groups can be formed, but this is very demanding in terms of methods, pedagogy, and the adequacy of the material to be used.

Step 4: Providing support
- It is suggested that the support of the person in charge of the programme should deal essentially with the training of the instructors or literacy teachers:
    - Support in adapting the proposed methods (if it is not reasonable to devise them) and the existing material to the expectations and needs expressed by the women and to their particular situation
    - Pedagogic support
    - Support in evaluating the progress and the results
- The support also involves the material aspects of the programme: lack of sufficient means is an additional handicap in an action which is in itself difficult. There are many problems which may lead the women to get bored and to lose interest quickly: classrooms lacking sufficient light, too little material, a badly chosen timetable, presence of small children. This is especially problematic if the women who are making the effort to read and write do not have the opportunity in their environment (posters, newspapers and other reading material) to use the skills they have acquired immediately.
- From previous experience, it appears to be important to give particular support to the village instructors responsible for the literacy programmes. Their great enthusiasm and devotion often evaporates because of lack of support. This support may take several forms: material, a presence or regular interest shown by the agents in charge of the programme at all levels or regular training which enables them to progress and appear as the real development agents.

Step 5: Evaluating the action
Certain methods are frequently used to evaluate literacy programmes, but there are a few methods for dealing with the maintenance of those achievements:
- A regional programme should be concerned with setting up an evaluation system which enables the agents to check that the programme has achieved its objectives and that it goes beyond the number of literate women (such as the permanent monitoring of the women who have learnt to read and write to identify the way they are using their knowledge: local production of written documents, increase in written exchanges).
- Indicators must be developed based on the objectives set out by the programme as well as on the women's expectations.
- Lastly, the evaluation should also deal with the written documents put into circulation by different regional operators: Are they accessible to literate women? Are they adapted to their concerns? To their centres of interest?

4.2 Tools

Different tools (pedagogic, audio-visual) can be used to understand women's issues. These tools do not differ from those used in the context of development actions:
- Meetings
- Individual and collective interviews
- Drama

- Songs
- Visual and audio-visual supports

Recent experiences demonstrate that visits and exchanges between people and groups are among the best methods to identify favourable conditions for actions: how in practical terms the action can be implemented; the sort of reaction it arouses and how it fits into the cultural and linguistic universe of the rural communities. Visiting a group that has been providing a husking service for many years is an invaluable tool which will teach the women more quickly and better than any development agent how to set a service up. The experienced group will be able to report on any difficulties and to analyse their action in a language that the new group will understand and identify with immediately.

It is always necessary to combine several tools, with particular attention to the following in the case of women: choosing carefully with them the best time to get them together; and where they are going to meet for a talk. Women are often concerned with their domestic and productive duties: mealtimes and time to go to the market, and the like, so they are not really free to think with the external operators. The latter should therefore adapt and not the reverse.

It is also useful, as far as the interviews and surveys are concerned, to look carefully at how the sample group of women is made up: women are not all the same. They are also not all equal in the face of the action envisaged. Many factors make them different: the family they belong to (a small household or important family), whether they have co-spouses to share the daily chores, young children or children who can already help them, cash income or not, freedom of movement owing to their age or good relationship with the head of family or, on the contrary, constraints in their daily schedule and little freedom to move around.

Certain tools (recorded cassettes, videos) are sometimes very useful in highlighting the different approaches men and women have to development problems. They enable the discussion to be handled from a less personal and emotional but more objective angle.

5. Recommended literature

Eade, D. (1999) *Development with women*. Oxfam Publishing, Oxford, UK

Everts, S. (1998) *Gender and technology: empowering women, engendering development*. Zed Books, London, UK.

Koopman, J. (1998) Gender and participation in agricultural development planning: key issues from 10 case studies. Sustainable Development Department, FAO. Web site: http://www.fao.org/WAICENT/FAOINFO/SUSTDEV/Wpdirect/Wpre0061.htm

Meinzen-Dick, R. et al. (1997) *Gender, property rights and natural resources*. IFPRI Food, Consumption & Nutrition Discussion paper #29.

Moock, J.L. (editor) (1986). *Understanding Africa's rural household and farming systems*. Westview Press, Boulder, Co, USA, 234p.

Ostergaard, L. (editor) (1994). *Gender and development: A practical guide*. Routledge, New York.

Otsyina, J.A. & Rosenberg, D. (1999) Rural development and women: what are the best approaches to communicating information? *Gender and Development*, 7 (2): 45–55

Pitcher, M.A. (1996). Conflicts and cooperation: gendered roles and responsibilities within cotton households in Northern Mozambique. *African Studies Review*, 39 (1996): 81–112.

Van Hook, M.P. (1994). The impact of economic and social changes on the roles of women in Botswana and Zimbabwe. *Affilia*, 9 (3): 288–307.

Wallace, T. & March, C. (editors) (1991). *Changing perception. Writings on Gender and Development.* Oxfam, Oxford. UK.

Chapter XIV : The non-agricultural sector
D. Pillot

The following chapters are related to the non-agricultural sector:

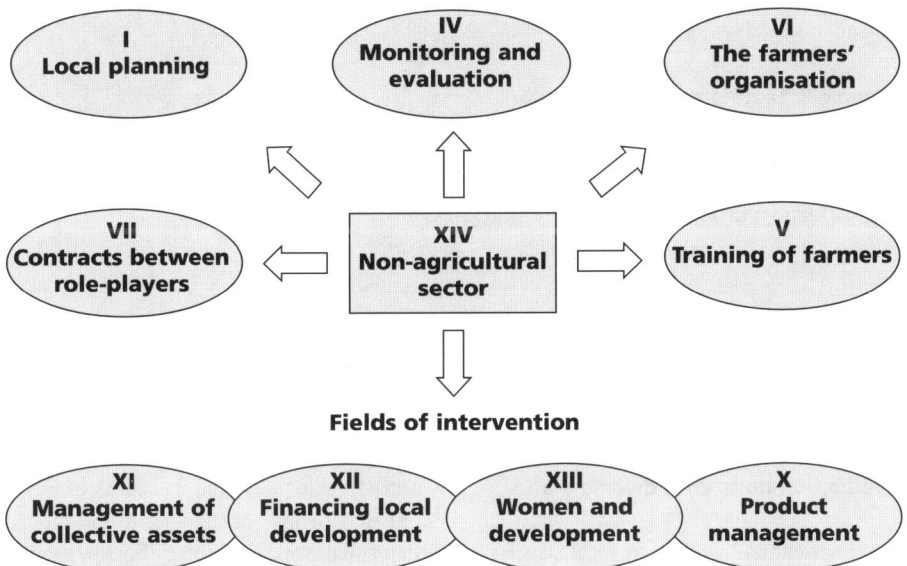

Fields of intervention

Rural development has often been understood as a purely agricultural development, whereas economic life is not only about agricultural production. Rural craftspersons, retailers and product processors contribute in the same way as farmers to the general well-being of society. In a development project it is essential, for many reasons, to take into account the non-agricultural sector.

- Firstly, for the benefit of agricultural development itself: farmers need carts, harnesses, hoes, baskets, and the like, as a means of production. These tools are sometimes produced locally, but local producers increasingly face competition not only from external producers, but also from foreign producers that benefit from a better reputation (or a better quality/price ratio). They also need to sell part of their produce; hence their need not only for active markets but also for dynamic retailers.
- Farmers also need their individual and communal equipment to be maintained. As the material is increasingly sophisticated, reference to specialists is indispensable.
- More and more often income from non-agricultural sources is essential to maintain farmers in rural areas and prevent an exodus to town.
- Finally, to be comprehensive, a balanced development plan must involve all the production role-players. The following examples illustrate all the non-agricultural activities that are involved in the development of two specific sectors, the onion and leather processing network.

> **Example 1: Onion production and marketing**
>
> The onion cropping system requires seeds, plant protection chemicals, fertilisers, and also the means to pump out irrigation water, erect fences, maintain hoes and manual implements, and so on.
>
> The stems have to be pressed, dried and stored, and the bulbs also have to be packed, dried and stored. This requires specific implements (manual press, on-farm dryer, storage silo).

> **Example 2: Leather processing network**
>
> Leather production depends on livestock production systems (water supply, marketing, animal traction, etc.). It also links up with slaughtering techniques, hide and skin quality, drying and tanning techniques, fine leather craft, shoe making, upholstery, and the like, for export or local consumption purposes.

1. The non-agricultural sector: diversity and constraints in development

1.1 Production units with diverse status

The non-agricultural sectors in local development include the following activities:
- Trade (wholesalers and retailers, boutiques, market stands, etc)
- Production of goods and services (metalworkers, blacksmiths, brick makers, carpenters, bakers, tanners, repairers, mechanics, eating houses, photographers, etc)

In rural communities, small service units and small-scale processing are obviously the most wide-spread categories. In small semi-urban centres, which are difficult to dissociate from the rural world, there are more important units with more regular and stable activities.

The configuration of the units pertaining to this sector is extremely diversified. A unit can be run by an individual, a family group, or an extra-family group on a co-operative basis or any other. It can be permanent, which is generally the case in small semi-urban rural areas, or seasonal depending on occupations linked to agricultural production, or temporary.

The means of production (the business, workshop, and tools) and the stocks of raw material may or may not belong to the producer. Often small workshopkeepers borrow their necessary working capital from the most important retailer to whom they give their goods in return.

The production units may use salaried labour, or conversely just the family workforce, possibly with the help of some apprentices. In certain countries, labour, which is sometimes abundant, is made up of apprentices only who, far from being paid, pay their boss for the training they are supposed to be receiving from him. This practice sometimes reaches such proportions (Benin, Togo) that it can be likened more to a phenomenon of social regulation than to a training system.

Small-scale industries and trade are sometimes linked to social castes and their specialities (blacksmiths in the Sahel countries, leather in Niger, pottery from the Twa of Rwanda), and are therefore steeped in culture and tradition.

1.2 Assets of the traditional non-agricultural sector

The assets of this sector illustrate the role that the sector plays, or aims to play in the development of rural populations.

1.2.1 Flexibility
The low capital investment and the almost total absence of rules and regulations in this field make the establishment of non-agricultural activities very flexible (at the outset).

1.2.2 Close links with agricultural production
This sector is closely linked to agricultural production:
- It enables agricultural surplus to be sold throughout the year, usually in small quantities, as and when the family needs a cash income.
- In the same way, it supplies families with the basics (oil, salt, soap).
- It ensures at a local level, the manufacture and maintenance of necessary tools and equipment.
- It thus contributes to decreasing production costs.
- It adds value to the agriculture products it has processed (in fact, some of these products can only be processed at this level).
- It relies on the trust bestowed on it by the rural population who, in turn, need the service.

1.2.3 Low production costs
Low production costs enable the sector to satisfy the domestic needs of the low income population (household equipment, home improvements).

1 2.4 Local labour involvement and capacity building
It can absorb and make valuable use of an important part of the available labour force thus contributing in a significant manner to stemming the rural exodus.

Owing to its varied systems of apprenticeship, it plays an important part (which may be vital in certain countries) in youth training.

1.2.5 Versatility, adaptability
Owing to its structure and dimension, the production unit is undoubtedly very flexible. This can be seen, or should be seen in relation to
- the market, which changes in nature and dimension;
- its products;
- employment (qualifications are less necessary);
- the techniques (sophistication is less necessary and therefore it is easier to grasp the technique);
- the possibility of using technologies, which, although obsolete in some contexts, are very efficient in others.

1.2.6 Increasingly favourable context
The institutional environment is becoming increasingly favourable to production units. Encouraged by the current trend, which in the matter of development tends to favour small-scale production units, it is destined by its dimension and flexibility to become more and more integrated into national and regional development policies and plans.

1.3 Limits of the social efficiency of the traditional sector

Although it undoubtedly provides a number of services to rural society, the traditional sector is marked by a number of handicaps, which limit progress in productivity or improvements in the quality of its services.
- The first of these handicaps is due to the narrowness of the market, which is limited in a virtually endemic manner by
  - low solvency in rural areas: smallholders' incomes (potential clients, by far the most normally) are by nature low and irregular smallholders are by far the greatest poten-

tial clients;
- inter-sector competition (this is one of the disadvantages of the ease with which small-scale production units can be set up. Many rural regions are saturated with craftspeople offering the same service or the same products.);
- competition from imports, urban businesses and smuggling activities (notably on border zones);
- lack of organisation regarding sales channels and the promotion of products.
• The second is due to the *progressive loss of technical expertise*.

> For example, it is not unusual to meet a carpenter who once knew how to turn wood, but who no longer masters this technique. In general, his/her personal ability is not to be blamed, and a simple technical training course would not improve matters. It is rather because he/she no longer has the opportunity to produce furniture with rounded feet and therefore he/she has gradually lost his/her competence in the most sophisticated techniques he/she possessed. This has been caused firstly by competition from imported products and secondly by the decline in his/her clients' purchasing power.

The apprenticeship systems are also sometimes to blame:
• Faced with an increasingly limited market, it is not objectively in the interest of a master to teach an apprentice all his/her skills as in the future he/she may compete with him/her (except if both are part of one family). For the master, an apprenticeship is more a means of controlling cheap labour rather than an opportunity to transfer all his/her skills. The tendency to increase the number of trainees jeopardises the latter's chance of getting proper training.
• Craftspersons are not necessarily good educators, and apprenticeship through mimicry has its limits. A young person trained through an apprenticeship will know, at best, how to reproduce what he/she has learnt, but will not have the necessary basic knowledge to imagine new products or create new forms of products necessitated by developments in the market,
• This leads to the third limitation of the traditional non-agricultural sector: *its poor capacity to innovate*, while the market and more generally the socio-economic environment are rapidly developing. Many factors have resulted in products, which come from elsewhere – be it a town or from abroad – penetrating the local markets and competing with local products. This has been accelerated by the integration of rural people into the market, the opening of roads and tracks, and improved means of transport. Competition is all the stronger since the foreign products have a valuable reputation. In general, local craftspersons decrease their production costs in order to supply the market with products that are cheaper than the imported ones. This contributes to the technical regression mentioned previously.
• A fourth limitation is that the production unit is generally *entrenched in a very strong client-orientated relationship*. The personality of the manager or retailer, his/her dependence network, the non-marketable service exchanges on which his/her activities are based determine his/her economic performance. On the other hand, these characteristics limit his/her economic space. It is difficult for a small-scale enterprise to change its source of raw material, or for an emerging retailer to stop selling to the same urban merchant, as they are prisoners, dependant on these relationships for credit, the children's education, and such like. Once again, these factors limit their ability to adapt to a rapidly growing market,
• Finally, *the lack of managerial skills* is the last constraint that can be identified. The concept of management is used here in its broader

sense. It is not a matter of finance or accounting. A rural craftsperson is happy to do without accounting, and, in any case, he/she cannot afford it. On the other hand, it is essential for him/her to know how to calculate costs, to draw up a correct estimate and to settle questions relating to provisions and organisation, which means in short to plan production. For example he/she has to take into consideration
- the necessary provision for the maintenance of his/her working tools, for investment and for the purchase of raw materials,
- avoiding continual maintenance (taking care of equipment),
- the organisation of production: layout of workstations, scheduling of operations, sales organisation,
- management and personnel training.

In the economic circumstances of small-scale production for a client-orientated market, this poor management capacity does not represent a real constraint. It is no longer the same, however, if the craftsperson or retailer has to quit the system and adapt to another.

2. Intervention and non-agricultural producers

In these circumstances, on what can one found the development of an intervention policy relating to the non-agricultural sector?

Experience firstly warns against the easy solution that consists of focussing only on the manufacturing aspect of the supply, manufacturing and sales cycle. More often, it is in the manufacturing process that the resources potential is the greatest and this can be used by the craftsperson. It is therefore in this area that it is less urgent to intervene.

However, owing to the characteristics of the sector one is led to focus on the individuals who are at the centre of the production.

One of the parameters and possibly the essential condition for the success of a small business is the personality of the manager. Creating or rejuvenating a small-scale business entails creating or improving units to serve as models of success. Hence the importance of identifying people whose competences, as managers of the actual or potential units, are evaluated not only according to their technical capacities, but also, if not more so, by their motivation and willingness to succeed.

Experience has also shown that one has to avoid producing everything locally. Some needs may be more easily provided for by urban craftspersons who are well equipped and benefit from a bigger market than rural producers.

Nevertheless, all positive interventions in small-scale units, whatever they may be, are based firstly on knowledge of the system and the surrounding environment. One cannot consider giving support to small-scale industries if it is disconnected from the problems affecting the entire system. From sound knowledge it is possible to identify needs, highlight opportunities, and choose priorities that will define the specific sort of support required. The interventions themselves will have a direct influence on the production of goods or services, or on their socio-economic environment (see next sketch).

Interventions fall into five main categories of actions:
- Market analysis and support for marketing products
- Credit and guarantee plans
- Advice and training in technical and management matters
- Technological information
- Support to professional organisations

The approach to supporting small-scale craft industries and SMMEs

Taking an interest in small-scale activities precludes mass action and therefore implies an elitist approach. How can one imagine, in fact, a project giving support to this sector not being forced to select units based on the best chances of progress improvement.

As mentioned earlier, such projects must focus on people, their personalities, the relevance of their achievements or their intentions regarding the social and economic context, their motivation to succeed or to grow.

However, supporting enterprises as they progress automatically means caring for their environment. This process implies of necessity an intervention in the business sector as a whole. Supporting managers is also about getting them to recognise that being organised in their profession is indispensable. It is therefore a question of finding with them the most appropriate means of organisation in the context; the action therefore serves all the other enterprises as it aims to improve their environment.

In these conditions, an elitist approach, far from being a defect, is a realistic option. Undoubtedly, the intensive experience gained by the operators in the daily running of the units is the best way to obtain information capable of defining, with the maximum of relevance, the actions to take in favour of a field of activity, a profession or a sector.

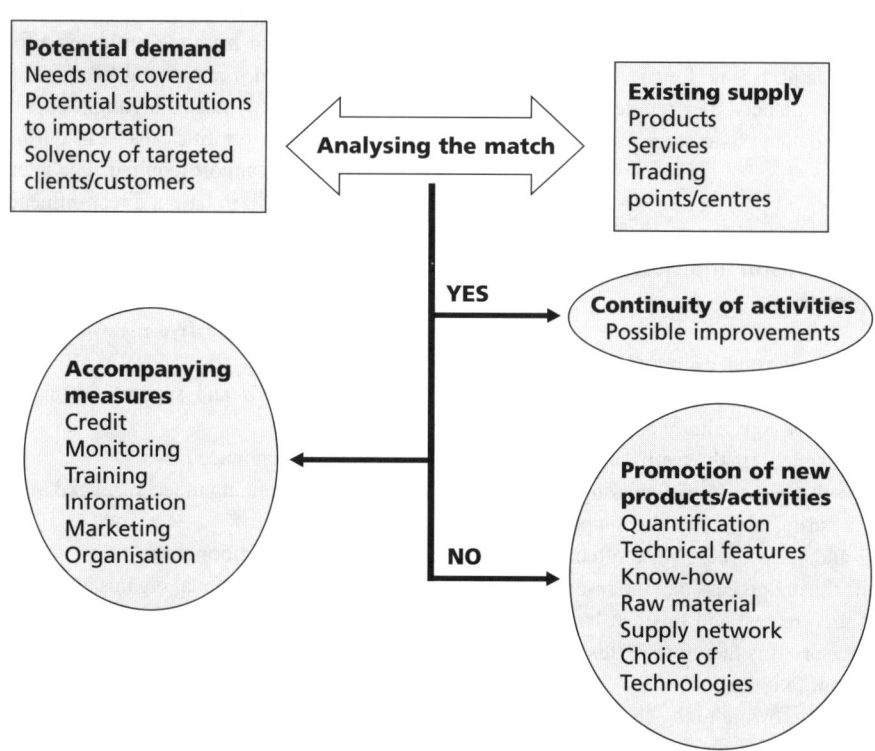

2.1 Market analysis

Support to small-scale enterprises necessarily implies and begins with studying the market:
- The study of the subjective market, as it is perceived by small enterprises and which is often no more than the sum of the orders made by the same well-known clients.
- The study of the potentially objective market (which is ignored but does exist), which takes account of the needs of the population and its solvency. Such a study must not be conducted from the outside; it must rather involve the craftspersons themselves as part of a training and awareness process.

Its objective is two fold:
- To identify, of course, new target markets for small-scale businesses
- But also, to train these businesses to change their habits in this matter, to avoid the passive attitude of waiting for the order, and to acquire a more aggressive commercial policy

It is therefore a matter of overcoming a "I produce and then I sell" strategy to become one of "I identify a market and I produce accordingly", which is a more offensive approach targeting the continual adaptation to supply and demand.

It is therefore necessary to find out the potential demand by
- identifying the overt needs and the possibilities of substitution for importations,
- studying the financial situation of the concerned population.

The analysis of the existing population will give correlated information on the degree of conformity between the actual supply and the potential demand.

From there it can be deduced if the products (or the activities) must be maintained, modified or if new products (or activities) should be launched.

If this is the case, one must
- quantify the products,
- define the technical characteristics,
- identify the corresponding production techniques (this often implies conducting experiments with the participation of craftspersons),
- specify the conditions of their manufacture: available raw materials, credit for supplies, adequacy of the equipment, producers' skills, accompanying measures (credit, training).

Possible markets for small-scale enterprises that should be investigated:
- National investment programmes (infrastructures, planning)
- Public or parastatal organisations
- Local development operations
- NGOs
- Big companies

These types of clients generally have important needs and constitute an appreciable distribution and extension network for the users (especially crop farmers). However, this market is more often subject to tenders and therefore becomes virtually inaccessible to craftspersons.
- The household and local agricultural institutions market (the needs of which have not been covered but which should be measured) is important but is clearly undermined by the weakness of its purchasing power.

2.2 Support for marketing procedures

The obvious consequence of market analysis is to support marketing, the organisation of sales and promotional activities. Below are the classic procedures that will vary in scale depending on the context:
- Direct customer prospecting
- Search for reliable intermediaries
- Creation of a label
- Use of the media
- Research on and use of informal networks
- Presentation at periodic local markets, fairs or other events, shows, and the like
- Presentations at permanent exhibition halls organised by communities or professional organisations
- Design of technical-commercial pamphlets
- Publishing catalogues

Small-scale enterprises often cannot provide the above on their own. Hence, the need for them to create a professional organisation and organise communal promotion services.

2.3 Credit

Credit is of vital importance for small enterprises, be it to create an activity or simply to develop it. The point of view of banks concerning loans to the small-scale business is well known: the credit demanded is often very low, therefore not profitable enough for the bank; the cost of managing the portfolio is disproportionate to the benefits; it is a high risk sector with uncertain guarantees.

As the small-scale enterprises are not eligible for credit, intervention projects in rural areas often have to be substitutes for the failing system and have to develop and implement more adapted and flexible systems.

Chapter XII (Financing local development) discusses this subject at length. Focussing on rural areas as a whole, the recommendations and ideas the chapter develops are also applicable to non-agricultural activities. This chapter adds some points concerning methods of helping (supportive measures for) credit beneficiaries when these are small-scale craftspersons or retailers. Experience has shown that for credit to be turned to profit, it is essential to help the beneficiaries in the creation or the extension of their activities. This follow-up procedure is indispensable to small enterprises and allows them to acquire the necessary skills and habits to manage their businesses. It also appears to the fund providers as an additional guarantee that they will recover the credit.

These supportive measures involve the following:
- A strict selection procedure for applications or projects based on market acceptance/interest, the personality of the entrepreneur, his motivations and competences
- Support for formulating the demand and putting together the commercial, technical and financial dossier that must be presented to the fund providers
- Support for the setting up of the professional activity which required the loan and for its close and individualised monitoring

Helping an emerging manager to obtain and reimburse credit is concrete and particularly efficient form of intervention that can be seen in terms of managerial training. In fact, he realises the importance of it when he discovers that, through this process, he can obtain the essential elements he needs to master the restricting process he is now involved in. As the advice and training are applied to matters that directly concern the manager, they are far better perceived and assimilated.

2.4 Technical training

In the first place, for the support procedures to be coherent, the technical training, which is only a part of the procedure to be implemented, must be based on products with a guaranteed market. One should avoid intervening in technical improvements that will not necessarily result in an increase in sales (an extended, new or wider market).

Following this logic, intervention should not be based on linear programmes, but rather on the *problems to be solved*, a project to be carried out, a potential market, and so on.

This approach excludes training for training's sake! It should be mentioned that such a resolution is far more uncomfortable for the trainer, who will be judged (or will judge him/herself) not on a simple evaluation of the skills acquired, but on concrete results, tangible improvements in production, demonstrated by an increased share of the market.

This approach also favours training in real production conditions in workshops as opposed to group training. In this way, both the manager and the personnel are made to feel involved.

Advanced training courses can obviously be taught in ad hoc structures (specific structures specialising in adult education, for example), but it should be stressed again, that one should avoid interventions which require the use of sophisticated equipment that small businesses will never have the opportunity to use in the future. This is too often the drawback of systems such as pilot or relay workshops.

Specific constraints inherent in the social structure and environment must also be addressed, such as the weather, the availability of the personnel involved, diversity on a cultural level (literacy for example), how basic the equipment is, and such like.

Technical training must result in improved production quality without increasing the costs excessively.

More useful training can be organised for certain skills or to complement other professions (panel beaters, mechanics, carpenters, etc.) or for self-equipment workshops (making their own equipment and machines).

2.4.1 Technical training and apprenticeship

The diverse nature of apprenticeship in sub-Saharan Africa (status, duration, financial conditions, organisation, future) will not be discussed; the common denominator remains the many endemic obstacles to be overcome in the setting up of the apprenticeships.

The following observations must orientate the action in the apprentice's favour:

The skills which enable small-scale enterprises to multiply are acquired essentially by in-house apprenticeships. Classical professional training structures are seldom adapted to the needs of this sector and, moreover, they are rare in rural areas. This situation, therefore, means greater interest should be taken in the means of improving the training apprentices receive in workshops.

The following are recommended:

- To improve, as a priority, the practices and skills of the manager himself, as the better a technique is mastered, the easier it is to teach someone else; moreover, in a workshop situation, mimicry is the natural method of passing on skills, thus an apprentice must have a real role model. The ideal situation is obviously that the workshop should be considered as a place of education where any production or management activity should be taken as an educational opportunity.
- To find ways of alternating. This consists of taking the apprentices out of their workshops for short periods of time to give them complementary training in line with their individual needs.
- In this respect, existing resources for training should be used and one should avoid creating new ones; this education by alternation must also be used by the traditional teaching systems as it favours open-mindedness and a system which adapts to the real needs of the productive sector.
- To study and show the apprentices the realistic opportunities for professional intervention

(setting themselves up on their own or in a group, being hired by existing workshops, creating new activities) and consequently to offer them support.

The example of the educational workshop at Camp Perrin, Haiti

Long-lasting equipment for rural areas.

The educational workshop at Camp Perrin is installed in the agricultural plain of Cayes, on the south peninsula of Haiti. The vocation of the centre is to make solid and reliable material to equip, as a priority, rural communities.

Training future entrepreneurs

The choice of apprentices is fundamental. One has to remember that the objectives of educational workshops are, on the one hand, to offer the rural world tools for its development and, on the other hand, to train the producers of these tools; it is not to train students or labourers. The choice of apprentices is made over a long trial period (about six months) and based on criteria such as good logical thinking and the ability to work in a team. The surprising thing is that initial technical knowledge is of little importance in the choice of candidates.

The training, which is of a very practical nature, revolves around current manufacturing. The newcomers, monitored by senior apprentices, do the more simple tasks; the more senior ones progressively undertake increasingly complex work.

The training focuses on experience, observation and work on the machines and not on scientific teaching. It is a matter of training local managers and not of processing technicians.

The workshop steward can only manage a limited number of apprentices (about 15); the training must therefore be efficient and fast. The choice of the candidates is important because young people must be recruited who are not only capable of assimilating technical training but who will also grow in maturity. In fact, the educational workshop does not train labourers but managers. The first quality required is therefore a sense of responsibility.

The training develops as time goes by: ten or so hours of calculations and geometry, and the same for workshop technology to complete the practical training and allow the future managers to confront the increasingly complex nature of the workshops in the region.

Creating small businesses

The final aim of the training programme is not to deliver a diploma or a classification, but to make the local managers efficient in their profession and therefore capable of managing their own workshop.

This process takes place slowly. With time certain trainees are mature enough to start their own independent workshop. Then they start going their own way: one goes into sharpening tools at regional markets; another starts a plough repairing business and a third establishes himself as a furniture manufacturer dealing also in wrought iron products, and so on. Each one leaves the main workshop with a manufactured good or piece of equipment.

Being trained and owning equipment is only a first step. Thereafter one must have a constant supply of raw materials, access to credit in order to extend or transform one's premises, and the like. The creation of an association seems important for everyone: whether it is to find the funds for the creation of a storehouse which supplies both raw material and equipment, to look for a competent manager, or create a credit fund to set up craftspersons or extend or reconvert existing workshops, it is better to be an association than an individual.

The educational workshop association has created a service to supply raw materials and equipment to the craftsmen and women in the Cayes area. This enables them to buy the raw materials at an affordable price (this is the case for a number of steel tools, for example).

The educational workshops have also created a credit service. The traditional credit institutions geared mainly to giving credit to commercial and consumer-orientated activities did not in fact give credit easily to emerging craftspersons.

Finally, the educational workshops have set up a development and research service to study prototypes and test them, before handing over their manufacturing to independent craftsmen and women. This service is also a place for craftspersons to think about the changes in manufacturing practices and relevant economic extensions.

Through the association, the delicate problem of competition has been solved. If ploughs are selling well, how can one prevent all the craftsmen and women setting themselves up in this line, whereas the market allows only one craftsperson to live from it? Every person specialises in one or several manufacturing techniques or services. The person who makes wheelbarrows cannot make silos. All of them do maintenance on agricultural material or vehicles, but only one can repair and make leaf springs. It is the sine qua non of peaceful co-existence and the economic viability of these small enterprises which enables them to continue to multiply today.

2.5 Partnerships with small rural industries/small urban industries

Forms of complementarity between small-scale production in rural areas and urban areas can be found. In certain professions, the first step in transforming a product can be done by a rural enterprise and the finishing touches by better equipped urban enterprises or by bigger firms.

One can see the benefit of this partnership formula which, by focussing on prefabrication in rural areas at the very source of the raw material, also improves it, in real production conditions, through technical training and by creating a real motivation through market support.

Here is one example of this kind in the field of utilitarian pottery, an action implemented by the ETTC (Ethiopian Tourist Trading Corporation):

Women make clay pots in their villages and fire them in the traditional way.

The ETTC, a public enterprise regrouping small-scale workshops, purchases them as they are, finishes them off (refires them up to 800°C in an electric oven and then enamels them) and markets them. The advantage of this formula is that it provides the potters with an easy and regular outlet. As for the ETTC, it cuts out time-consuming operations, optimises the use of its ovens, increases its production capacity while decreasing certain general expenses.

At the same time, some women from the same villages come to work in the ETTC's workshop to learn new techniques that they can use in their home environment. Finally, there is a permanent link between the ETTC and the villages; it helps to select the products, orientate the manufacturing and improve their quality to meet market standards.

A similar operation is underway with a group of craftspersons working with bamboo.

2.6 Technological information

It is obvious that craftsmen and women need access to technological information and that this is particularly difficult in rural areas owing to several factors:
- Chronic lack of technological information at local level (be it in rural or urban areas)
- The difficulty for an emerging manager to access the little there is
- His/her understandable difficulty in formulating his/her needs in that matter and making decisions concerning certain fundamental technological choices (trials, equipment, optimal dimension of his/her business; and often, how to avoid being the victim of his/her own dreams)
- The risk he/she is taking of being the victim of unscrupulous suppliers, and so forth

All support procedures in favour of the producer must create methods and means that take into account this information dimension and its particular constraints.

2.6.1 Intervention and technological information

Experience has shown that technological information and documentation services seldom have the impact hoped for by the initiators. In the worst cases, they are abandoned by the people to whom they were supposedly addressed.

This situation is the result of an approach regarding the communication of technological information that does not really take into account the operational aspects of the mission.

Compensating for this implies that a system of communicating technological information
- cannot be seen as a finality,
- must have sufficient means to make it work,
- cannot be dissociated from a support-advice structure,
- must have the capacity to reformulate, to adapt the information to the needs of the beneficiaries, and if needs be, to edit the books and adapt them to the local context,
- must be an open structure which leads to it being visited because of its efficiency and professionalism,

---

In Chad, in the context of a project to support micro-enterprises co-financed by the World Bank, the vocation of the technological information service is to respond to the needs of small enterprises.

The technological information service is based on the concept of technology:

"Technology is the study of a set of techniques and the relationship they establish in view of a definite improvement, in a human, economic or geographical environment at any given time."

Technological information does not limit itself to information concerning objects or mechanisms; it is also about leading individuals to understand thoroughly the specific content and methods of the techniques they wish to implement:
- Market study
- Supply (suppliers, skills)
- Product conception (engineering file)
- Creativity techniques
- Marketing techniques, and so on.

The information/documentation service is open to the general public (contractors, students, etc.) to create a dynamic and synergic environment and to encourage the public to get to know small enterprises better.

To avoid being isolated and to make intervention more professional, it is supported by a development-support institution (GRET), which enables the information/documentation service to use its network of experts and knowledge acquired from similar projects that have given support to small businesses. The transfer of technology is also programmed between Senegal and Chad concerning grain-processing equipment (husking machines for rice and millet).

In the long term, it is planned to hand back the system of technological information to a local structure (NGO). Its aim is to offer support advice to small enterprises. This type of service, which is similar to a public service, can only work if it receives subsidies. In the current situation, the beneficiaries of the services do not have the financial capacity to pay the real cost of the technological information they need.

In the case of Chad, one can see how a long-lasting system of technological information remains dependent on a set of active partnerships which need to be defined and established so that it can meet its initial objectives.

- must be open to the exterior, especially to the professional sector which can supply practical information ready to be used,
- must have a sort of global technico-socio-economic approach corresponding to identified needs,
- must play the role of technological memory, that is, it must capitalise the acquisitions of developed actions in the country concerned, and the successful experiences in other places, in other countries.

3. Conclusion

Action in favour of the non-agricultural sector in rural areas requires a great deal from the operator who must combine different levels of intervention. The action must concentrate on the production units, while paying particular attention to the environment in which these units are situated. It must associate the technical aspects (the tools, the procedures) with the economic aspects (organisation, market management).

The role-players themselves are in very different positions. Their interests are sometimes contradictory. It is always important to identify the different categories of craftspersons or retailers with whom one wishes to work. Identifying the markets and products is also a delicate matter; the interested parties should always be very much involved. The diagnostic phase of the situation is therefore essential.

Finally, the action itself, in whatever form it takes – training, technological information or organisation – must, as for the agricultural sector, be a flexible procedure centring on the role-players and regular evaluations must be undertaken with them.

4. Recommended literature

In general, the non-agricultural sector of rural development is not the subject of many studies and publications, as is the case with the urban traditional sector, and the informal sector.

Brautigam, D. (1997). Substituting for the State: Institutions and industrial development in Eastern Nigeria. *World Development*, 25 (7): 1063–1080.

Flora B.C. & Flora, J.L. (1993). Entrepreneurial social infrastructure: a necessary ingredient. *Annals of American Academy*, 1993: 48– 58.

Van der Ploeg, J. & Long, A. (editors) 1994. *Born from within: Practices and perspectives of endogenous rural development*. Van Gorcum, Assen, The Netherlands.

Whiteside, M. (1998). *Living farms. Encouraging sustainable smallholders in Southern Africa*. Earthscan Publications, London, UK, 217p.